The Nature of Quantum Paradoxes

Italian Studies in the Foundations and Philosophy of Modern Physics

Fundamental Theories of Physics

A New International Book Series on The Fundamental Theories of Physics: Their Clarification, Development and Application

Editor: ALWYN VAN DER MERWE
University of Denver, U.S.A.

The Nature of Quantum Paradoxes

*Italian Studies in the
Foundations and Philosophy
of Modern Physics*

edited by

Gino Tarozzi
*Institute of Philosophical Sciences and Education,
University of Urbino, Italy*

and

Alwyn van der Merwe
*Department of Physics,
University of Denver, U.S.A.*

KLUWER ACADEMIC PUBLISHERS
DORDRECHT / BOSTON / LONDON

Library of Congress Cataloging in Publication Data

The nature of quantum paradoxes: Italian studies in the
  foundations and philosophy of modern physics.
  edited by Gino Tarozzi and Alwyn van der Merwe.
      p.   cm. -- (Fundamental theories of physics)
  Includes index.
  ISBN-13: 978-94-010-7826-9   e-ISBN-13: 978-94-009-2947-0
  DOI: 10.1007/ 978-94-009-2947-0

  1. Quantum theory--Congresses. 2. Physics--Philosophy-
-Congresses.   I. Tarozzi, G. II. Van der Merwe, Alwyn.
III. Series.
QC173.96.Q815 1988
530.1'2--dc19   88-3068
                                                     CIP

ISBN 978-94-010-7826-9

Published by Kluwer Academic Publishers,
P.O. Box 17, 3300 AA Dordrecht, The Netherlands.

Kluwer Academic Publishers incorporates
the publishing programmes of
D. Reidel, Martinus Nijhoff, Dr W. Junk and MTP Press.

Sold and distributed in the U.S.A. and Canada
by Kluwer Academic Publishers,
101 Philip Drive, Norwell, MA 02061, U.S.A.

In all other countries, sold and distributed
by Kluwer Academic Publishers Group,
P.O. Box 322, 3300 AH Dordrecht, The Netherlands.

CONTENTS

PART 5.   LOGIC AND PROBABILITY IN QUANTUM MECHANICS

PART 6.   HISTORICAL DEVELOPMENTS OF THE EINSTEIN-BOHR CONTROVERSY

PREFACE

For three days in April of 1985, Cesena (Italy) was the scene of a national conference which was convened, by the Assessorato alla Cultura of this town under the auspices of the Società Italiana di Logica e Filosofia delle Scienze (SILFS), in order to celebrate two historical milestones: the centenary of the birth of Niels Bohr, who was to become the leader of the orthodox, or Copenhagen, interpretation of quantum theory, and the fiftieth anniversary of the publication of the most influential challenge to this interpretation which was contained in the well-known paper coauthored by Einstein, Podolsky, and Rosen.

The proceedings of the Cesena meeting, which are collected in the present volume, are intended to provide an exhaustive and panoramic view of the most recent investigations carried out by Italian scientists and philosophers engaged in research on the foundations of quantum physics. What emerges is a critical review of, and alternative approaches to, the orthodox interpretation of the Copenhagen school.

Enrico Fermi, the acknowledged founder of modern Italian physics, early on expressed doubts about the final validity of the Copenhagen interpretation, criticizing its "tendency to refrain from understanding things." But sustained interest in the study of conceptual problems in microphysics arose in Italy only around the mid-sixties, after creative thinkers overcame the mental barriers created by the neo-idealist philosophy of Benedetto Croce and Giovanni Gentile, which viewed the Copenhagen thesis regarding the insuperability of quantum paradoxes and contradictions as the surest grounds for renouncing a scientific world view. The last twenty years since then have seen remarkably perceptive Italian works of research on the quantum-mechanical theory of measurement, the axiomatization of the quantum formalism, Bell-type theorems, and realistic local theories, all of which have greatly contributed to the renewed international debate that surrounds the foundations of physics and is aimed at understanding the nature and clarifying the origin of quantum paradoxes.

The editors owe a debt of gratitude to the institutions and individuals whose labor and cooperation made the Cesena conference possible. In particular, we should like to express our warmest thanks to Professor Alberto Pasquinelli, President of the SILFS, to Dr. Giordano Conti, assessore alla Cultura del Comune di Cesena, and to Dr. Franco Pollini and Professor Franco Selleri, members of the organizing committee of the conference.

Gino Tarozzi
Alwyn van der Merwe

INTRODUCTION. THE ITALIAN DEBATE ON QUANTUM PARADOXES

G. Tarozzi

Institute of Philosophical Sciences and Education
University of Urbino, Italy

1. THE REACTION AGAINST THE NEO-IDEALIST PHILOSOPHY OF PHYSICS

The debate in Italy on the foundations of quantum mechanics and the discussion of the conceptual problems raised by this theory began, in relatively recent times. It was only in the late sixties that Italian physicists and epistemologists started to discuss, at a considerably high level of theoretical awareness and formal elaboration, the issues connected with the analysis of physical theories, the critical understanding of their basic concepts and principles, the philosophical implications of these concepts and principles, and the clarification of the logical functions of theories, of the open questions that they are expected to be able to solve, of the new conceptual problems they create, of the relationship between different theories and the question of their compatibility.

The main reason for this late start can be attributed to the ideological climate created by the neo-idealist philosophy that dominated Italian culture for the entire first half of this century and even later.

The strongly negative influence of idealism on the research into the methodology and philosophy of science has been widely discussed and universally acknowledged in the case of formal logic and the foundations of mathematics. It is well known that Benedetto Croce, leading exponent, along with Giovanni Gentile, of Italian neo-idealism, disparagingly dismissed mathematical logic as "something to be laughed at .... when regarded as *a science of thought* .... truly worthy of the minds that created it," in his *Logica come scienza del concetto puro* [1].

As to the foundations of physics, the situation was more complex. The neo-idealist philosophers, though they had a very restrictive attitude to the cognitive value of science, did not deny its practical value and stressed its essentially pragmatic nature. And it was physics that, more than any other empirical science (especially in the light of the fundamental early 20th century discoveries on the structure of atomic phenomena), seemed capable of satisfying the previous requirements of practical utility and pragmatic value.

1

G. Tarozzi and A. van der Merwe (eds.), The Nature of Quantum Paradoxes, 1–50.
© *1988 by Kluwer Academic Publishers.*

From this perspective, experimental studies into the applications and technological impact of physical theories were promoted in Italy, whereas the research into their conceptual foundations, regarded by Croce as "pseudoconcepts" absolutely devoid of universality and by Gentile as too abstract and dogmatic, did not receive the slightest attention.

This was the principal cause that led to the open questions of quantum physics being almost completely neglected for a long time in Italy, while it was being heatedly debated in other European countries and in the United States.

The most critically discerning neo-idealist philosophers did not restrict themselves to denying the cognitive value of physical sciences; they also focussed their attention on the paradoxes and conceptual problems of microphysics. In these contradictions and unsolved questions, they saw the most evident demonstration of the limits of scientific knowledge and of its inadequacy to grasp a more complex and deeper reality than that of the natural world and, moreover, the proof of the need to appeal to this spiritual reality for a proper understanding of the same natural phenomena. The basic thesis of the Copenhagen interpretation, assuming the centrality of the role of the human observer, which made it possible to legitimate the insuperability of the wave-particle dualism and to avoid the investigation of a consistent physico-mathematical solution to the measurement problem, appeared thus as a confirmation of one of the fundamental assumptions of idealism according to which the properties of the physical world do not exist independently of the knowledge that we, as beings endowed with a mind and consciousness, have of them.

In de Ruggiero's treatise of history of philosophy, published in Italy by Laterza in the thirties, the "new atomism" is viewed as decisive proof of the active role of the human mind within the physical world and of the definitive overcoming of the mechanistic conception of classical physics in favour of an explicitly idealistic and finalist perspective:

"It is not really the indeterminacy of the atom as such but rather its capacity to assume diverse determinations according to the nature of the energy field that gives cause for reflection of an idealistic nature. We perceive the presence of levels of action influencing from the top down the structure of the physical world, of teleological anticipations, ordinating and distributing the material components. While in the old atomistic conception, the mind was completely outside the description of nature and could not be introduced into it except through the double miracle of divine creation and human knowledge, in the new conception the mind appears to be more intrinsic to the picture and better distributed in its effect on the various orders and levels of cosmic life. Ancient mechanism did not know history; the mind could be regarded only as *mens momentanea*, i.e. as a negation of those characters which the mind assumes as its most distinctive and which are connected with the names of process and development. The new atomism attributes a wider role to these *ideal* values."[2]

Another very explicit claim as to the idealistic nature of quantum theory can be found in the introduction written in 1934 by Giovanni Gentile, Jr. to the Italian translation of a book by the English astrophysicist James Jeans. In this introduction, Gentile, Jr., son of the more famous neo-idealist philosopher and politician and together with Enrico Fermi and Ettore Majorana, one of the first university professors of theoretical physics in Italy, expresses his full agreement with the Copenhagen interpretation of quantum mechanics and a critical attitude towards the point of view of the main founders of the theory, such as Planck and Einstein, whom he accused of naive realism:

"The person who goes in for the study of contemporary scientific thought will be able to distinguish, as is well known, two trends of opinion: one of the physicists searching for the *ultimate reality* (and this group includes men like Planck and Einstein); the other of physicists who have a more critical consciousness of the methods of scientific work and have decidedly abandoned the traditional approach of classical physics. The latter physicists do not, to tell the truth, like *theorizing*. They are rightly content to have *faith in their ideas* and to work with their methods."[3]

Gentile shares the thesis, implied in the Copenhagen interpretation, restricting the concepts of physical theories to the only directly observable quantities. Nevertheless, unlike Heisenberg or other exponents of the Copenhagen school, who always maintained a strictly operationistic and phenomenistic point of view requiring a direct dependence of the theoretical description on empirical procedures, the Italian physicist considers the identification of the observables with the only legitimate scientific concepts as a confirmation of the subjectivistic conception of immanentistic idealism:

"In relation to the scientific point of view on the problem of reality, Jeans, following the ideas of Heisenberg, summarizes the situation in this way: There is *thought*, there are *then*, *sense data*, from which after the necessary corrections have been made, we go back to the *observables*, i.e. the knowledge of the quantities which are directly observable. These can then have behind them -- as is natural to suppose -- other unobservable quantities: the world beyond our experiences. What really exists, however, what creates reality, is only the observable and, according to Heisenberg, only this. The other reality, that unobservable, can be object of faith and hypothesis, according to taste, but not of science. Science must be concerned only with observables, all the rest is chimera or a mathematical framework. ... For my part, I wish only to emphasize that the observables account for the necessity *immanent* in thought that subject is always present to object. Otherwise, how could there be science? ... In the conclusion, Jeans raises the question whether *science supports idealism* and finds many reasons for answering in the affirmative."[4]

The inhibiting effect of neo-idealistic philosophy on the research into the foundations of physics was thus exerted in two main directions: Firstly it discouraged theoretical research in favor of the study of experimental applications, in the name of the merely practical value of scientific activity; secondly, it appealed to some philosophical consequences of the Copenhagen interpretation to support subjectivism. The epistemological status of the orthodox interpretation could be used, by legitimating the paradoxes of microphysics, to confirm the idealist conceptions, since even the most advanced scientific theory seemed forced to recognize the impossibility of eliminating the observer viewed as a subject endowed with consciousness, from the context of its theoretical and empirical laws.

After the second World War, the development of epistemological and methodological studies according to the perspective of logical empiricism's scientific philosophy due mainly to Ludovico Geymonat and his school, led to a gradual abandonment of anti-scientific, idealistic metaphysics. The work of Geymonat, however, did not have as its result the conclusive overcoming of idealism, since it was prevalently concerned with opposing Croce's and Gentile's philosophy but, in its effort to reconcile logical empiricism with dialectic materialism, completely endorsed Hegel's conception of dialectics. And it is by appealing to this very conception, as we shall see in the next section, that Geymonat and his school have proposed to solve the main conceptual problems of quantum theory.

A deeper, more penetrating criticism of idealistic philosophy was made in the sixties by Alberto Pasquinelli, who pointed out that the same logical possibility of a theory of scientific knowledge was absolutely incompatible with Hegel's idealistic metaphysics. His book, *Nuovi principi di epistemologia* published in 1964, marked the real superseding of neo-idealistic philosophy in Italy and the first systematic exposition of a scientific philosophy consistent with the logical empiricist conceptions of Rudolph Carnap, whom he had studied under at Chicago university. As its title suggests, this publication was about general epistemology, but it was more than this, and although it did not deal specifically with questions in the philosophy of physics, it nevertheless represented a methodological reference point in the development of the research into the foundations of quantum theory for at least two reasons.

The first is the priority attributed to the empirical over the formal sciences, contrasting the standard view of the Geymonat school, which saw in mathematics the paradigmatic model of scientific knowledge. Maintaining the instrumental or "auxiliary" nature of mathematics -- starting from the thesis of the analyticity of the laws of logic and appealing to the reducibility to these latter of mathematical laws, implied by logicism -- Pasquinelli thus completely identifies epistemology with the critical analysis of empirical knowledge provided by physical or "real" sciences. The second reason is the in-depth discussion of the epistemological contribution of Albert Einstein, often accused, by the supporters of the orthodox interpretation, of deterministic mechanism, naive realism, and generally of scanty philosophical awareness. Pasquinelli dwells on more than one occasion not only on the philosophical implications of the scientific work but also on the specifically critical and methodological contributions of Einstein, whom he considers "the man of science who elaborated these issues in perhaps the *most open-minded and deep manner*, arriving at a formulation of an explicit 'epistemological creed' totally in accordance with his own determinant scientific work."[5]

The assertion of scientific over metaphysical philosophy created the basis for a critical discussion of the epistemological foundations of quantum mechanics. The need for a revision of its conceptual foundations turned into a criticism of the justificationistic philosophy of the orthodox interpretation.

The reaction of Italian physicists against this interpretation took three main directions corresponding to the principal theoretical presuppositions of the Copenhagen school, namely:

- the contradictory solution given through the complementary principle to the wave-particle dualism, conceived on the basis of this principle as an insuperable dilemma;
- the inadmissible interference of the mind or consciousness of the human observer with the laws of physics, implied by the von Neumann theory of measurement;
- the myth of the completeness of quantum theory based on the unconditional acceptance of the von Neumann theorem (generalized in those very sixties by Jauch and Piron) against the same mathematical possibility of any causal generalization of the existing theory through the introduction of additional parameters (or hidden variables).

In this process of critical revision of the orthodox interpretation, which began in Italy in the sixties three fundamental dates can be singled out.

In 1962, Daneri, Loinger, and Prosperi proposed a first version, which was to be subsequently improved of their realistic theory of measurement [6] as an alternative to the subjectivistic theory of von Neumann. Unlike the latter, which appealed to the consciousness of the observer, conceived of as an entity not describable by the laws of physics but capable at the same time of interfering with their course, the theory of measurement of the Italian physicists sought to provide, as we shall see in section 6., a consistent physical explanation of the measuring process in terms of a physical interaction between the measured micro-object and the measuring macro-instrument.

In 1969, Franco Selleri advanced his realistic interpretation of the wave function founded on the assumption of the existence of physical objects devoid of the basic properties possessed by any other physical systems, such as energy and momentum, and characterizable only through relational properties.[7]

In 1970, Capasso, Fortunato and Selleri discussed the epistemological significance of superseding the theorem of von Neumann and the new prospects opened in those very years by the discovery of the Bell theorem and the consequent possibility of discriminating experimentally between local hidden-variable theories and quantum mechanics.[8]

At the same time, one is faced in the philosophical context, with Evandro Agazzi's critical re-examination of the rigidly operationistic and phenomenistic conception of physical theory, which leads to some of the basic assumptions of this conception being overcome. In his book *Filosofia della Fisica*, Agazzi proves the artificious nature of the distinction between observative and theoretical terms and shows the need to abandon the strictly operationistic interpretation of scientific concepts, reducing them to their mere measuring procedures, stressing how this interpretation, as we shall see in section 2, lies at the origin of the most serious contradictions of the orthodox interpretation of quantum mechanics. The basic idea on which the overcoming of the operationistic point of view is based can be summarized in Agazzi's statement that "a physical concept does not denote a *single* operation (or a *single set of operations, arising from an operation but which is not identifiable with it*)."[9] This idea, later formalized by the author in a proposal of intensional semantics[10], was to be developed by M.L. Dalla Chiara and G. Toraldo di Francia in their formal approach to physical theories.[11]

2. THE EPISTEMOLOGICAL OBJECTIONS TO COMPLEMENTARITY

One of the most penetrating critical analyses of the epistemological status of the complementarity principle, which constitutes, as is well known, the official interpretation of the wave-particle dualism put forward by the Copenhagen school, is contained in the book by Agazzi, just mentioned. This author highlighted the presence of two interpretations of this principle which are profoundly different from a conceptual point of view but nevertheless often are confused by the supporters of the Copenhagen interpretation.

The first, due mainly to Pauli but then *de facto* followed by many others, views complementarity simply as a synonym for the indeterminacy relations, i.e., as a prohibition of the simultaneous use of two classical concepts like position and momentum or time and energy. According to this latter meaning, the principle of complementarity is prior to Heisenberg's indeterminacy relations and corresponds essentially to the original formulation of this principle as a renunciation of the possibility of a description of microphysical phenomena which was at once a causal one -- where causal to Bohr always means "in accordance with the fundamental physical conservation laws of energy and momentum" -- and a space-time one. In both cases -- whether the use of two classical concepts is prohibited or whether the incompatibility of two classical descriptions is maintained -- one is faced with a *restrictive* interpretation which does not make any attempt to clarify the dualistic nature of quantum phenomena, restricting itself to pointing out the inadequacy of the concept of classical particles as objects for which the two properties of position and momentum can be defined simultaneously or to which a certain precise value of the energy at a well-defined time can be attributed.

The second interpretation, more closely, according to Agazzi, connected to Bohr's idea, views complementarity as the necessity to appeal, in order to explain microphenomena, to two different *classically incompatible* "images" or descriptions, the undulatory, on which Maxwell's electromagnetism was based, and the corpuscular which Newtonian mechanics was founded on. According to this perspective, complementarity is therefore synonymous with the wave-particle dualism or, rather, *dilemma* because, on the one hand, the recourse to both images seems indispensable to explain the behavior of micro-objects and, on the other, these images are experimentally incompatible.

Now, while the first interpretation of complementarity as indeterminacy seems to be quite obvious and not to produce particular problems, the second, related to the wave-particle dualism, gives rise to considerable epistemological difficulties. In the latter case, not only is the necessity to have recourse to both images or descriptions maintained, but so is the contradictoriness deriving from the application of both descriptions to the same physical situation.

At the root of the problem, there is for Agazzi the insufficient emancipation of the Copenhagen interpretation from the classical concepts, due to the physicists of this school following too strict a version of operationism which led them to identify completely the scientific concepts with their measuring procedures. The necessity to apply classical concepts to microphenomena was thus made inevitable by the fact that, in order to investigate the properties of micro-objects, one makes use, in one's observations or measurements, of macroscopic instruments describable by means of classical laws and concepts.

Agazzi, however, shows how the use of the instruments does not determine in the strict sense the investigated physical concepts, pointing out, for example, that electromagnetic phenomena can be and are *de facto* often investigated through mechanical instruments without this fact preventing the acknowledgement of the peculiarity of these phenomena and the introduction of new concepts such as that of charge, current, and induction, which are all clearly nonmechanical even if revealed on the basis of their mechanical effects registered by mechanical instruments.

To remove these difficulties, he proposes his thesis of the

contextualistic nature of the meaning of physical concepts. According to this thesis, which is discussed in detail in Agazzi's contribution to this book, one single physical concept, such as that of "material particle," is the object of different characterizations depending on the different levels or contexts in which it appears: Thus, in the classical context it is viewed as an object endowed with *both* well-determined position *and* momentum, while in the quantum context, it loses the contemporaneous possession of these properties but assumes new properties such as spin. The difficulties of the Copenhagen interpretation regarding the wave-particle dualism might find a solution if one assumes that the classical concepts, considered at a formal level, appear as the element of a new semantic combination in which the contradiction disappears since it is not formally connected to the concepts themselves but to their classical denotation.

Agazzi believes, however, that his thesis does not provide a conclusive solution to the problem of the nature of the microphysical object and stresses the need for the introduction of concepts that are *new* not only because they are the result of a new combination of classical concepts but also because they are able to replace in whole or in part those classical components with *something really new.*

In the very year in which he was making these epistemological reflections on the nature of the wave-particle dualism, an interpretation of the wave function based on the introduction of a new concept and destined perhaps, as we shall see in Section 5, to solve the Copenhagen dilemma of the dual nature of quantum phenomena was advanced in Italy by Franco Selleri.

Another detailed philosophical analysis of the complementarity principle may be found in Ludovico Geymonat's *Storia del pensiero filosofico e scientifico,* in a chapter written by Silvano Tagliagambe[12], and in the introduction, also by the latter author, to the book *L'interpretazione materialistica della meccanica quantistica* discussing the relationship between physics and philosophy in the USSR.[13]

Tagliagambe discusses complementarity from a mainly historical point of view, proposing to reconstruct its origin in Bohr's thought and the subsequent debate in the USSR that led to the attempt at a materialistic reformulation of this principle. The author's analysis of the influence on Bohr's early philosophy of Hoffding's conceptions, though it did not contain any basic novelty with respect to the theses expressed in Jammer's book *The Conceptual Development of Quantum Mechanics,* shows in greater detail the very close connection between the qualitative dialectics of existentialism and Bohr's view of the wave-particle dualism as an insuperable dilemma due to the limitations of our macroscopic categories being incapable of enabling us to fully understand the properties of the world of micro-objects.

Of special interest is also Tagliagambe's discussion of the relationship between Bohr and the Soviet physicist Fock, who attempted to disassociate complementarity from the negative connotations of existentialistic dialectics and to overcome the wave-particle dilemma by reinterpreting it as a consequence not of limitations in principle which our mental categories are subjected to but of the contingent limitations due to the unavoidable use of macroscopic instruments. Fock started from Bohr's observation that the process of interaction between the measuring apparatus and the measured object cannot be eliminated or arbitrarily reduced, because every interaction always implies the exchange of at least one quantum h, and concluded from this that the experimental physicist observes only a global process involving both the instrument and the micro-object and never the *intrinsic* properties of the latter. It is therefore possible that some instruments make the micro-object appear as a wave, whereas others make it appear as a particle: The undulatory and corpuscular properties of the microscopic world would thus be *ways*, mediated by the measuring apparata, by which microscopic systems *reveal themselves* at the macroscopic level.

This is the concept of *relativity to the observational means*, already partially anticipated by the logical empiricist physicist P. Franck in his attempt to reinterpret complementarity in terms of our instrumental limitations. Fock and the Soviet philosophers of science, followed in this operation by Tagliagambe and Geymonat, intend to provide a clarification in the positive sense of the nature of dualism by replacing the existentialistic contraposition of waves and particles with a *synthetic* view whereby the thesis and the antithesis, i.e., in this case, wave and particle, would be overcome by means of a new, but not well-defined concept. According to Tagliagambe, it is precisely this approach to the problem that enables us to solve "one of the *presumed paradoxes* of quantum mechanics, that is, that of the corpuscular-undulatory dualism of microphenomena."[14]

The concept of the relativity to the observational means is also applied by Tagliagambe to what he believes is the solution to the Einstein-Podolsky-Rosen (EPR) paradox, which can be briefly summarized by the statement that properties of micro-objects do not exist but only relational properties between micro-objects and the instruments used for observing them. The relational conception of the properties of the quantum state typical of the orthodox interpretation is in this way extended from the explanation of the ordinary measuring processes, whereby it can be in some way justified as a consequence of the *practical* inseparability between measuring instrument and measured object, to the explanation of the EPR cases by assuming the existence of a not very plausible inseparability *in principle* of the instrument from an an unmeasured object located at an arbitrarily large distance from it.

The proposing in Italy in the mid-seventies of the Soviet philosophy of physics seems to us a doubtful cultural operation, and one that was, at best, behind the times, as we have stressed elsewhere.[15].

The defence by the Soviet physicists and epistemologists of some of the materialistic and also realistic conceptions, even though of a typically metaphysical realism, had already been endorsed by Valerio Tonini who, immediately after the second World War, published and discussed papers by these Soviet authors in his review *La Nuova Critica*. In this period, when the Copenhagen "paradigm" prevailed almost everywhere without meeting any opposition, the rejection of the most dogmatic and subjectivistic implications of the orthodox interpretation played instead an essential role in keeping the debate on the foundations of quantum mechanics open.

This substantially positive opinion of the Soviet philosophy of physics has recently been asserted once again by Tonini according to whom

> "in spite of some ideological and polemic excesses, it must be admitted that the views of the Soviet scientists, as a whole, provides us with a precise and legitimate epistemological approach, valid not only in a strictly materialistic context but also from a general epistemological point of view, being *classically realist* ..."(16)

A similar attempt to conciliate quantum mechanical description and a realist conception was also made in those earlier years by Antonio Pignedoli, who was one of the first Italian physicists to oppose the subjectivistic connotations of the Copenhagen philosophy, stressing how they are in contrast with the very nature of scientific research, since they would lead to an unacceptable form of solipsism:

> "... the particle's being cannot coincide with its being perceived because then, no probabilistic law would have general validity either, but *every observer would create his own probabilistic physics.*"(17)

This realistic view lies at the basis of Pignedoli's interpretation of the indeterminacy principle according to which the prohibition expressed by this principle that a micro-object cannot simultaneously possess well-defined position and momentum has purely gnoseological significance and is inapplicable at the ontological level since it refers only to a limitation of our knowledge and not to the existence of physical situations:

> "It nevertheless appears rather arduous to admit, for example, that a photon, between the elementary act of emission and the one of absorption, has not followed a given sequence of states ...: these states, even if they are not determinate and not physically determinable, they must be thought of as philosophically existent."(18)

Some philosophical objections to the metaphysical nature of the Soviet (and Geymonat-Tagliagambe) "materialistic" interpretation of quantum mechanics have already been discussed by the present author(19). There is, however, a more serious criticism regarding the problem of the physical foundation of the principle of relativity to the observational means. This principle, since it establishes in rather precise terms the impossibility of constructing an experimental device capable of revealing both the undulatory and the corpuscular properties of micro-objects at the same time, constitutes the only non-metaphysical assumption of such an interpretation, and, precisely because of its empirical nature it can be refuted by the following *gedanken* experiment.

Let us consider the physical situation of the double-slit experiment where a very weak source, instead of quanta of light as in the standard case, emits *massive particles* such as electrons which, after passing through a first screen provided with two slits are detected by a second screen behind the slits. This experiment represents one of the starting points of several formulations of the wave-particle dualism, as we shall see in detail later: As is known, on the second screen is observed the classical interference pattern typical of the undulatory phenomena which, however, vanishes as soon as one tries to attribute to the micro-object its corpuscular individuality, by putting a detector at each slit to establish which one the particle passes through.

The interpretation of the experiment, in the light of the concept of the relativity to the observational means, is that the micro-object reveals *either* corpuscular or undulatory properties, depending on the measuring device used, i.e. *double slit* either *with* or *without* detector at the slits, but *never both* properties at the same time. Yet this is precisely what our experiment does show.

Let us suppose that we replace the ordinary detector at each slit with a "semi-transparent" particle revealing apparatus which has the characteristic of detecting the particle with a probability K , disturbing it or not detecting and allowing it to pass through without perturbation with a probability $1-K$. Let us suppose also that on the second detecting screen there is a battery of counters each of which is connected to both semi-transparent apparata in such a way as to perform coincidence measurements that make it possible to detect the passage of the particle through one of the two slits and then the arrival of the particle on the second screen.

In this way, there are two series of events: the first with probability $1-K$, where no coincidence occurs and the micro-object is revealed only on the second screen and contributes in this case to the creation of the interference pattern characteristic of the undulatory image; the second, with probability K, where coincidences do occur and the micro-object arrives on the second screen after its passage through a well-defined slit has been established but without, in this latter case, producing interference as happens for any particle. One is thus faced with a single experimental situation in which the undulatory and corpuscular properties of micro-objects reveal themselves at the same time and are, moreover, perfectly distinguishable.

In our view, this experiment does not necessarily contradict Bohr's complementarity principle, since it is not able to isolate purely undulatory properties of micro-objects, as the experiments on quantum waves, which we shall discuss in Section 5, attempt to do, but constitutes a conclusive objection to all the proposals to relativize to our means of observation the undulatory and corpuscular properties of micro-objects which, as we have just seen, can instead, at least in principle, be measured contemporaneously with a single experimental device.

3. QUANTUM LOGIC

In view of the serious conceptual difficulties of the orthodox interpretation of the wave-particle dualism, as well as the inadequacy of the reformulation of the complementary principle in terms of the concept of relativity to the observational means, one is faced with two alternatives.

The first, as we have already seen, consists in the search for new concepts representing a real and radical emancipation from the classical concepts of particle and wave on which to base a different and more satisfactory interpretation of the quantum mechanical wave function. This solution will be discussed in Section 5.

The second alternative is based on the rejection of the dualistic nature of micro-phenomena in favor of a strictly corpuscular interpretation -- the insuperable empirical objections to a purely undulatory interpretation, that only Schrödinger historically attempted to uphold are well known -- where some formal presuppositions about the logical structure of quantum theory have been modified. This kind of solution, as we shall see better later, can be viewed within the framework of an attempt to extend, to the problems of the foundations of physics, the approach that proved highly successful in the field of the foundations of mathematics, making it possible to solve the main paradoxes of classical mathematics, but which was to prove not quite so satisfactory when applied to the paradoxes of microphysics. It is not by chance that, in the introduction to their book on the formal analysis of physical theories, Dalla Chiara and Toraldo di Francia, the leading Italian exponents of the quantum logical approach, along with Beltrametti and Cassinelli, assert that their research program can be set in the framework of attempts to solve one of the twenty-three problems, listed by Hilbert during the international Brussels congress of mathematicians in 1900, which was formulated as follows:

> "Durch die Untersuchungen über die Grundlagen der Geometrie wird uns die Aufgabe nahe gelegt, *nach diesem Vorbilde diejenigen physikalischer Disziplinen axiomatisch zu behandeln, in denen schon heute die Mathematik eine hervorragende Rolle spielt* ..."[20]

This reference to the foundations of geometry is particularly significant also because both the quantum-logical approach and the quantum-probabilistic one regard the abandonment of a fundamental law of classical logic or probability as a process perfectly equivalent from a conceptual point of view to the disavowal of the postulates of parallels in the nonEuclidean geometries adopted by the relativistic theories.

The starting point of these formal analyses corresponds

therefore to the attempt to give an empirical foundation to their rejection of at least one basic law of logic or of probability calculus, demonstrating that, just as Euclid's parallel postulate is no longer valid for the great distances of astrophysics, so one or more logical or probabilistic law(s) ceases to be valid when applied to the very small distances of microphysics.

Such a possibility of founding on empirical grounds a new nonclassical logic or probability calculus has usually been based on the analysis of the famous double-slit experiment, a particular variant of which we have already discussed. In this peculiar physical situation, which, as Feynman emphasized, is formulated in such a manner as to contain all the mysteries of quantum theory, one can easily show, by a simple formal analyis, how one can reach paradoxical conclusions starting from three different kinds of hypotheses:
(a) the validity of the axioms of classical probability calculus;
(b) the validity of the laws of classical logic;
(c) the logical individuality of micro-objects corresponding to the rejection of dualism in favor of a merely corpuscular interpretation.

If A and B are the slits of the first screen and X the second screen where the particles are revealed, X_p and A_p (B_p) will correpond, respectively, to the events "the particle has arrived in a region of the revealing screen X" and "the particle has passed through the slit A (B)."

Now, the hypothesis c) implies the validity of the two following relations:

$$A_p \vee B_p \tag{1}$$

$$\neg (A_p \wedge B_p) \tag{2}$$

For relation (1) we can write

$$X_p = X_p \wedge (A_p \vee B_p) \tag{3}$$

which, on application of the principle of the distributivity of conjunction with respect to disjunction, implied by the hypothesis (b), becomes:

$$X_p = (X_p \wedge A_p) \vee (X_p \wedge B_p). \tag{4}$$

Moreover, since A_p and B_p denote disjunct events because of relation (2), one obtains for the additivity postulate of the probability of these events, implied by (a):

$$P(X_p) = P(X_p \wedge A_p) + P(X_p \wedge B_p), \tag{5}$$

in contrast with the predictions of quantum mechanics confirmed by experimental results. According to this theory, the wave function that describes the particle in the region x of the second screen is given by:

$$\psi(x) = \psi_A(x) + \psi_B(x) \tag{6}$$

where $\psi_A(x)$ and $\psi_B(x)$ are complex quantities whose squared modulus corresponds, respectively, to

$$|\psi_A(x)|^2 = P(X_p \wedge A_p) \; ; \; |\psi_B(x)|^2 = P(X_p \wedge B_p) \tag{7}$$

while $|\psi(x)|^2$ represents similarly the probability $P(X_p)$ that, in this case, will therefore be:

$$P(X_p) = |\psi_A(x) + \psi_B(x)|^2 = P(X_p \wedge A_p) + P(X_p \wedge B_p) + I_t, \tag{8}$$

where I_t indicates the known interference term given by:

$$I_t = \psi_A{}^*(x) \; \psi_B(x) + \psi_B{}^*(x) \; \psi_A(x), \tag{9}$$

whose presence allows the experimental data to be reproduced.

A similar analysis of this microphysical paradox was proposed by Maria Luisa Dalla Chiara[21] who stressed how the quantum-logical solution consists in the abandonment of one of the hypotheses of kind (b), namely, the distributivity principle of classical logic. The rejection of this principle would make it impossible to pass from relation (3) to (4), avoiding in this way the contradictory conclusion of (5).

The double-slit experiment would thus represent the *experimentum crucis* proving the falsity of a principle of ordinary logic generally considered to be a *priori* and making necessary a conceptual revolution comparable with the physical applications of non-Euclidean geometries. From this perspective, logic should therefore be considered, in the same way as geometry, an empirical science. Just as Euclidean geometry proved to be a special case of the more general, non-Euclidean geometries, so Aristotelian logic would be a special case of the more general quantum logic.

While maintaining that objections do not exist in principle to the introduction of a system weaker than that of classical logic, owing to the basically conventional nature of the choice of any formal system, it does not appear to us to be philosophically correct to trace this necessity back to a specific empirical situation as is the double-slit experiment.

It should also be noted that in the previous analysis there is in our view an assumption that is much less plausible than that of the validity of the distributivity principle, namely, hypothesis (c) relative to the logical individuality of the micro-object. Such an assumption means essentially that the micro-object is a classical particle, as clearly appears from the introduction of A_p, B_p, $A_p \vee B_p$, and from the rejection of $A_p \wedge B_p$; if, on the contrary, the micro-object were, according to Schroedinger's view, a purely undulatory phenomenon, $A_p \wedge B_p$ should of course always be taken into consideration. Nevertheless, it is precisely the undulatory aspect which is usually totally denied by quantum logicians: "The world consists of particles, not of waves nor of wave-particles," according to a well-known statement by Putnam, one of the most authoritative exponents of this approach to the foundations of physics.

The main researches in the field of quantum-logic, however, are due to the Italian physicists, E. Beltrametti and G. Cassinelli, who have made a fundamental contribution to the axiomatization of quantum theory.

These authors are not looking for a direct experimental foundation for quantum logic but restrict themselves to considering the appeal to it as a necessity imposed by the mathematical structure of the theory. They conclude their basic review article on the logical and mathematical structures of quantum mechanics by stressing that

> "important issues such as ... the role to be acknowledged to quantum logic, with the more extreme attitudes, which, on the belief that no *a priori* universally valid logic can exist, consider quantum logic as a new, autonomous logic supplanting, on the grounds of experience, the classical logic, go beyond the purpose of this paper."[22]

Though we cannot here go into a detailed analysis of the important contribution constituted by the works of these authors, we would like to discuss briefly a limitation to this formulation of quantum axiomatic, which is that this axiomatic system is not able to account for the superposition principle, one of the two basic formal features that distinguish the mathematical structure of quantum mechanics from the structure of classical and relativistic theories. As has been stressed by Dirac,

> ".. quantum mechanics is based on two basic principles, one negative in character, the indeterminacy principle, and one positive in character, the superposition principle. Besides these two principles, there is the meaning of the squared modulus of the wave-function as probability density."[23]

On the basis of the definition of superposition given by Beltrametti and Cassinelli, we say that

> "the state β, belonging to the set S of the states of any given logic, is a superposition of the states of the subset S of S, if all the propositions giving with certainty the result *yes* for all the states of S behave in the same way for the state β."[24]

This definition is similar to that of Varadarajan, who had, however, defined the concept of superposition for the false propositions, that is, for those that give with certainty the result *no*.

Let us now consider the quantum mechanical description of two correlated physical systems given by the singlet state

$$| \psi_S > = (1/\sqrt{2}) \{ |u_1^+ > |u_2^- > - |u_1^- > |u_2^+ > \} \tag{10}$$

corresponding to the superposition of the two states

$$|u_1^+ > |u_2^- > \quad \text{and} \quad |u_1^- > |u_2^+ > \tag{11}$$

to one of which $| \psi_S >$ is reduced after the process of measurement.

If we now consider the proposition "The observable J^2 has not a well-defined value," we can see that such a proposition is true for each of the states (11) but false for the state (10), for which J^2 always assumes the exact value *zero*, as we shall see in more detail later.

This objection could not be overcome even if one tried to define, according to Varadarajan, the superposition for the propositions having with certainty the result *no*. In this case, if we consider the proposition "J^2 is equal to zero," we can easily observe that it is false for states (11) but true for their superposition (10).

The inadequacy of this quantum logical definition of the concept of superposition appears related to the fact that the superposition of two more quantum states does not constitute only the *ordinary set union* of the properties of the physical substates which compose it but something qualitatively new, precisely because of this positive character of the superposition principle stressed by Dirac. We cannot, however, neglect to mention the fact that the irreproducibility of the superposition of the states of type (10) could be viewed not so much as a formal limitation of Beltrametti's and Cassinelli's quantum logical approach but rather as an indicator of the anomalous character of this kind of superposition of non-factorable states, which, as is known and as we shall see in Section 8, lie at the roots of the most serious paradoxes of quantum theory.

4. QUANTUM PROBABILITY

As emerged from the above discussion of the double-slit experiment, another possibility of avoiding the contradictory conclusion expressed by relation (5) is provided by the abandonment of the postulate of additivity of probabilities, implied by the hypothesis (a). This, for example, is the point of view adopted by Jordan, Fine, and Feynman, who proposed to replace this postulate with the typically quantum mechanical one of the additivity of the probability amplitudes. It should be noted, however, that for the authors just mentioned the rejection of the additivity postulate appears more like a consequence of the assumption of the nonvalidity of the relation (2) -- typical of the dualistic interpretation which does not accept the idea that the micro-object passes through one slit only, as is assumed in the merely corpuscular interpretation -- than an indication of the need to modify the basic rules of probability calculus.

A different solution in terms of a rigorously defined concept of quantum probability has been proposed by Luigi Accardi[25] according to whom relation (5) is not yet contradictory *per se* since it contains values of joint probabilities that cannot be measured experimentally in microphysics. The paradox can easily be reformulated, however, by replacing the joint with the conditional probabilities, which always have empirical significance and which, according to an idea that Accardi attributes directly to Bohr, would represent the fundamental object of quantum theory.

The replacement can be made through recourse to the Bayes identity, which, in our case, is expressed by

$$P(X_p/A_p) = P(X_p \wedge A_p)/P(A_p); \; P(X_p/B_p) = P(X_p \wedge B_p)/P(B_p) \qquad (12)$$

where $P(X_p/A_p)$, $P(X_p/B_p)$ denote the conditional probability that a micro-object arrives at X, when it has passed through A or B; $P(X_p \wedge A_p)$, $P(X_p \wedge B_p)$ represent the joint probabilities of the arrival of the particle in X and of its passage through A or B; and $P(A_p)$, $P(B_p)$ indicate the *a priori* probability that the micro-object passes through the slit A or B, respectively, satisfying the relation: $P(A_p) + P(B_p) = 1$.

From relations (12) and (5), we obtain the theorem of composite probabilities, given by

$$P(X_p) = P(A_p) . P(X_p/A_p) + P(B_p) . P(X_p/B_p), \qquad (13)$$

which is in open conflict with experimental data.

Now, since the passage from (5) to (13) is based on the application of the Bayes identity, it is the latter postulate which Accardi proposed to eliminate. The abandonment of the Bayes axiom, defined as "the hidden postulate of classical probability," implies the introduction in the statistical description of micro-systems of non-Kolmogorovian models which are the object of a systematic and ample discussion in Accardi's contribution to the present volume.

According to the latter author, the abandonment of the Bayes postulate is preferable to the rejection of the distributivity principle of quantum logic. He rightly maintains the weak foundation of quantum logical presuppositions that micro-objects do not follow the laws of classical logic ·only when they are not observed, which would lead one to base the rejection of distributivity on the grounds of unobservable physical situations. In the case of the refutation of the Bayes identity, one would be faced with a completely different situation, since it could be "proved only in the terms of experimentally measurable quantities." Accardi's claim that the adoption of his concept of quantum probability "is not a question of postulates" but of experiments and could therefore represent a choice "free of interpretations" does not seem justified to us.

We have just seen how, in his analysis of the double-slit experiment, the author holds that relation (5) that can be obtained without recourse to the Bayes identity, does not yet give rise to any contradiction, in contrast with the considerations proposed in the previous section. To obviate the contrast between relations (5) and the experimentally confirmed (8), Accardi deliberately avoids representing the joint probabilities $P(X_p \wedge A_p)$, $P(X_p \wedge B_p)$ with the square modulus of the wave-functions $\psi_A(x)$ and $\psi_B(x)$ as established by relation (7). Nevertheless, such a procedure can be justified only on the grounds of an explicitly frequency interpretation of the probability calculus, which corresponds to subdividing the particles arriving on the detecting screen into two ensembles: that of the particles coming from slit A and that of particles coming from slit B, whose cardinalities

$$N(X_p \wedge A_p) \text{ and } N(X_p \wedge B_p),$$ ()

would not be experimentally measurable, but whose relative frequencies

$$P(X_p \wedge A_p), \ P(X_p \wedge B_p),$$ ()

would continue to satisfy the additivity postulate.

In this way, the validity of relation (5) is preserved without, however, it being still possible to write $P(X_p \wedge A_p)$, $P(X_p \wedge B_p)$, considered now as relative frequencies, in the form of $| \psi_A(x) |^2$ and $| \psi_B(x) |^2$, as established by relation (7), since, according to the orthodox interpretation of quantum mechanics, the wave functions $\psi_A(x)$ and $\psi_B(x)$ describe individual physical systems and not statistical ensembles.

Nevertheless, we do not fully understand why, after re-interpreting the individual probabilities $P(X_p \wedge A_p)$ and $P(X_p \wedge B_p)$ in terms of relative frequencies, Accardi does not extend this statistical interpretation to the wave function by indentifying, as for instance in the Popper-Landé-Margenau statistical interpretation of quantum mechanics, his squared modulus with the relative frequency of a statistical distribution of particles and continues, instead, to consider it, in accordance with Born's orthodox interpretation, as the expression of the individual probability to find a given particle in a given space-time region.

A quantum probabilistic approach to the foundations of microphysics would be able, according to Accardi, to solve also the Einstein-Bell contradiction. His solution is based on the attempt to prove that the EPR paradox (even in its modern formulations) has its logical roots in the application of the Bayes postulate to physical situations where it cannot be applied:

"The argument according to which quantum theory should imply the existence of superluminal signals stems from the fact that one has to deduce a conclusion on the physical world ... from a demonstrably wrong matehmatical premise".[26]

To this extent, Accardi has proposed the following quantum mechanical formulation of the Bayes postulate:

QMB - Let 1, 2 be states of a physical system

corresponding to the orthonormal basis (ψ_1, ψ_2) of the quantum mechanical state space of a system. If on an ensemble of systems we know with certainty that the fraction P_1 is in the state 1 and the fraction P_2 is in the state 2, then the statistical description of the ensemble is given by the mixture

$$W = P(1) \; P\psi_1 + P(2) \; P\psi_2 \tag{14}$$

where $P\psi_1$ and $P\psi_2$ denote the rank-one projections onto the states ψ_1 and ψ_2, respectively.

The equivalence of the QMB with Bayes' postulate would be guaranteed by the fact that such a definition is nothing but a transcription of the theorem of composite probabilities (13) in which the conditional probabilities $P(X/1)$ and $P(X/2)$ have been replaced by their quantum mechanical description. On these grounds, Accardi has critically discussed and reformulated a "modern" proof of the EPR paradox given by Selleri and Tarozzi[27], arriving at the conclusion that it contained an implicit appeal to the QMB, and that the rejection of such a postulate could therefore have allowed a reconciliation between quantum mechanics and local realism.

The argument started from Bohm's version of the EPR paradox: A system Σ is given which is capable of decaying into spin 1/2 particles S_1 and S_2. Now if $|u_1^\pm\rangle$ and $|u_2^\pm\rangle$ are spin state vectors of the systems S_1 and S_2, respectively, corresponding to the third component $\pm 1/2$, the state vector for $S_1 + S_2$ is the singlet state $|\psi_S\rangle$ given by relation (10). A very large number N of such decays $\Sigma \Rightarrow S_1 + S_2$ is considered, repeating on each pair of the decay product the following reasoning:

(1) At time t_0 a measurement of the third component σ_3 is performed on S_1. Suppose +1/2 is obtained. In other cases, −1/2 is obtained with equal probability as follows from the quantum mechanical state (10).

(2) We are then sure that a future $(t>t_0)$ measurement of σ_3 on S_2 will give −1/2, as predicted by quantum mechanics and confirmed by experiments.

(3) But at time $t = t_0$, when S_1 interacts with an instrument, nothing can happen to particle S_2 which can be as far away as one wishes from S_1. All that is true for S_2 at time $t \geq t_0$ must have been true even before, namely at $t<t_0$.

(4) The only state vector describing the outcome of a measurement of the third spin component of a S_2 as certain and equal to −1/2 is $|u_2^-\rangle$. As a consequence of the previous point, this must be the description of S_2 *both before and after* the time t_0.

(5) But quantum mechanics predicts that the third component of the total spin of the two particles S_1 and S_2 must be zero before t_0, and that it remains so even after the measurement of σ_3 on S_1 at $t=t_0$.

(6) The only quantum mechanical state vector for $S_1 + S_2$ which describes S_2 as $|u_2^-\rangle$ and which gives zero for the third component of the two particles S_1 and S_2 is $|u_1^+\rangle \ |u_2^-\rangle$. This is therefore their state vector before and after time t_0.

(7) Repeating the previous reasoning for every one of the N pairs of particles S_1 and S_2, we conclude that the statistical ensemble which they form is described, even before any measurement on S_1 and/or S_2 is performed, as a mixture, with equal statistical weights (1/2), of $|u_1^+\rangle \ |u_2^-\rangle$ and $|u_1^-\rangle \ |u_2^+\rangle$, corresponding in Accardi's notation to the density matrix (14), where ψ_1 and ψ_2 are given in this case by $|u_1^+\rangle \ |u_2^-\rangle$ and $|u_1^-\rangle \ |u_2^+\rangle$, respectively.

(8) The latter conclusion contradicts, in an observable manner, the initial quantum mechanical assumption that at any time before t_0, the description of $S_1 + S_2$ is provided by the singlet state (10).

This means, according to Selleri and Tarozzi, that at least one of the previous points contains assumptions foreign to, and incompatible with, quantum mechanics. But (1), (2) and 5) are strict consequences of quantum formalism, (7) and (8) are completely derived by the first six points, and, moreover, (4) and (6) are consequences both of point (3) and of quantum mechanics. They conclude, therefore, that quantum mechanics is incompatible with the locality hypothesis introduced in point (3).

Contrary to this conclusion, Accardi has maintained that another "hidden" hypothesis foreign to quantum mechanics is contained in this derivation of the EPR paradox. To support his claim, he has proposed the following modified formulation[28] of points (4) to (7) of Selleri and Tarozzi's proof:

(4') Using the argument at point (2), we conclude that, before and after t_0, the state of the system $S_1 + S_2$ was (+ , -).

(5') But, according to quantum theory, if at the time t, the system $S_1 + S_2$ is in the state (+ , -), then in the mathematical model, it is described by the density matrix $P\psi_1$, where ψ_1 corresponds to the state $|u_1^+\rangle \ |u_2^-\rangle$.

(6') As a consequence of point (1), we know with certainty that in the ensemble of N systems, approximately one half are in the state (+ , -) and the other half, by applying to it the same reasoning, is in the state (-, +).

(7') From point (6') and from the QMB, we deduce that at any time before or after t_0 the statistical description of $S_1 + S_2$ is given by the density matrix (14) -- where ψ_1 and ψ_2 correspond, respectively, to the states $|u_1^+\rangle \ |u_2^-\rangle$ and $|u_1^-\rangle \ |u_2^+\rangle$ -- which, as previously stressed in point (8), is in contrast with the quantum mechanical description of this system.

It must be admitted that Accardi's criticism of the Selleri and Tarozzi proof is consistent as long as he maintains the introduction of an additional implicit assumption for the derivation of their contradictory conclusion of point (8) but it seems totally unacceptable when he identifies this further assumption with an hypothesis foreign to, and incompatible with, quantum mechanics. In actual fact, at step (4), the two authors introduced the typical quantum mechanical hypothesis, shared and upheld by all the exponents of the orthodox interpretation, of the completeness of quantum formalism by assuming that, since the only state vector describing in this theory the result of a measurement of the third spin component of S_2 as certain and equal to $-1/2$ is $|u_2^-\rangle$, this must be the mathematical description of S_2.

It has, however, recently been shown that the contradictory conclusions of (8) can be obtained even if one replaces the hypothesis of point (4) with the assumption of the incompleteness of quantum formalism [29]. The fact that the same result can be deduced from both these two opposite assumptions (completeness and lack of completeness) demonstrates that the question of the completeness of quantum theory is totally irrelevant as far the EPR paradox is concerned. From this it follows that Accardi's criticism is in no way applicable to the conclusions of the Selleri and Tarozzi proof.

It must be added, moreover, that in the reconstruction of the EPR paradox given by this author, a contradictory conclusion, in contrast with quantum mechanics, is already obtained in point (4') and the remaining steps from (5') to (7'), containing the appeal to the QMB, therefore appears totally redundant. As a matter of fact, if the state of the composite system S_1+S_2 is $(+,-)$ even before the time in which any measurement on S_1 and/or S_2 is performed, such a system cannot be described by the singlet state (10) as prescribed by quantum mechanics.

In this way, without any recourse to the Bayes postulate, one is led to the conclusion either that quantum mechanics is an incomplete physical theory incapable of describing composite systems like S_1+S_2 before a measurement is carried out on one or both of them, or that these composite systems must be described even before any measuring operation takes place by the mixture (14) and not by the state vector (10). In both cases, this conclusion implies an observable contrast between quantum mechanics and Einstein locality for even the appeal to the incompleteness of quantum formalism cannot, as previously stressed, avoid the derivation of the Einstein-Bell contradiction.

5. THE TESTABILITY OF THE NEW REALIST INTERPRETATION
OF THE WAVE FUNCTION

A third kind of solution to the paradox of the double-slit experiment based on the abandonment of hypothesis (c) and also of the consequent merely corpuscular approach to the wave-particle dualism has been advanced by Selleri and Tarozzi[30]. They have proposed an interpretation of this experiment which appears to allow the presence of the interference pattern to be explained but without thereby introducing a radical change in the basic laws of classical logic and probability. Their analysis was founded on the acceptance of the

dualistic nature of quantum phenomena but, in contrast with
complementarity according to which the undulatory and corpuscular
descriptions are mutually exclusive in the same experimental
situation, they assumed instead, following Einstein's and de Broglie's
point of view, that such descriptions complete each other and are both
necessary to explain the same physical situation.

On the basis of the latter interpretation, relation (2) of the
reasoning in Section 3 was rejected by maintaining that the
propositions A_p and B_p corresponded actually to the more complex
events

$$A_p = A_1 \wedge (A_2 \wedge B_2) \quad , \quad B_p = B_1 \wedge (A_2 \wedge B_2) \tag{15}$$

where A_1, B_1 indicate the passage of the particle through one of the
two slits and $A_2 \wedge B_2$ indicate, instead, the passage through both
slits of the quantum wave associated with the particle. The presence
of the interference and its wiping out could thus be explained as the
result of the action of the quantum wave on each particle, which
action could be exerted only until the wave was disturbed. On the
grounds of this interpretation, it was possible to explain those
physical situations which, as we shall see, give rise to one of the
most serious paradoxes of the theory of measurement where there is
only one instrument located behind one of the two slits and one
considers only the cases in which the arrival of a particle on the
second screen is registered without, however, the instrument's having
detected the passage through the slit of any micro-object.

The disappearance of the interference is viewed by Selleri and
Tarozzi in this case not as a consequence of the observer's knowledge
acquired by *not observing* the occurrence of a certain physical event,
as implied by the quantum logical interpretation, but as the result of
the interaction of the quantum wave with the measuring instrument.
According to the description proposed, the events A_p and B_p are thus
transformed into

$$A'_p = A_1 \wedge (A_2 \wedge B'_2) \quad , \quad B'_p = B_1 \wedge (A'_2 \wedge B_2) \tag{16}$$

where B'_2 (A'_2) represents the wave that passed through B (A) and was
absorbed by the instrument located behind B (A). From the point of
view of the experimental situation for the second screen, there is no
difference between the events

$$A'_p = A_1 \wedge (A_2 \wedge B'_2) \quad , \quad A''_p = A_1 \wedge A_2 \tag{17}$$

the second of which represents the cases where slit B is closed. By a
reasoning similar to that which led us to relation (5), we can
conclude:

$$P(X_p) = P(X_p \wedge A'_p) + P(X_p \wedge B'_p); \tag{18}$$

$$P(X_p) = P(X_p \wedge A''_p) + P(X_p \wedge B''_p) \tag{19}$$

where no interference can occur.

One could object to this model that even events like

$$\gamma = A_2 \wedge B_2$$

are logically present (the particle is absorbed by the first screen and the quantum wave passed through both screens), the physical significance of which seems rather obscure since it appears to defy the possibility of any experimental verification.

It is precisely to the problem of giving a physical meaning to the γ-type events, usually considered empirically untestable, that Selleri's hypothesis of quantum waves provides an answer.

The concept of quantum wave can be regarded as the attempt to make a theoretical synthesis of three different views of the wave-particle dualism historically advanced by the principal founders of quantum theory.

It has in the first place a close analogy with conceptions of Einstein, who, though he reintroduced into physics a corpuscular theory of radiation through his famous hypothesis of the quanta of light, believed that the interference and diffraction phenomena could not be explained on merely corpuscular grounds but also required the presence of a wave to accompany and guide the quanta in their motion. Nevertheless, the fact that all the energy was concentrated in the quantum, and that the wave associated with it was consequently devoid of these fundamental properties led Einstein to introduce the term of *Gespensterfeld* (ghost field) to denote a concept which seemed to him logically necessary but which lacked empirical support.

When de Broglie, with his wave mechanics, then extended the dualism to matter in an attempt to overcome the contradiction due to the "existence" of entities devoid of the properties characterizing any other physical object, he found no way out to guarantee the reality of his "pilot wave," other than to attribute to it an extremely small portion of energy which would have been almost entirely concentrated in the particle.

The fact that no one has yet been able to reveal this very small portion of energy of that which de Broglie very significantly called the "empty wave," seems to justify the objection that the de Broglie waves represent a philosophical assumption on which a metaphysical solution to the wave-particle dualism could be based, rather than a genuine physical hypothesis. Such an objection is, for example, contained in Jauch's Galilean dialogue on the reality of quanta, in which de Broglie's point of view is defended by Simplicius.

A third conception, which came very near to that of quantum wave, was introduced by Bohr, Kramers and Slater in the framework of their attempt to reformulate a purely undulatory theory of radiation in opposition to Einstein's corpuscular hypothesis of light quanta. This was the concept of *virtual wave*, to which these authors attributed the fundamental characteristics of carrying neither energy nor momentum and of producing only stimulated transitions in the atoms that the wave interacted with.

The atomic transitions were thus alleged to occur in open

violation of the laws of conservation of the these physical quantities. In fact, according to the theory of Bohr, Kramers and Slater, any given atom could pass from one energy level to another without there being for this reason exchanges of energy with the electromagnetic field. The concept of virtual wave was, however, soon abandoned as a result of the refutation of the theory of these authors by the Compton and Simon and the Bothe and Geiger experiments, which confirmed the Einstein corpuscular hypothesis of radiation.

Although they elaborated very similar views, none of the aforementioned authors succeeded in coming to the concept of quantum wave: Neither Einstein who, having made the most important contribution to defining the concept of energy and transforming it into the basic notion of modern physics, found it contradictory to assert the existence of objects devoid of this fundamental property; nor de Broglie who, having created wave mechanics, was unable to conceive waves without energy and momentum and had not found any solution other than to attribute to the waves an *untestable* quantity of the previous properties; nor Bohr, Kramers and Slater whose virtual waves were introduced as an alternative to the corpuscular hypothesis of light quanta and who, faced with the failure of their undulatory radiation theory, soon replaced them with Born's merely probabilistic interpretation of the wave function.

Selleri's paper of 1969 marks the conceptual turning point with respect to the interpretations of the wave function discussed earlier.[31] He started from the Einstein-de Broglie realistic conception that waves and particles have an objective existence and from the observation that the experiments carried out in this field until then proved beyond any reasonable doubt that all the energy, the momentum and the charge are closely associated with the particle. He thus posed the question: What is an entity existing but not having any observable property associated with it? To this question Selleri believed there were two possible answers.

The first was that physical quantities are *mainly* associated with particles, but that a portion of every quantity, so small that it has escaped all experimental observations, is associated with the waves. He believed however, that this first answer, which, as we have already seen, coincides with de Broglie's view, constituted an "unappealing way out."

The second answer that appeared to Selleri a "better solution to this epistemological problem" was that, "even if devoid of any physical quantity associated with it, the wave function can *still* give rise to observable physical phenomena"[32]. He pointed out, in this regard, that we do not measure energies, momenta, or similar physical quantities only but also *probabilities*, as in the case, for example, of the average life of an unstable physical system. According to Selleri, therefore, the wave function could "acquire reality independently of the associated particles if it could give rise to changes in the transition probabilities of the system it interacts with"[33].

On the basis of this new idea, he proposed the first version of his experiment for detecting physical properties of quantum waves. He considered a piece of matter composed of unstable entities (nuclei, atoms, or excited molecules) traversed by a flux of neutrinos. The experiment then consisted in measuring the average life of these entities under these conditions and comparing them with the average of the same entities in the absence of any particle flux: If any difference was observed, the only logical explanation was, according to him, that such difference

"is due to the action of the action of the wave function, since neutrinos are particles that interact very weakly and only a few of them, in the best of cases, can have interacted with the piece of matter."[34]

In his remarks on Selleri's paper, Louis de Broglie valued this experimental proposal as an important attempt at

"obtaining an interpretation of wave mechanics more satisfactory than that presently adopted and as a confirmation of the ideas that had guided me when in 1923-24 I proposed the basic conceptions of wave mechanics,"[35]

but did not appreciate the essential novelty of the new idea and tended to view it in terms of his old conception of the pilot wave:

"The experiment you propose to prove the existence of the wave ψ will be of great interest to prove the existence of this very weak (très faible) wave, which carries the particles ..."[36]

Some years later, Selleri perfected his original idea in the experimental proposal[37] shown in Fig. 1, where S is no longer a source of neutrinos but of photons and LGT is not any piece of matter composed of unstable entities, but a laser-gain tube, and where there was the addition of two photomultipliers P_1 and P_2 and the semi-reflecting mirror SRM. The latter behaves in the same way as the double slit: The particle is propagated in one direction only (depending on whether it is transmitted or reflected by the mirror), whereas the wave is both transmitted and reflected, according to Selleri's hypothesis.

Selleri then proposed to concentrate on the cases in which the photomultiplier P_1 located along the reflected beam reveals the presence of a photon: this means that in the transmitted beam only the quantum wave is present. This latter can, however, reveal its presence by generating the stimulated emission of photons (which can be detected by P_2), on passing through a laser-gain tube LGT, whose molecules are at an excited level corresponding to the wavelength of the incident wave.

In this way, the coincidences between the detection of P_1 and P_2 would reveal the propagation of a "zero-energy undulatory phenomenon" transmitted by SRM. The space-time propagation of these entities could

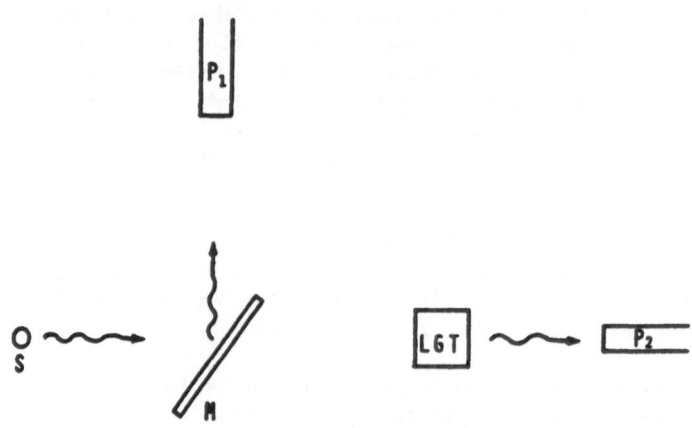

Fig. 1. Set-up of Selleri's experiment

be studied by verifying whether the P_1 P_2 coincidences disappear when an obstacle is placed in the transmitted beam in front of LGT. A positive result of this experiment would finally show that "something not having energy and momentum but which can produce transition (of probability) propagates in space and time."

Two variations of Selleri's experiment have been proposed by the present author[38]: The first represents a complete test of our realistic interpretation of the wave function discussed at the beginning of this section, while the second tries to demonstrate the possibility of a simultaneous presence of the two properties of quantum waves to generate stimulated emission, according to Selleri's hypothesis, and to produce interference, highlighting how even both these properties are not complementary in Bohr's sense.

Other Italian authors, often in collaboration with physicists of the de Broglie school, have advanced further experimental proposals to reveal the properties of quantum waves, which will be discussed briefly in Section 7, since, as we shall see, they do not simply test the interpretation of the wave function but also a basic postulate of the quantum-mechanical theory of measurement. We should like, now, to take a look at some of the epistemological and philosophical implications connected with the testability of this new realistic interpretation of the wave function.

In the first place, it should be noted that the hypothesis of quantum waves appears to satisfy the need for new concepts in microphysics, as pointed out, for example, by Agazzi, since it introduces an important conceptual novelty in quantum theory. Quantum waves lose their classical properties of energy and momentum but acquire the new nonclassical ability to modify the transition probability of particles and their probability distribution, in the same way as the quantum mechanical concept of the particle is subjected to precise restrictions in relation to simultaneous possession of its classical properties of position and momentum but is characterized by the new observable properties of spin.

Another important reason for the novelty of the quantum wave concept lies in the fact that it rejects the symmetric nature of wave-particle dualism, maintained by the other realistic interpretations of the wave function: Indeed, though it disagrees with a merely corpuscular interpretation, a kind of "ontological priority" of particles with respect to waves is acknowledged in the sense that quantum waves cannot be characterized through the fundamental monadic properties possessed by particles, but only through (diadic) relational properties with the latter. This means that quantum waves should belong to a weaker level of physical reality which contains objects that are carriers of exclusively relational predicates in the language of quantum theory.

It is this latter aspect that is probably the most interesting from the epistemological point of view. It can be put in the framework of a conception of empirical realism, the fundamental thesis of which is the notion of shifting the reality concept from the object to its predictable properties as a guarantee of the reality of the object itself. From this perspective, the reality of an object is not founded so much on the possession of properties defined for it as on the independence of the subject (or of his mind or consciousness) of the properties that are predictable without interacting with the object itself. It is, therefore, starting from the existence of this kind of (independent) properties that we define an object as real and not *vice versa*: For empirical realism, it is the objective reality of its predictable properties that logically precedes the reality of the object.

Now, the concept of quantum wave is very closely connected to what we have just said since it is not on the its direct possession of properties that we can base the reality of this object. Nevertheless, the existence of properties such as interference and, in Selleri's hypothesis, the stimulated emission of particles, which cannot be attributed to a single class of objects but which seem to be relational properties between two different kinds of entity, i.e., particles and quantum waves, in turn constitutes a guarantee of the reality of the latter.

The remarkable philosophical novelty of the concept of quantum wave is, in final analysis, connected with the fact that the experiments that have been advanced in order to detect its physical properties do not test the predictions of quantum formalism, but *discriminate experimentally between two different philosophical interpretations* of the main theoretical concept of quantum theory: Bohr's complementarity and this new realist interpretation of the wave-particle dualism.

6. THE ALTERNATIVE THEORIES OF MEASUREMENT

The reaction against the philosophy of the Copenhagen school in

Italy also took the form of an attack against the orthodox solution to
the paradoxes of the theory of measurement. The problem of
measurement, as is known, has its logical roots in the co-existence,
within the mathematical structure of quantum theory, of two different,
incompatible kinds of temporal evolution of the wave function -- on
the one hand, the Schrödinger deterministic equation of motion, which
governs the temporal evolution of the state of unobserved, microscopic
and/or macroscopic physical systems and, on the other, the
probabilistic description of the observations or measuring operations
performed to determine the observable properties of these physical
systems -- but without there being given any unambiguous
specification of precise, mutually exclusive conditions for these two
different kinds of evolution to occur.

The appeal to the these two kinds of description (deterministic
and probabilistic), seems inevitable, however, in the same way as it
is necessary to use both the undulatory and corpuscular
interpretations to explain all the empirical properties of
micro-objects. Indeed, every time the measured micro-object is in an
initial state that is a superposition of two or more microscopic
states, describing the measuring process by means of the Schroedinger
equation as a physical interaction between micro-object and measuring
instrument would imply, because of the linearity of this equation,
that the final state of the global system "micro-object + instrument,"
is still given by a superposition, in this case, of different
macroscopic states. Such a possibility would mean that an observer
controlling the measuring instrument, for example, would see the
latter registering simultaneously all the different results predicted
by formalism, which is obviously absurd.

To avoid such a conclusion, an axiom was introduced in quantum
theory to establish that the Schroedinger equation does not apply to
the description of the results of our measurements or observations and
that the observed system is in a well-defined state after every
measuring process. This was the so-called "reduction of the wave
function", a completely *ad hoc* postulate that the orthodox
interpretation tried to explain in two completely different ways.

The first solution of the problem, advanced by Bohr, consisted
in the renunciation of any attempt to develop a mathematically
consistent theory of measurement and was therefore founded on a purely
qualitative interpretation of the reduction postulate, viewed as a
consequence of the inapplicability of the quantum formalism to the
description of the measuring apparatus on account of the macroscopic
nature of the latter and the resultant necessity to provide a
description of it entirely based on classical concepts.

The second solution, due to von Neumann, took the form of a
rigorous mathematical theory of measurement, where the instrument, in
the same way as any other physical object, was describable by quantum
formalism and the reduction postulate was interpreted as a consequence
of the intervention of the observer's consciousness.

The clearly subjectivistic nature of the von Neumann theory of
measurement was definitively shown by the present author[39] who
proved the incompatibility of this theory of measurement with a very

general macrorealistic hypothesis, that was first discussed by Lewis, Schlick, and Carnap ("if all minds disappeared from the Universe, stars still go on in their courses"), and which constitutes a reformulation, endowed with meaning, of the theses of metaphysical realism refuted by Hume.

According to our argument, which will be discussed here also in relation to the Bohr solution to the measurement problem, let us suppose, that a 100,000 megaton nuclear missile is launched from a base on Earth towards a planet of our solar system. The instant the missile hits its target, it produces such a violent explosion as to disturb the ordinary orbit of the planet, which will in turn cause a general displacement (however small) of the entire planetary system. More precisely, if the set of observables $P = F(q_{ia}, p_{ia}; t_a)$ corresponds to the ordinary configuration of the planetary system at time t_a, then $P' = F(q'_{ia}, p'_{ia}; t_a)$ will correspond to the perturbed case. The possibility of such an event would not, however, represent an obstacle to the validity of the Lewis-Carnap realist hypothesis, since the laws describing the temporal evolution of the considered classical observables or the average value of the corresponding macroscopic quantum observables P would enable us to predict the different planetary configuration given by P' in the case of the nuclear explosion on the planet. According to this realist hypothesis, the stars would continue on their perfectly predictable orbits independently of our observation.

Let us now suppose that the missile has a device consisting of a spherical cavity, divided into two parts S_1 and S_2, inside which there is an excited atom decaying from its fundamental state through the emission of a photon γ. The absorption of γ in S_2 occurs without any macroscopic effect, whereas absorption in S_1 generates an electric pulse that activates, at time $t_b < t_a$, a disintegration system that prevents the missile from reaching the surface of the planet. In this case, then, a microscopic process inside the missile determines two different macroscopic events described by P and P', and, since such a microscopic process is described in quantum formalism by a state of superposition of the type

$$|\psi_1 > = a \ |A_1> \ + b \ | A_2>, \tag{20}$$

where the states A_1 and A_2 indicate the absorption of γ in S_1 and S_2, respectively, and $|a|^2$ and $|b|^2$ the probabilities of these absorptions, the planetary system will be described by a similar wave function

$$|\psi_2 > = a \ |A_1> \ |P> + b \ |A_2> \ |P'>. \tag{21}$$

We are thus faced with a very odd description according to which the planetary system would be partially disturbed and partially not, in an undefined state of superposition between its ordinary and its perturbed configuration.

Now, according to the Bohr interpretation, to obtain the reduction of the wave function $|\psi_2 >$ to one of the well-defined final states

$$a \ |A_1> \ |P> \ \text{and} \ b \ |A_2> \ |P'>, \tag{22}$$

the presence of a measuring instrument M is required to register at a given instant t_c the presence or the absence of a given celestial body in the position predicted by one of the two different configurations P_1 and P_2. This is clearly an unsatisfactory explanation since it is founded on the assumption that there is a *difference in principle* between a measuring instrument M that can give rise to the reduction process and any other macroscopic system, such as one of the very same celestial bodies, that would not be capable of generating this process.

Upholding the existence of such a difference is equivalent to asserting that the interaction of an electron with a mass of liquid hydrogen and the measurement of the momentum of that electron in a bubble chamber are conceptually different events, thus attributing to the measuring apparatus a special epistemological status with respect to other physical systems, without, however, any plausible reason being given to justify such a privileged status. Nevertheless, Bohr's solution has the advantage of not reflecting a subjectivistic view and succeeds in providing a sort of physical explanation to the reduction process of the superposition () to one of the final states of relation (), which is compatible with the Lewis-Carnap realistic hypothesis: Even in the absence of any human observer, the intervention of the measuring instrument M, by provoking the reduction of the wave function, enables the stars to continue in their (ordinary or perturbed) courses.

In his theory of measurement, von Neumann, unlike Bohr, assumes, on the basis of his well-known completeness theorem, that quantum theory is able to describe all the properties of the physical world and therefore also those of the measuring apparatuses. This of course implies that, after the measuring process, one will obtain yet another superposition state

$$|\psi_3> = a \ |A_1> \ |P> \ |M_1> + b \ |A_2> \ |P'> \ |M_2> \tag{23}$$

where $|M_1>$ and $|M_2>$ denote the quantum states associated with our instrument M.

One might then be tempted to introduce another instrument X observing "A + P + M", and this chain could be extended *ad infinitum* without ever obtaining any reduction process. This means that, according to the von Neumann theory, in the absence of an observer outside the physical world, the stars could not continue in their courses. Such a conclusion has proved beyond any residual doubt the subjectivistic nature of the von Neumann theory of measurement, contrasting even with the only very general version of realism considered not to be metaphysical by the logical empiricists.

It is in this framework that the theory of measurement of Daneri, Loinger, and Prosperi (DLP) can be viewed[40]. According to these Italian physicists, in fact, this theory represents an attempt to overcome the von Neumann theory -- as well as certain "rather extremist philosophical interpretations ... that would deprive of all their objectivity our knowledge of the physical world" -- by appealing to Bohr's conception of the measuring process.

The central thesis of the DLP theory is based on Bohr's idea of the macroscopic nature of the measuring instrument. By "measuring apparatus of a certain magnitude" they mean "a macroscopic system so devised that its *macroscopic state* is modified by the effect of the interaction with the *observed object*."[41]

To explain how a very small perturbation produced by a single micro-object during the measuring process can alter the measuring instrument, DLP assume that the latter is not an ordinary macroscopic object but a special system in a state of metastable thermodynamic equilibrium which, due to this very small perturbation, passes to another more stable state of equilibrium. There must, therefore, exist different possible states of final equilibrium, and to each of them there must correspond an eigenfunction of the magnitude relative to the measured object.

This hypothesis of the existence of different possible states of equilibrium, corresponding to the same macroscopic value of energy, implies for a macroscopic system the necessity to introduce a second constant of motion in addition to the energy of the system itself. The values assumed by this constant determine the decomposition of the energy shell into subvarieties V_0, V_1, ..., V_k which may be called "channels." Every channel breaks down into "cells," corresponding to the different possible macroscopic states that are compatible with the values given for the energy and the other constants of motion. The authors assume that in each channel the conditions of ergodicity are satisfied and a state of equilibrium, where the microcanonical probability is very close to unity, exists which consequently the system will strive to reach as fast as possible.

The effect of the interaction between microsystems and macroscopic apparatus is that of a transition of the state vector of the latter from an initial channel V_0 to a new one V_k, which depends on the particular state, among the ones of an orthonormal system, in which the microsystem is. The state vector of the micro-object acting as a measuring apparatus is subjected to a spontaneous modification in such a manner as to coincide with the state of equilibrium belonging to channel V_k at the end of the measuring process.

We can therefore conclude that the original idea of the DLP theory of measurement is not so much the classical nature of the measuring instrument as its characteristic of metastable thermodynamic system: As we have seen, the reduction of the wave function is not produced in this theory by the interaction of macro-instrument with micro-object but rather by an ergodic process occurring in the instrument after the interaction with the micro-object has taken place.

The DLP theory would thus seem to be able "to overcome the objection that quantum theory is capable of explaining how micro-objects behave but not how their behaviour can be observed," appearing as "an indispensable completion and a natural crowning of the basic structure of present-day quantum mechanics."

This point of view was fully shared by Leon Rosenfeld, one of the leading exponents of the Copenhagen present interpretation, who declared that these "Italian physicists" with their "very thorough and elegant discussion of the measuring process ... have conclusively established the full consistency of the algorithm (of quantum mechanics), leaving no loophole for extravagant speculations."[42] Rosenfeld stressed the presence of an essential continuity between Bohr's approach and this theory of measurement, which to him represented the translation into mathematical formalism of the qualitative physical arguments of the Danish physicist.

This latter theory was severely criticized by Jeffrey Bub, who pointed out, instead, the absolute incompatibility of the DLP theory with the basic theoretical principles of the Copenhagen interpretation, showing how this theory was "basically opposed to Bohr's ideas"[43] because it applied classical concepts at the macroscopic level on the grounds of purely quantum theoretical arguments. Hence, classical mechanics was regarded as an "approximation to a quantum theory of macrosystems" and as such replaceable, in principle, by quantum theory, in opposition to Bohr's ideas that the conceptual framework provided by classical mechanics is indispensable for the description of quantum phenomena and therefore has logical priority over any application of quantum mechanics. Besides this criticism, Bub also highlighted how the theory failed in its very purpose to give an objective description of the macro-world since, by explaining the reduction of the wave function as a consequence of the interaction of instrument with micro-object, it did not succeed in removing the possibility of the paradoxical superposition of macro-states occurring in the case of an ordinary interaction of a microscopic with a macroscopic system that is not a measuring instrument.

Bub's criticism of the absence of continuity between Bohr's interpretation of the measuring process and the DLP theory was endorsed by Dalla Chiara and Toraldo di Francia in their formal approach to the problem of measurement that we shall deal with in the next section.

A much more serious criticism was advanced against the DLP theory of measurement by Eugene Wigner. In fact, the latter showed how this thoery is not able to account for the so-called negative result measurements, in which the reduction of the wave function occurs without any detection process by the measuring apparatus and, consequently, without any microscopic modification of the state of the apparatus[44]. The existence of such physical situations was stressed for the first time by Renninger through the proposal of the following thought experiment[45].

Let us consider a source of photons isotropically emitting these particles in all directions and which is partially surrounded by a hemispherical screen E_1 with center P and radius R_1 which subtends a solid angle Ω about P and completely surrounded by a spherical screen E_2 with radius $R_2 > R_1$. The photons emitted by P can be absorbed by one of the two detecting screens, either E_1 or E_2, with probabilities given, respectively, by

$$\omega_1 = \Omega \, / \, 4\pi \quad , \quad \omega_2 = (4\pi - \Omega) \, / \, 4\pi \tag{24}$$

We can therefore write for the initial state $|\psi_0\rangle$ at time t_0 the following wave function

$$|\psi_0\rangle = \sqrt{\omega_1} \, |\varphi_A\rangle + \sqrt{\omega_2} \, |\varphi_B\rangle \tag{25}$$

In the next instant, $t_1 = R_1/c$, two different events may occur, the probability of each of which is given by ω_1 and ω_2, respectively. The first possibility is the detection of the photon by E_1, which implies that ω_2 vanishes and ω_1 becomes certainty. According to the postulate of the reduction of the wave function, the new physical situation will be described by

$$|\psi_1\rangle = |\varphi_A\rangle \tag{26}$$

In this case, the reduction process can be explained on the basis of the DLP theory of measurement, as a consequence of the interaction between the detected photon and the macroscopic measuring apparatus E_1, which, as result of such interaction, will evolve towards the state of thermodynamic equilibrium given by the well-defined state vector $|\varphi_A\rangle$. This explanation is of course far more satisfactory than the one provided by the von Neumann theory where, to justify the transition from (25) to (26), one is forced to appeal to a metaphysical entity such as the consciousness of the observer.

The second possibility is that at time t_1, there is no detection of the emitted photon by the screen E_1. This implies that $\omega_1=0$ and $\omega_2=1$ and, again on account of the reduction postulate, that the physical situation will be described by

$$|\psi_1\rangle = |\varphi_B\rangle \tag{27}$$

In this second case, however, the change in mathematical description can in no way be put in relation to any physical process of interaction between the emitted photon and the measuring apparatus E_1 but seems to be merely a consequence of the information that a human observer has obtained about a given physical situation by *not* observing the occurrence of a given phenomenon. This means that, to explain the transition from (25 to (27), we can no longer appeal to a theory of measurement which, like the DLP theory, considers the measuring apparatus as a macroscopic object in a metastable state evolving towards stability as a consequence of the interaction with the measured micro-object. The Renninger paradox appears, on the other hand, to support, in an apparently conclusive way, the spiritualistic von Neumann theory of measurement, which seems to be the only one that explains how, even in the *absence* of any interaction between measuring apparatus and measured system, a change in the observer's knowledge can, via a modification in the mathematical description, produce change in the physical situation.

7. LOGICAL AND EPISTEMOLOGICAL ANALYSES OF THE MEASUREMENT PARADOXES

A very thorough formal discussion of the measurement process has been recently given by Maria Luisa Dalla Chiara and Giuliano Toraldo di Francia who have highlighted the existence of close analogies between the proposed solutions to the problem of measurement of quantum theory and the solutions that have been put forward in the past by logicians and mathematicians in relation to the paradoxes of set theory[46]. The two authors have shown how the well-known contradiction connected with the wave function reduction postulate can be obtained starting from two different groups of hypotheses: On the one hand, the axioms of quantum theory (QT) and, on the other, what they define as "special hypotheses."

The first of these special hypotheses is a) that the measuring apparatus A is a physical system describable in the same way as any other physical object by QT, while the second hypothesis, which essentially follows from the first, is b) that the interaction between the measuring apparatus A and the measured system Σ can be described by the Schrödinger equation.

The authors assume, therefore, with their special hypothesis, that QT describes in the same way "*normal* physical systems," such as any microsystem Σ and "*strange* systems" such as are for them A and Σ + A. Both hypotheses seem very reasonable because the Schrödinger equation, it is perhaps well to remember, is correctly applied to describe the interaction both between two microscopic systems and that between a microscopic and a macroscopic system, such as a measuring instrument would seem to be.

They then point out how the ways out of the paradox are "founded on different choices of the *guilty* hypothesis to be rejected (or at least weakened)," which have, from the formal point of view, close analogies with two corresponding hypotheses that are used in set theory (ST) and from which Russel's paradox can be derived:

"(a) certain *strange* collections which are objects of the metatheory of ST (such as the collection of all sets described by the theory) are also *objects* of the theory ...; (b) all axioms of ST hold, in the same way, for *all* the objects of the theory ..."

Now, according to Dalla Chiara and Toraldo, there are three possible solutions in the case of both QT and ST: (i) rejecting (a) and accepting (b); (ii) accepting a) and rejecting (b); (iii) rejecting (a) *and* (b).

The first possibility, in the case of ST, corresponds to Zermelo's solution and, in the case of QT, to Bohr's approach. The basic idea behind both is that the *strange* collections, such as, in ST, that of all sets described by the theory and, in QT, that of the (macroscopic and therefore classical) measuring instruments "are *essentially only metatheoretical.*" The third possibility, leaving aside the second for the moment, is represented, for ST, by Russel's solution limiting the universe of sets and, at the same time, relativizing the axioms to specific classes of them and, for QT, by the DLP theory and similar approaches such as that of Jordan, Ludwig, and Prigogine. According to these, the measuring instruments must be treated as a quantum system, and the reduction process must be described as a physical relation, and yet both the instrument and this relation are not directly describable by QT but require "a more sophisticated physical theory" able to account also for irreversible processes. Dalla Chiara and Toraldo's formal analysis of the measurement problem thus makes it possible to clearly distinguish Bohr's solution from DLP's, which the latter authors tended erroneously to link, despite Bub's criticisms.

The second possibility corresponds to the solution proposed by von Neumann both in relation to the ST paradoxes (in this case, better known as the von Neumann-Bernays-Gödel solution) and in relation to the paradoxes of the quantum mechanical theory of measurement. According to von Neumann, the measuring instrument is a legitimate object of QT in that, although it is a macrosystem, it is made up of elementary particles and, as such, must be describable by this theory

The two authors emphasize the fact that von Neumann advanced the same type of *logical operation* in the case of both ST and QT. His attempt can therefore be viewed in the context of the attempt to solve the physical paradoxes from a purely formal point of view, as we have seen to be the case in the quantum logical approaches. Dalla Chiara and Toraldo then try to propose a purely *logical interpretation* of von Neumann's thesis "completely free of any subjectivistic and spiritualistic connotations," expressed as follows:

> "If the apparatus-observer A is an object of the theory, (...) then it *cannot realize* the reduction of the wave function. This is feasible only by another A' which is *external* with respect to the universe of the theory."[47]

The conclusion is thus reached that "*any apparatus which realizes the reduction of the wave function is necessarily only a metatheoretical object*," which according to the authors "does not sound as a subjectivistic claim."

Such a conclusion, in our view, can be shared only if one acknowledges the essential incompleteness of the quantum mechanical description of the physical world, which in any case Dalla Chiara and Toraldo do not at all exclude. According to the latter, a theorem of completeness for any given physical theory T -- which in their logical analysis, they identify with a pair <FS,K>, consisting of a *formal system* FS and the class K of all the models of FS -- seems to be unprovable for reasons both purely logical, connected to Gödel's incompleteness theorem -- on account of which there exist some mathematical sentences of T true for every physical model of T but not provable in T -- and because K does not, as in the standard case, represent "the class of all possible worlds" but only the class of all physical worlds that verify the theory. In this latter case, a formal property of completeness of T would mean that one would be able, solely on the grounds of T, to predict "whatever empirically happens in all parts of the physical world."[48]

Nevertheless, this thesis of the incompleteness of QT appears diametrally opposed to the von Neumann myth of completeness, based on the intransigent defence of his famous theorem even against a mathematical possibility of any completion of the existing theory, and much closer to the conception of Einstein who until 1927 had maintained the necessity for such a completion. Moreover, this logical reinterpretation of the von Neumann theory of measurement does not seem to differ very much from Bohr's solution, although, according to the former, the measuring instrument is a metatheoretical object *only* when it produces the reduction of the wave function, whereas for the latter author any given instrument, unlike other ordinary macrosystems which are instead describable by QT, is *always* a metatheoretical object. Hence, the rather implausible difference in principle between ordinary macroscopic systems and measuring apparata typical of Bohr's interpretation is shifted to the even less understandable difference between two different kinds of instruments: those that produce the reduction and cannot be described by QT and those that do *not* produce the reduction and can be described by QT. It seems, however, that such a difference can be explained only by distinguishing between observed and unobserved instruments, thus appealing again to the thesis of the centrality of the observer in the standard subjectivistic theory of von Neumann.

A theory of measurement having significant analogies with that of DLP has recently been advanced by Marcello Cini. Its central assumption is that the reduction of the wave function must not be considered as an additional axiom external to quantum theory but rather as an approximate consequence of the standard law of temporal evolution provided by the Schrödinger equation[49]. This theory also has as its starting point the criticism of von Neumann's subjectivistic approach and Bohr's idea that the interaction of the micro-object with the measuring instrument is the only source of information for the physical properties of the former. It differs, however, from the DLP theory in that it assumes that the measuring instrument is a macroscopic system, not so much because it is subject to an irreversible evolution characteristic of thermodynamic systems, but rather because it is a classical object whose observables have the property of being commutable with each other, unlike quantum observables, which are not endowed with this property of universal commutativity. Jauch's theory of measurement was also based on the latter idea and, as has elsewhere been shown[50], implied the paradoxical existence of a "frozen universe," in which nothing happens and which is totally contradicted by the extremely rapid processes of amplification occurring in the real physical world. Cini's approach does not therefore seem to constitute any decisive progress with respect to the DLP approach, both on account of the objections that can be raised against Jauch's theory and because it, too, is quite unable to explain the paradox of negative result measurement.

A solution to this paradox, based on the hypothesis of quantum waves, has recently been advanced by the present author[51] who has discussed a modified version of Selleri's experiment where the distance between SRM and P_1 is greater than that between SRM and P_2, and where the photomultipliers P_1 and P_2 are arranged in such a manner as to measure delayed coincidences. We thus obtain a physical situation which is conceptually equivalent to Renninger's thought experiment and the mathematical description of which is given by the wave function (25), the only difference being that ω_1 and ω_2 correspond, in this new physical situation, to the probabilities, both equal to 1/2, of detection of the photon by P_1 and P_2, respectively.

In this modified situation, the amplifier LGT, which is situated in the path of the transmitted beam of radiation, behaves either like an ideal duplicator in the case of the incident particle or, in the case of the wave without particle, because of the process of stimulated emission of light implied by the realist interpretation of the wave function, like a source of a single photon. Since it is also possible, at least in principle, on the basis of the pulse height of the photomultiplier P_2, to discriminate experimentally between the detection of the twin photons resulting from the duplication and that of the single photon produced by the amplified quantum wave, we are thus able to explain the reduction of the wave function as a consequence of these two different processes of physical interaction.

From this perspective, it would therefore be possible to overcome the paradox connected with the two different kinds of reduction involved in the orthodox interpretation of Renninger's experiment: The one corresponding to a physical interaction between measuring instrument and measured object, the other to a psychophysical, or purely mental, event due to the change of our knowledge relative to the probabilities of the occurrence of certain events. This type of solution would not, however, allow the DLP theory of measurement to be preserved as such, since, in the light of this theory, there would be no process of physical interaction between macroscopic instrument and microscopic object when the latter corresponds to a zero-energy undulatory phenomenon.

This impossibility of explaining the paradox of negative result measurement is due to the fact that the DLP theory provides a realistic interpretation of quantum mechanics that is restricted only to the concept of measurement viewed as a real process of interaction between two different physical objects and is not extended to the other basic quantum mechanical concept of wave function. The paradox can, on the other hand, be explained both on the basis of the, in our view unsatisfactory, subjectivistic and spiritualistic approach of von Neumann, who considers measurement as a psychophysical interaction between (observing) subject and (observed) object and the function ψ describing the observer's knowledge, and on the basis of the hypothesis of quantum waves interpreting realistically not only the concept of measurement but also the concept of wave function ψ. In fact, once we have established the correspondence between the function ψ and the real physical properties of micro-objects, we can explain even the process of reduction of $|\psi_0\rangle$ given by (25) to $|\varphi_B\rangle$ as the result of a physical interaction between two real physical objects: The instrument E_1 and the quantum wave without the associated particle, which will subsequently be detected by E_2.

Two different attempts to falsify experimentally the postulate of the reduction of the wave function which, as we have seen, is at the origin of the most serious paradoxes of the theory of measurement has been advanced on the one hand by Augusto Garuccio, Vittorio Rapisarda and Jean-Pierre Vigier[52] and, on the other hand, by the present author[53]. The common idea behind these experiments is the possibility of establishing the path followed by a photon propagating inside an interferometer, revealing at the same time the interference pattern on the final detecting screen, in contrast with what is prescribed by the reduction postulate. The difference lies in the fact that the Garuccio-Rapisarda-Vigier experiment can be performed as an alternative to Selleri's experiment, i.e., only in the case where quantum waves are not able to produce stimulated emission of light, whereas the proposal of the present author can be performed whether the result of Selleri's experiment is successful or not.

Experiments of this kind will perhaps enable us to go beyond the very same, previously discussed, solution of the paradox of negative result measurement, since they are designed to demonstrate that when the change in the mathematical description envisaged by the theory is not the result, as is the case of the Renninger experiment, of a process of physical interaction but can only be related with a modification of the observer's knowledge, such a change in the mathematical description must correspond to a wrong prediction of the theory itself. And this is what would be proved by the persistence of the interference even when the observer knows which path is followed by each micro-object in the experimental device. In this way, the opportunity of solving empirically even the controversial problem of measurement based, as we have seen, on two different levels of testability of quantum mechanics, is created. At the first level, the solution would be consistent with the predictions of quantum mechanics but not with the Copenhagen interpretation, whereas, at the second level, it would require even formal, though -- on account of the possibility of reformulating this theory without introducing the reduction postulate -- not radical, modifications of its mathematical structure.

8. QUANTUM MECHANICAL ACTION AT A DISTANCE VERSUS REALISTIC LOCAL THEORIES

While the Italian debate on the paradoxes of quantum mechanics discussed up to now is basically concerned with the *interpretation* of this theory -- although many authors, as we have seen, have proposed solutions of a formal nature -- there is a third conceptual problem that cannot be exhausted in this epistemological perspective. This problem is not in fact connected with the possibility of assuming an alternative philosophical interpretation of the fundamental concepts and principles of quantum theory, but with the very compatibility of its formalism with the prescriptions implied by the postulates of relativity associated with the hypothesis of the validity of certain philosophical conceptions. We are referring to the question raised by the Einstein-Bell contradiction between quantum mechanical description and the principle of local reality (often called Einstein locality) or realistic local theories.

We have seen in Section 4 how the origin of the problem from the formal point of view is closely connected with the logical and empirical incompatibility between two types of quantum mechanical descriptions: one based on the second-type (non-factorable) state vectors and the other on first-type (factorable) state vectors. The first to realize that quantum mechanics uses these two different types of descriptions for two separate systems was Furry in 1936. It was, however, Bohm and Aharonov, followed in this approach by Jauch, who proposed the elimination of the second-type state vector as the cause of the EPR type paradoxes, attempting to refound quantum formalism without this "inconvenient" mathematical entity.

It is in the framework of these investigations that some important contributions by Vincenzo Capasso, Donato Fortunato, and Franco Selleri can be viewed. These authors introduced the concept of *sensible observable*[54], a term used to define those physical magnitudes whose expectation value on a state of the second type such as (10) is observably different from the one for any mixture of first-type state vectors such as (11). Fortunato and Selleri, in collaboration with Augusto Garuccio[55], also demonstrated that if $|\eta\rangle$ is a second-type state vector for two correlated physical systems, S_1 and S_2, then the projection operator $|\eta\rangle \langle\eta|$ is a sensible observable for the overall system $S_1 + S_2$. Starting from their concept of sensible observable, they showed the failure of the Furry (or Bohm-Aharonov) hypothesis to be a purely formal solution to the Einstein-Bell problem.

Let us consider the inequality that can be deduced from the sensible observable $|\eta\rangle \langle\eta|$ when $|\eta\rangle$ is the singlet state (10) of two spin-1/2 systems:

$$K = -P(\hat{\imath}\hat{\imath}) - P(\hat{\jmath}\hat{\jmath}) - P(\hat{R}\hat{R}) \leq 1, \qquad (28)$$

where $\hat{\imath}$, $\hat{\jmath}$, $\hat{R}$ are three unitary vectors along the three orthogonal axes x, y, z. The singlet state gives for any $\hat{a}$,

$$P(\hat{a}\hat{a}) = -1, \text{ so that } K = 3 \qquad (29)$$

The description of two correlated spin-1/2 physical systems through state vectors of first type therefore leads two a strong violation of quantum predictions, which has never, however, been confirmed by experiments.

The weakness of Furry's hypothesis is even more evident as soon as the state vectors of first type are applied to the description of ordinary classical situations. In this regard, Garuccio, Scalera, and Selleri[56] have suggested a classical model consisting of a statistical ensemble of pairs of spheres propagating in opposite directions with constant velocity. All the spheres are spinning, and in each pair the two rotations take place around opposite directions, which in the statistical ensemble may have an isotropic distribution. These authors then consider two experimental apparata A_1 and A_2, located on the path of the oppositely moving spheres in such a way as not to disturb their motion but of recording the *sign* of the spin projection on a certain direction a. On account of the opposite rotations, if A_1 records +1, then A_2 shall record -1, and *vice versa*. Therefore, the correlation function $P(aa)$, the average of the products of the correlated results obtained by A_1 and A_2, is necessarily -1, and this is true for all possible choices of a. If we refer to relation (29), we see that this classical model implies $K = 3$, and it is thus absolutely not describable through state vectors of the first type implying, according to (28), $K \leq 1$.

This conclusion that a modified, or more precisely "reduced," quantum mechanics cannot even reproduce the ordinary properties of the elementary classical model just described seems the more significant because it confirms the illusoriness of the formal solutions to microphysical paradoxes based on a modification of the mathematical structure of quantum theory or of its domain of application. Faced with the impossibility of arriving at this kind of solution to the Einstein-Bell contradiction, Italian research into this field have taken two main directions.

On the one hand, the relationship between the prescriptions of the Bell theorem and the formal description of two separate physical systems provided by theories other than quantum mechanics has been investigated. In this connection, Selleri and the present author have proposed a theorem showing that even some explicitly nonlocal theories, allowing the propagation of instantaneous action at a distance, always satisfy Bell's prescriptions[57]. We shall not deal with this theorem here except to comment on two consequences of our extension of the domain of validity of the Bell inequality. One concerns the distinction between the concept of locality directly following from Einstein's postulates of relativity theory and the more general idea of separability satisfied by classical or semiclassical theories, such as Newtonian or Hamiltonian mechanics and certain stochastic reformulations of quantum theory. Despite the actions at a distance implied by the latter theories, this concept of separability made it possible to separate the properties of the investigated physical objects from those of the instruments employed for their observation, in contrast with the quantum mechanical inseparability. The other consequence is that the only kind of mathematical description of physical systems capable of violating the Bell inequality is the quantum mechanical theory of spin, a typically quantum observable, which has no classical analog and whose commutation rules have thus been given heuristically without one being able to invoke the correspondence principle to deduce them from Poisson-bracket type classical relations.

Another approach of Italian research concentrated on the analysis of the hypothesis used to prove the Bell inequality. As is well known, Bell's famous proof of 1964[58] was valid for any *deterministic* local theory of hidden variables and was only later generalized to *probabilistic* theories by Clauser and Horne[59] whose theorem was founded on an additional assumption called the "*factorability* hypothesis," which consisted roughly speaking in the identification of the physical notion of locality with the axiom of statistical independence of standard probability calculus.

A realistic local model capable of violating this additional hypothesis has been suggested by Selleri and the present author[60] who have thus shown the insufficient generality of Clauser and Horne's probabilistic version of the Bell theorem. An idea of our model can be given by the following example. Let us consider N pairs of identical twins, with an arbitrarily large space-time interval separating the twins of each pair in such a way that, from the locality principle, any interaction with one of the two individuals can in no way disturb the state of the other. Let us now suppose we are interested in the two observables height h and weight w, defining each individual as *tall* when he taller than 1.80 meters and *heavy* when he is heavier than 80 kilograms.

We then call $p(h)$ and $p(w)$ ($q(h)$ and $q(w)$) the probabilities that the first (second) twin is tall and heavy respectively. We also know, and this is confirmed by our observations, that when the first twin is tall (heavy), it is highly probable that the second, who is identical to the first, is heavy (tall) and *vice versa*. We can thus define the conditional probability $P(h/w)$ ($P(w/h)$) that the second twin is heavy (tall) when the first is tall (heavy), the value of which will of course be different from that of the individual probability $q(w)$ ($q(h)$) that the second twin is heavy (tall). There now only remains for us to define the value of the joint probability $P(h \wedge w)$, of finding the first individual tall and the second heavy. According to Clauser and Horne, as a consequence of their identification of the physical notion of locality (or separability) with the mathematical axiom of statistic independence, this value will correspond to the product of the individual probabilities $p(h)$ and $q(w)$, which is openly in contrast with experience showing, instead, that the value of $P(h \wedge w)$ is given by the product of the value of $p(h)$ with that of the conditional probability $P(h/w)$. The latter way of writing the joint probability is not in contrast with the locality principle but merely indicates that we are dealing with correlated, not independent, events, demonstrating that there exist *probabilistic local* physical models where *separate but not statistically independent* events can be defined.

This result has been interpreted by some authors, such as Angelidis[61] and Popper[62], as proof that the Bell theorem is not a prescription necessarily satisfied by any probabilistic local theory and consequently as evidence for the groundlessness of its "universality claim," according to which "all possible local theories (of emission and propagation of particles in opposite directions) lead to statistical predictions that differ from the predictions of quantum formalism."[63]

A probabilistic refoundation of the Bell theorem not based on the factorability hypothesis has been attempted by Selleri and the present author[64] who, in different but converging ways, have obtained considerably general results[65]. The basic feature shared by both approaches lies in our use of a probabilistic principle of reality requiring neither the appeal to the hidden variable hypothesis nor to the notion of the predictability with certainty required by the original EPR criterion. Without dwelling on these works, the main recent developments of which are amply discussed in other contributions to this book and in particular in Selleri's paper, we shall make only some brief remarks on the nature of the reality principle introduced in the probabilistic version of the Bell theorem, which, in one of its more general forms, has been expressed as follows:

> "When, without interacting with a physical system S, we can predict at a given instant t_1 that we will find the property P -- i.e, that the physical quantity R defined on S has different possible values $\{r_1, r_2, \ldots, r_n\}$, with probabilities given respectively by $\{p_1, p_2, \ldots, p_n\}$ -- as a result of a subsequent observation or measurement performed on S at time $t_2 > t_1$, then we maintain that P is a real property of S and can therefore be attributed to it even at a time $t_0 < t_1 < t_2$."[66]

In the first place, we can observe that, as already mentioned, this new principle of reality makes it possible to supplant the EPR deterministic criterion of reality in two ways, both by eliminating the idealized notion of predictability with certainty and by not necessarily putting the existence of predictable properties in relation some hypothetical *elements of sub-microscopic reality*, generally known as hidden variables, which in turn would have determined the previous predictable, observable properties. Our assumption that the predictable properties are *themselves real* does not correspond to denying the epistemological value of the theory of hidden variables but rather to a search for a more general foundation for a realist conception, one that does away at the roots with the objection advanced by the supporters of the orthodox interpretation to the (presumed) metaphysical nature of the EPR reality criterion and also of some successive reformulations of it.

This new reality principle conserves, moreover, the concept of the *independence* of the observer (or of his mind or consciousness), as already occurred in the case of the theses of metaphysical realism, on the basis of which reality was identified as that something that "would exist even if we and every sensible creature were absent or annihilated" according to a famous formulation that Hume considered as meaningless. Nevertheless, quite unlike metaphysical realism, this independence does not regard the empirical objects as such but their predictable properties, such as in the case of the Lewis-Carnap realist hypothesis, discussed in Section 6, which constitutes the *meaningful reformulation* of the metaphysical realism refuted by Hume.

This shifting of the notion of reality from the object to its predictable properties[67] corresponds to endorsing the logical empiricist refutation of the identification of the concept of reality with that of a special kind of property of empirical objects endowed with a privileged status, already partially anticipated by the Kantian criticism of existence as a predicate. For, such an identification may produce an improper and often wrong use of the reality principle, consisting in its direct application to physical systems instead of to their predictable properties, as occurred for example in the cases of D'Espagnat, according to whom

> "the existence of a system as such (that is, independently of other systems) is to be considered one of its properties and it too should therefore also follow the EPR reality principle."[68]

This probabilistic re-foundation of the Bell inequality makes it possible, by virtue of the extreme generality of the reality principle employed, to overcome the false alternative between the two solutions discussed below, usually presented as the only possible way out of its (presumed) violations.

The first consists in negating the reality of the micro-world, in the sense that no kind of physical property directly attributable to micro-objects would exist. This solution sees the Bell inequality experiments as the definitive experimental confirmation of the orthodox interpretation of quantum mechanics and of Bohr's reply to the EPR argument. In this reply, Bohr did not question Einstein's locality condition, while he refused the EPR epistemological "criterion" of physical reality, maintaining that there do not exist properties directly attributable to micro-objects but only relational properties between micro-objects and the instruments that we use to observe them.

As I have already tried to show[69], this is a philosophically unsound solution: if predictability, which is a stronger requirement for scientific objectivity, did not represent a guarantee of physical reality, then neither could the experimental result $B=2\sqrt{2}$, confirming quantum mechanical predictions and, according to this alternative, disproving the reality principle, be considered a (real) property of micro-systems. The very experiment that should prove the falsity of the reality principle would thus be meaningless. For, on the basis of a single experiment or single class of experiments, one cannot falsify a philosophical principle on whose assumption the interpretation of every experiment depends and whose abandonment would not only involve a renouncement of objectivity but also the falling away of the very inter-subjective evidence and value of scientific knowledge.

The second assumes the existence of instantaneous superluminal signals, whose efficiency does not decrease as the distance increases, between different space-time regions. Some authors, such as for example Cufaro-Petroni in his contribution to the present volume, have tried to prove that such a possibility on the hand makes it possible to preserve a realist conception based on the EPR reality principle and on the other does not contradict the theory of relativity. However, such a solution is compatible with the relativity theory only if one is prepared to introduce new assumptions such as "tachyons," Dirac's ether, time-symmetric theories, the telegraphy into the past and similar *ad hoc* hypothesis which imply, amongst their consequences, a reversal between cause and effect, an interaction between objects that no longer exist in the present physical world, a universe regulated by an absolute determinism in which nothing happens because everything has already been written once for all in the space-time continuum.

Moreover, as has been stressed, if locality were experimentally refuted, the premise of the statement expressing the principle of reality would be false -- that is, the requirement of predictability without disturbance would no longer be satisfied because the measurement performed on the first of the two correlated systems might, through an instantaneous action at a distance, modify *even* the state of the second system -- and the same principle would therefore be *trivially true*. Accordingly, any violations of the locality condition would imply the inapplicability of the reality principle to the realm of microphysics.

The effects produced by a refutation of locality would therefore create havoc not only in the conceptual foundations of modern science, which has been based on this principle ever since Bacon's rule of *dissectio naturae*, and especially since Einstein's criticism of simultaneity, but also in a realist conception of nature which, like the EPR reality principle, or similar probabilistic generalizations of it, expressly avoids the recourse to old-fashioned holistic and organicist ontologies which oppose a complex and irreducible totality to its constituent parts. It should also be noted that the existence of superluminal signals that do not decrease in efficiency with increasing distance would be in contrast with the other presently known kinds of physical interactions which *all*, without exception, share the property of decreasing in intensity as distance increases. For the gravitational and electromagnetic interactions generate potential energies that decrease inversely proportionally to distance while nuclear and weak interactions decrease exponentially and therefore in an even more rapid manner.

Finally, as recently shown by Marshall, Santos and Selleri (MSS), there is a further physical objection to all the interpretations which view the current tests of the Bell theorem as a decisive experimental falsification of realistic local theories[70]. This is that the experiments carried out up to now with atomic cascades have been analyzed in the light of some *additional hypotheses*. The main of these hypotheses was proposed by Clauser, Horne, Shimony and Holt (CHSH) and consisted in the statement: "if a pair of photons comes out of two polarizers, the probability of their joint detection is independent of the orientation of the axes of the polarizers."[71] This was later reformulated by Clauser and Horne (CH) as follows: "for every atomic emission, the probability of detection when the polarizer is in place is not greater than the probability of detection when the polarizer is removed."[72] Now, according to MSS, there is no apparent reason for sustaining that the previous typically quantum-mechanical hypotheses apply even to realistic local theories. As a matter of fact, once one has accepted the idea that present quantum theory predicts only statistical results and that a more detailed theoretical description is possible if one takes into account the individual properties of physical systems, the detection process can be described in a different way for different objects. This natural assumption is sufficient to violate both the CHSH and the CH hypotheses.

Such a criticism of the additional hypotheses introduced into the analysis of the experiments on the Bell theorem does not, however, mean, as recently highlighted by Garuccio and Selleri, that the empirical contrast between quantum mechanics and realistic local theory has fallen away but simply that it has been shifted to a different level[73].

Whatever the results of the experiments on the Bell theorem, the latter seems bound to lead to crucial developments for the conceptual foundations of contemporary physics. If the locality principle is found to be violated, the existence of instantaneous influences between objects separated by arbitrarily large distances will have to be admitted, this being in contrast not only with the postulates of special relativity but even with the mathematical structure of nonlocal theories such as non-relativistic, classical mechanics and de Broglie's and Bohm's hidden variable theories. Theoretical researches into "tachyonic" effects have recently stressed that the existence of such presumed superluminal actions can be considered to be compatible with the formalism of the theory of relativity even if it appears to be in open contrast with Einstein's interpretation of this formalism in that it is based on the assumption of an inversion of cause and effect or on similar ad hoc hypotheses like those mentioned earlier.

If, on the other hand, Einstein locality succeeds in overcoming the experimental tests on the Bell theorem, then it is quantum formalism -- and not only its Copenhagen interpretation -- that will have to be radically modified, as already stressed:

> "If a change in quantum theory will take place in the future in order to get rid of nonlocality, this will probably not be a *minor* change. As we saw ... state vectors of the second type are responsible for nonlocal effects. Their elimination implies a drastic modification of the superposition principle, that is, of the linear nature of quantum laws. This would, however, imply very probably an automatic resolution of the measurement problem (the reduction of the wave packet which is the passage from a superposition to a mixture of states would no longer be necessary) ..."[74]

9. CONCLUSIONS

The research of the last two decades carried on in Italy in the foundations and philosophy of quantum mechanics has considerably contributed to the clarification of the nature of the paradoxes of this theory and indicated at the same time the possibility of solving its basic conceptual problems, from at least three viewpoints.

In the first place, the common origin of quantum paradoxes has been highlighted. We have seen that all these paradoxes have their logical and epistemological roots in three different kinds of unsolved duality within the present theory -- waves vs. particles, deterministic vs. probabilistic description, state vectors of the first kind vs. state vectors of the second kind.

Secondly, the possibility has been shown of their resolution through a realistic interpretation of quantum theory that involves not only its basic theoretical concepts such as that of wave function in the case of Selleri's quantum wave hypothesis and that of measurement, as in the case of the DLP theory, but also the theory as a whole, establishing its compatibility (or incompatibility) with a general principle of physical reality such as that of EPR or its subsequent probabilistic generalizations.

Finally, it has been stressed how such a proposed solution is based on the following three, different kinds of experiments on the validity of the present theory:

- experimental tests of the interpretation of the wave function, which regard the philosophical interpretation of one of the fundamental theoretical concepts of quantum mechanics and the result of which would not question the validity of quantum formalism but only that of the principle of complementarity on which the orthodox interpretation is based;
- experimental tests of the postulate of the reduction of the wave function which implies, instead, the possibility of refuting the mathematical formulations of quantum mechanics based, like the orthodox one, on this postulate and the need to appeal to alternative formulations, such as, for example, the so-called statistical interpretation of Popper, Landé and Margenau;
- experimental tests of the (superluminal) quantum correlations, the result of which would question the validity of the superposition principle and the very, linear, mathematical structure of quantum formalism.

The search for an empirical solution to quantum paradoxes seems, therefore, to have led to the individuation of three different *levels of testability* for a given physical theory which range from the level of the philosophical interpretation of its basic concepts which is thus no longer matter of an arbitrary metaphysical choice but becomes a genuine empirical problem, to that of its mathematical formulation based on additional postulates which can be eliminated in alternative formulations, as far as the level of its fundamental physico-mathematical principles, the refutation of which implies in turn an actual experimental falsification, in Popper's sense, of the existing theory.

REFERENCES

1.   B. Croce, *Logica come scienza del concetto puro*, Laterza, Bari

(1908); p.90.

2. G. de Ruggiero, *Storia della filosofia*, Laterza, Bari (1934); Vol.X, pp.203-204.

3. G. Gentile, Jr., Introduction to J. Jeans, *I nuovi orizzonti della scienza*, Sansoni, Firenze (1934).

4. *Ibid.*, pp.ix-xi.

5. A. Pasquinelli, *Nuovi principi di epistemologia*, Feltrinelli, Milano (1964); p.57.

6. A. Daneri, A. Loinger, G.M. Prosperi, "Quantum theory of measurement and ergodicity conditions," *Nucl. Phys. 33*, 297 (1962).

7. F. Selleri, "On the wave-function of quantum mechanics", *Lett. Nuovo Cimento 1*, 908-910 (1969).

8. V. Capasso, D. Fortunato, F. Selleri, "Von Neumann's theorem and hidden-variable models", *Riv. Nuovo Cimento 2*, 149-199 (1970).

9. E. Agazzi, *Temi e problemi di filosofia della fisica*, Manfredi, Milano (1969); p.50.

10. E. Agazzi, "The concept of empirical data. Proposal for an intentional semantics of empirical theories", in M. Przelecki *et al.*, eds., *Formal Methods in the Methodology of Empirical Sciences*, Reidel, Dordrecht (1976).

11. M.L. Dalla Chiara, G. Toraldo di Francia, "A formal analysis of physical theories," in G. Toraldo di Francia, ed., *Problems in the Foundations of Physics*, North Holland, Amsterdam (1979).

12. S. Tagliagambe, "Il dibattito sull'interpretazione della meccanica quantitistica", in L. Geymonat, *Storia del pensiero filosofico e scientifico*, Garzanti, Milano (1972); Vol.VI, pp.729-761.

13. S. Tagliagambe, ed., *L'interpretazione materialistica della meccanica quantistica. Fisica e filosofia in URSS*, Feltrinelli, Milano (1972).

14. *Ibid.*, p.115.

15. G. Tarozzi, "Realistic interpretation of physical theories", *Memorie dell'Accademia Nazionale di Scienze, Lettere ed Arti di Modena 20*, 49-62 (1978).

16. V. Tonini, "Il testamento scientifico di Einstein e la filosofia della fisica oggi", *La Nuova Critica*, 50-51 (1979).

17. A. Pignedoli, *Alcune teorie meccaniche superiori*, CEDAM, Padova (1969); pp.14-15.

18. *Ibid.*

19. G. Tarozzi, "Realisme d'Einstein et méchanique quantique: un cas de contradiction entre une théorie physique et une hypothèse philosophique clairement définie", *Revue de Synthèse 101-102*, 125-158 (1981).

20. M.L. Dalla Chiara, G. Toraldo di Francia, *La Teoria Fisica*, Boringhieri, Torino (1982); p.10.

21. M.L. Dalla Chiara, "A general approach to non-distributive logics," *Studia Logica 25*, 139-162 (1976).

22. E.G. Beltrametti, G. Cassinelli, "Logical and mathematical structures of quantum mechanics", *Riv. Nuovo Cimento 6*, 321-404 (1976); p.321.

23. P.A.M. Dirac, *The Principles of Quantum Mechanics*, Clarendon Press, Oxford (1958).

24. E.G. Beltrametti, G. Cassinelli, "Logical and mathematical structures of quantum mechanics", *Riv. Nuovo Cimento 6*, 321-404 (1976); p.354.

25. L. Accardi, "The probabilistic roots of quantum mechanical paradoxes", in S. Diner et al., eds., *The Wave-Particle Dualism*, Reidel, Dordrecht (1984); pp. 297-330.

26. *Ibid.*, p.317.

27. F. Selleri, G. Tarozzi, "Quantum mechanics, reality and separability", *Riv. Nuovo Cimento 4*, 1 (1981).

28. L. Accardi, "The probabilistic roots of quantum mechanical paradoxes", in S. Diner et al., eds., *The Wave-Particle Dualism*, Reidel, Dordrecht (1984); pp. 318-319.

29. F. Selleri, G. Tarozzi, "Why quantum mechanics is incompatible with Einstein locality", *Phys.Lett.*, *119*, 101 (1986).

30. F. Selleri, G. Tarozzi, "Is non-distributivity for microsystems empirically founded?", *Nuovo Cimento B 43*, 31 (1978).

31. F. Selleri, "On the wave-function of quantum mechanics", *Lett. Nuovo Cimento 1*, 908-910 (1969).

32. F. Selleri, "Realism and the wave-function of quantum mechanics", in *Foundations of Quantum Mechanics*, Academic Press, New York (1971); pp. 398-406.

33. *Ibid.*

34. *Ibid.*

35. L. de Broglie, Letter to F. Selleri, dated April 11, 1969.

36. *Ibid.*.

37. F. Selleri, "Gespenstelfelder", in S. Diner et al., eds., *The Wave-Particle Dualism*, Reidel, Dordrecht (1984).

38. G. Tarozzi, "Two proposals for testing physical properties of quantum waves", *Lett. Nuovo Cimento 35*, 53-59 (1982); "From ghost to real waves: a proposed solution to the wave-particle dilemma", in S. Diner et al., *The Wave-Particle Dualism*, Reidel, Dordrecht (1984).

39. G. Tarozzi, "The principle of empiricism and quantum theory", *Epistemologia 3*, 13-28 (1980).

40. A. Daneri, A. Loinger, G.M. Prosperi, "Quantum theory of measurement and ergodicity conditions," *Nucl. Phys. 33*, 297 (1962).

41. G. M. Prosperi, *Teoria quantistica della misurazione*, in *Enciclopedia della scienza e della Tecnica*, Mondadori, Milano (1963); Vol.VI, p.863.

42. L. Rosenfeld, "The measuring process in quantum mechanics" *Suppl.Progr.Theor.Phys.*, 212 (1965).

43. J.Bub, "The Daneri-Loinger-Prosperi quantum theory of measurement", *Nuovo Cimento, 57B*, 503 (1968). The reactions to the DLP theory are discussed in more detail from a historical perspective in M. Jammer, *The Philosophy of Quantum Mechanics*, Wiley, New York (1973); pp.492-493.

44. E.P. Wigner, "The problem of measurement", *Am.J.Phys.*, *31*, 6 (1963).

45. M. Renninger, "Messungen ohne Störung des Messobjects", *Zeit.Phys.*, *158*, 417 (1960).

46. M.L. Dalla Chiara, G. Toraldo di Francia, "A formal analysis of physical theories," in G. Toraldo di Francia, ed., *Problems in the Foundations of Physics*, North Holland, Amsterdam (1979); p.134.

47. *Ibid.*, p.110.

48. *Ibid.*, p.101.

49. M. Cini, "Quantum theory of measurement without wave packet collapse," in G. Tarozzi, and A. Van der Merwe, eds., *Open questions in Quantum Physics*, Reidel, Dordrecht (1985); p.185.

50. G. Tarozzi, "Teoria e strumento in microfisica", *Epistemologia 8*, 83 (1985).

51. G. Tarozzi, "A unified experiment for testing both the interpretation and the reduction postulate of the quantum mechanical wave function", in G. Tarozzi, and A. Van der Merwe, eds., *Open Questions in Quantum Physics* (Reidel, Dordrecht, 1985).

52. A. Garuccio, V. Rapisarda, J.P. Vigier, "New experimental set-up for the detection of de Broglie's waves", *Phys. Lett. 90 A*, 17 (1982). An earlier but less stringent experimental proposal is discussed by A. Garuccio, K.R. Popper, J.P. Vigier, in "Possible direct physical detection of de Broglie's waves", *Phys. Lett 86 A*, 397 (1981).

53. G. Tarozzi, "Experimental tests of the properties of the quantum mechanical wave function", *Lett. Nuovo Cimento 42*, 439-442 (1985).

54. V. Capasso, D. Fortunato, F. Selleri, "Sensitive observables of quantum mechanics", *Int. J. Theor. Phys.* 7, 319 (1973); D. Fortunato, F. Selleri, "Sensitive observables on infinite-dimensional Hilbert spaces," *Int. J. Theor. Phys. 15*, 333 (1976).

55. D. Fortunato, A. Garuccio, F. Selleri, "Observable consequences from second-type state vectors of quantum mechanics," *Int. J. Theor. Phys. 16*, 1 (1977).

56. A. Garuccio, G. Scalera, F. Selleri, "On local causality and the quantum-mechanical state vector", *Lett. Nuovo Cimento 18*, 26 (1977).

57. F. Selleri, G. Tarozzi, "Nonlocal theories satisfying Bell's inequality", *Nuovo Cimento B 48*, 120 (1978); "Extension of the domain of validity of Bell's inequality", *Epist. Lett. 21*, 1 (1978).

58. J. S. Bell, "On the Einstein-Podolsky-Rosen paradox", *Physics 1*, 195 (1964).

59. J. F. Clauser, M. A. Horne, "Experimental consequences of objective local theories", *Phys.Rev.D 10*, 526 (1974).

60 F. Selleri, G. Tarozzi, "Is Clauser and Horne's factorability a necessary requirement for a probabilistic local theory?", *Lett. Nuovo Cimento 29*, 533 (1980).

61. T. D. Angelidis, "Bell's theorem: Does the Clauser-Horne inequality hold for all local theories?", *Phys.Rev.Lett. 51*, 1819 (1980); "Does the Bell inequality hold for all local theories?"

in G. Tarozzi, and A. Van der Merwe, eds., *Open questions in Quantum Physics*, Reidel, Dordrecht (1985); p.51.

62. K. R. Popper, "Realism in quantum mechanics and the new version of the EPR experiment", in G. Tarozzi, and A. Van der Merwe, eds., *Open questions in Quantum Physics*, Reidel, Dordrecht (1985); p.3.

63. *Ibid.*, p.11.

64. F. Selleri, G.Tarozzi, "A probabilistic generalization of the concept of physical reality", *Spec. Sci. Tech. 6*, 55 (1983).

65. F. Selleri, "On the consequences of Einstein locality", *Found. Phys. 8*, 103 (1978); G. Tarozzi, "Local realism and Bell's theorem without the hidden-variable hypothesis", *Atti Accad. Scienze Torino 108*, 119 (1981).

66. G. Tarozzi, "Physical reality: from the metaphysical notion to its empirical definition", in Bitsakis, E., ed., *The Concept of Physical Reality*, (Zacharopoulos, Athens, 1983), p. 197.

67. G. Tarozzi, "The conceptual development of the E.P.R. argument", *Mem Accad. Naz.Scienze, Lettere ed Arti di Modena 21*, 353 (1979); "Realism as a meaningful philosophical hypothesis", *Atti Accad. Scienze Bologna, Rendiconti 7*, 89 (1980).

68. B. D'Espagnat, *I fondamenti concettuali della teoria quantistica*, Bibliopolis, Napoli (1977); p.114.

69. G. Tarozzi, *Discussion* following Popper's opening lecture, in G. Tarozzi, and A. Van der Merwe, eds., *Open questions in Quantum Physics*, Reidel, Dordrecht (1985); pp. 31-32.

70. T.W. Marshall, E. Santos, F. Selleri, "Local realism has not been refuted by atomic cascade experiments", *Phys.Lett. 98A*, 5 (1983).

71. J.F. Clauser, M. A. Horne, A. Shimony, R. A. Holt, "Proposed experiment to test local hidden-variable theories", *Phys. Rev. Lett. 23*, 880 (1969).

72. J. F. Clauser, M. A. Horne, "Experimental consequences of objective local theories", *Phys.Rev.D 10*, 526 (1974).

73. A. Garuccio, F. Selleri, "Enhanced photon detection in EPR type experiments", *Phys. Lett. 103 A* 99 (1984).

74. F. Selleri, G. Tarozzi, "Quantum mechanics, reality and separability", *Riv. Nuovo Cimento 4*, 1 (1981); p.50.

PART 1

THE EPISTEMOLOGICAL ROOTS OF QUANTUM
PARADOXES

WAVES, PARTICLES, AND COMPLEMENTARITY

Evandro Agazzi

Séminaire de Philosophie, Université de Fribourg Suisse
Fribourg, Switzerland

ABSTRACT

It is shown that the complementarity principle does not really eliminate the conceptual difficulties of the coexistence of wave-like and particle-like features in quantum phenomena. In order to overcome them a purely formal consideration of quantum sentences must first be applied, which provides a consistent implicit contextual definition of the concepts involved. This is enough for providing exact meanings of these concepts and for eliminating contradictions. At the same time it opens the possibility of constructing models of the microworld that may be endowed with physical meaning though not being intuitive.

1.THE WAVE PICTURE AND THE PARTICLE PICTURE

Classical physics, i.e., the physics which received its definitive form at the end of the nineteenth century, provided a sort of all-encompassing picture of natural phenomena, within which different concepts could be referred to either of two sides of a fundamental bipartition. On the one side were the material substances (that is, essentially, the atoms of the different chemical elements, considered as unchangeable entities, and the molecules of the various "substances" proper, obtained from the composition of such atoms); on the other hand were fields and different forms of radiation (light, heat radiation, electricity, and magnetism). The general attitude was to consider material substances as the real natural basis of physical phenomena juxtaposed to fields and radiation, which were intended rather as mental frameworks and logical models aimed at representing the evolution of the former: This was especially true after the attempt to "substantiate" the field by means of the concept of ether met with difficulties.

A particular series of characteristics was attributed to each of these categories: Matter was viewed as being constituted of particles and hence as having a discrete corpuscular structure and as being localizable within a limited region of space. Fields and radiation, on the

53

G. Tarozzi and A. van der Merwe (eds.), The Nature of Quantum Paradoxes, 53–74.
© 1988 by Kluwer Academic Publishers.

other hand, were taken to have a wave-like continuous nature and were thought of as energy carriers extending over the whole of space.

In this situation it was altogether natural that, whenever new phenomena presented themselves, physicists tried to understand them by incorporating them into one of these two conceptual frameworks.

It is well known, however, that from the very beginning of the twentieth century it became more and more difficult to handle physical entities in this way. All things showed themselves to be, at one and the same time, both particles and fields, both matter and radiation. All things appeared to possess the continuous structure and well known wave-like properties of the field, as well as the discrete and just as well known corpuscle-like properties of particles.

Due to this ambiguity, the representation of the physical world became uncertain: Indeed this overlapping of the corpuscular and wave-like images gives rise to real conceptual difficulties.

2.THE CORRESPONDENCE PRINCIPLE

It is true that real difficulties were already implied by the "quantum postulate", but in fact they only became apparent after a certain time, namely after the "correspondence" principle had ceased to act as a safety valve. (The principle itself is the simple assumption that the quantum theory, or at any rate its formalism, contains classical mechanics as a limiting case.)

Availing oneself of the principle, the interpretation of new phenomena could always be attempted by applying classical concepts and theories, except for the ad hoc introduction of "quantum conditions." This principle afforded a valuable and versatile tool for the development of early quantum theory, but, as time went by, it showed itself to be more and more difficult to reconcile with another of Bohr's ideas. This other idea is that which gave rise to the "principle of complementarity," which implies the fundamental irreducibility of quantum mechanics to classical mechanics. The principle of complementarity thus also led to the setting aside of the belief that quantum mechanics is a "rational generalization" of classical mechanics (as Bohr reasoned), as well as the belief that, on the contrary, the former is a "restriction" of the latter (in line with the thought of Sommerfeld and Ehrenfest, who emphasized the restrictive character of quantal conditions).

At this point, there were two alternatives: either to give up classical concepts altogether, or to try to adapt them to the new situation in a way more effective than that afforded by the correspondence principle. The second route was preferred, both on practical grounds (essentially because the theoretical tools available for the development of the new mechanics were precisely those of the old, for instance the use of the Hamiltonian function), and for reasons of principle (it was not clear how new concepts were to be formulated and correlated with classical concepts, which were still essential for the description of measuring devices and their results).

3.THE COMPLEMENTARITY PRINCIPLE

The most peculiar thing about this principle is the fact that it
has never been stated in a clear and unequivocal way, not even by its
own originator, Niels Bohr. Indeed, someone has remarked, somewhat ma-
liciously but not unreasonably, that one of the reasons for "the persistence of the faith in complementarity" irrespective of consistent
objections, was to be found in the very vagueness of the fundamental
principles of such belief.[1]

If one considers certain of Bohr's remarks, the principle seems
essentially to involve a "description" of phenomena.[2] But on other
occasions, Bohr speaks of "pictures":

"Evidence obtained under different experimental conditions
cannot be comprehended into a single picture, but must be
regarded as *complementary*, in the sense that only the totality of the phenomena exhausts the possible information about
the objects."[3]

The vagueness which Bohr, perhaps not unintentionally, failed to
dissipate with his terms "description" and "picture" was in part removed by others. Pauli, for instance, does not mention complementary
descriptions but rather complementary classical concepts:

"If the possibility of utilizing one classical concept is
related by way of mutual exclusiveness to the possibility
of utilizing another, we, along with Bohr, refer to these
two concepts as being complementary (for instance, the coordinates and momenta of a particle)."[4]

It is not difficult to see that the concept of complementarity is
here interpreted differently: We are not dealing, as in Bohr's case,
with two mutually exclusive classical descriptions of atomic phenomena,
but with two concepts which, belonging to the same "classical" description (in the above example, the corpuscular description), cannot be
utilized at the same time. We are thus faced witn a faithful transcription of Heisenberg's uncertainty principle rather than a transcription
of Bohr's complementarity principle. Others have reformulated this principle in yet another fashion (for instance C.F. von Weiszsäcker).

It is important to emphasize these differences, not only because
they give an ides of the conceptual difficulties which invariably
accompanied this famous principle, but also because they can help to
explain why there are still so many contrasting views concerning its
evaluation. Historically, moreover, it is precisely this ambiguity that
has helped to conceal the logical difficulties encountered by this principle. It is indeed clear that the principle, considered as the simultaneous proposal of two contrasting "pictures," two "descriptions," or
two embryonic "theories" pertaining to microphysical reality, could not
escape a severe judgement, such as the trenchant one once expressed by
Schrödinger.

"There is," he says," another concept, the complementarity
concept, which Niels Bohr and his disciples defend and of
which everybody makes use. I must confess I don't understand
it. To me, it is nothing but an evasion. Not a voluntary one.
In fact one ends up admitting that we have two theories, two
images of matter, which don't agree with each other, so that
at times we must utilize one, at times the other. Once, seven-
ty-odd years ago, when such an event occurred, one concluded
that the investigation was still unfinished, since it was considered absolutely impossible to utilize two different concepts in connection with one phenomenon or with the constitu-
tion of a body. Now they have invented the word "complementarity," and this seems to me to be the attempt to try and justify this use of two different concepts, as if it were not
necessary to find eventually a single concept, a complete
image which can be understood. The word 'complementarity' invariably reminds me of Goethe's words: 'Because where the
concepts are missing, at the right moment a word appears.'"[5]

How can one answer such arguments? In order to prevent accusations
about the "provisional" nature of the theory, and at the same time to
defend it from the charge of inconsistency, the backers of complementarity found valuable support in Heisenberg's uncertainty principle: Although it is true (and historically verifiable) that Bohr was not led
to the idea of complementarity by the knowledge of Heisenberg's principle, it is just as true that he found in it a kind of psychological
confirmation and logical justification. In fact, the uncertainty principle clarified for him the price one has to pay for the adoption of
complementary but irreconciliable notions, and at the same time showed
that one would never be brought to contradiction, since one could not
hope to treat the two complementary and irreconciliable aspects of a
given phenomenon simultaneously.

This justification has been repeated for decades by the supporters
of the "Copenhagen school," but its apparent strength derives from the
very confusion between the two different aspects of the complementarity
notion mentioned above. It is one thing if the newly discovered microphysical reality *prevents* us from simultaneously using, with an accuracy
exceeding a given limit, two *mutually compatible* concepts (e.g., position and velocity) the meanings of which come from classical mechanics:
It is quite another matter if we admit that the new situation appears
to *oblige* ua simultaneously to utilize two *mutually incompatible* concepts (e.g., wave and particle). The first case does not pose problems
of logical compatibility or intrinsic consistency; at most it raises a
question of the adequacy of the single classical concepts in their application to quantal situations. The second case, however, actually poses
problems of intrinsic consistency, the solving of which is not aided by

the uncertainty principle, for various reasons. First, the principle is
nothing but the realization of a situation of the former kind, which has
no clear-cut connection with that of the latter; second, the logical
compatibility of the concepts cannot be guaranteed by the simple fact
that they do not come into direct confrontation on the experimental
level due to uncertainty in measurements.

4.THE INTUITIVENESS OF QUANTUM PHYSICS

What do we really mean when we say that the concepts of classical
physics only have limited applicability to the new reality of quanta?
We mean, no doubt, that their use is not wholly sufficient to bring us
to understand such a reality; and from this point of view, the principle
of complementarity offers no further help. Indeed, in the Bohr formulation we have been considering up to now, the principle contains an implicit assertion, namely that only the totality of phenomena covers the
possibility of information on the subject. This assertion is not so much
obvious as altogether trivial, since no one denies that a knowledge of
quantum facts must take into account all phenomena, including the aspects
we can understand using a wave-like model, and those we can understand
with a corpuscular model. The problem lies elsewhere, namely in the fact
that we must understand $h\ o\ w$ one type of phenomenon is to be reconciled with the other. It is not sufficient to state that they are complementary, since simply to use the term "complementary" is, as
Schrödinger argued, to hide behind a word when the concepts are lacking.

A justification of the following kind is often encountered: Without
recourse to waves and particles, we cannot obtain an image of the microscopic world, and we cannot understand it without having an image of it.
Therefore, these images must be perceived; and the only way to do this is
by recognizing that they are complementary, while generally being considered mutually exclusive. Here again we note that calling them by a new
name (i.e., "complementary") does not suffice to insure the reconcilia-
bility of such images; and moreover, we may say that what we have here
is a typical kind of self-punishment for the choice made. It is well
known, in fact, that the supporters of complementarity are compelled to
admit that, just because of the simultaneous presence of these concepts
and these complementary but contrasting images, quantum physics is not
intuitive.[6]

Here the situation becomes almost paradoxical. One begins by repeatedly stating that we can only understand what can be expressed in classical images, such as waves and corpuscles, and on this basis we justify
the use of such images even in the realm of microphysics. But at a certain point we realize that the very presence of such images makes it
impossible to obtain an image of the whole of quantum physics; and this,
strictly speaking, amounts to making quantum physics itself incomprehen-
sible.

5.A LOGICAL ANALYSIS OF COMPLEMENTARITY

Faced with the choice between wave and particle, scientists- first divided themselves into two parties, one of which argued exclusively for the particle picture, while the other argued exclusively for the wave picture. From a logical point of view, this taking of sides is quite correct: When two statements are mutually exclusive, one must inevitably admit that one of them is false, and this agrees with logic, according to which *only one* is true. However, in the present case, neither of the two opposing points of view can be completely eliminated, since both are completely successful in certain cases, though both fail whenever one tries to extend them in such a way that they explain all known facts.

Logic pure and simple would then require one to declare both models false, while institutional physics has chosen the opposite view, declaring both true, though "complementary." This, however, rather than solve the problem of wave-particle dualism, is simply a recognition of the fact that this dualism cannot be eliminated; and, by *stating* that there is no real contradiction, one suggests that the problem is not in need of a solution. How can the majority of today's physicists be satisfied with a position which, from the point of view of logic, is so very weak?

The answer to this question is hidden but not difficult. Let us begin by considering a very intuitive fact of which formal logic offers a rigorous demonstration: If a set of expressions can be given an "interpretation" in terms of some universe of objects, in such a way that the expressions are all simultaneously true (i.e., if they admit any "model" whatsoever), then the set of expressions is logically consistent. This means not only that such expressions are not mutually contradictory, but also that their logical consequences cannot contradict one another.

Now, it has been established that those expressions of quantum mechanics which have a corpuscular "aspect," so to speak, are true with respect to essentially the same universe of objects as are the expressions which have a wave-like "aspect"; and such a universe can be viewed, generally speaking, as the set of subatomic physical entities. One can then conclude that the expressions of each type constitute a non-contradictory set, since they are all true with respect to the same universe of objects.

The people who formulated the principle of complementarity must have been aware of something of this kind, albeit not in such an explicit form. The agreement of the quantal formalism with previous experimental results, its compatibility with new results, and the great variety of phenomena it encompassed, gradually helped dispel doubts concerning its consistency, in spite of the absence of a really satisfactory "interpretation."

The strong point of the principle of complementarity therefore lay in an implicit explanation of the abstract guarantees of compatibility, which consisted in the fact that all the statements involved were somehow "true of the same objects."

The weak point, however, consisted in the belief that this simple fact might itself constitute an *explicit indication* of the compatibility of the various expressions, and, even worse, that this might mean that *the two models*, particle-like and wave-like, were compatible. One must not forget that the different expressions of quantum theory can all be considered true of the same objects only if they are viewed from the outset as *purely formal statements*. If we interpret them on the contrary, as referring to, for example, waves (that is, assuming that subatomic entities are waves), we know that some of them are false, while, if they are taken as referring to particles, others are false.

The real situation then appears to be the following: The existence of a universe of objects for which all statements are true is an *a priori* guarantee of compatibility. It does not however offer any indication of *how* such a compatibility is realized. The irreconcilability of wave and particle "interpretations" is not removed by a knowledge of the consistency of the formalism, since this guarantee of consistency is of a purely formal nature. It means, at most, that there exists at least *one* universal interpretation of the various expressions which is satisfactory from a logical point of view, but it does not mean that any interpretation *whatsoever* (in particular those already existing) is satisfactory. On the contrary, we can predict that an interpretation capable of explicitly reconciling all the expressions of the quantum formalism (if someday it should present itself) will certainly be *neither* corpuscular *nor* wave-like (instead of being *both* corpuscular *and* wave-like). This is not to deny that such an interpretation might be expected to contain some aspects *analogous* to the corpuscular model and others analogous to the wave model, though the analogies would in any case be very partial.[8] But the envisaged interpretation would never constitute a true "reconciliation" of the two models, since each of them necessarily contains its "negative" aspects as well. It would simply be something new.

6. THE PROBLEM OF RESORTING TO INTERPRETATIONS

Is it really necessary however to obtain such an interpretation? This is not an easy question to answer, since it is not at all clear what it should mean in physics.

In mathematical logic (as we have already mentioned and as will be recalled later) the notion of interpretation is relatively clear, but in physics such clarity has yet to be attained. From our best evidence, i.e., from the writings of the greatest theoreticians, we might conclude that an interpretation is more or less an "image," an "intuitive model," or the possibility of "visualizing" phenomena. We might thus consider the above question as asking whether an interpretation of the quantum formalism, viewed in this light, is at all necessary.

We have earlier denied such a necessity, and, if we look at the situation more closely, we should see that the backers of the complementarity principle had already opened the way for this renunciation. One cannot help but attribute a considerable heuristic value to this principle, since, due to the lack of better tools, it suggests that one should try to understand subatomic phenomena by exploiting certain conceptual models that are already available. The weak aspect of the principle consists in its going beyond its heuristic function and presenting itself as a logical justification for the adoption of contrasting models, by means of the purely verbal device of declaring them "complementary." On the other hand we have already observed that this is accompanied by a kind of "self-punishment," which is contained in the repeated statements of the supporters of complementarity, according to whom the simultaneous presence of corpuscular and wave-like properties informs us that we must forever give up the hope of forming an intuitive image of microphysical events. This amounts to an admission that it is not necessary to possess an "interpretation" or a "model" of such events.

The logically most correct way to maintain that the equations and laws of quantum mechanics do not suggest a model of reality would be to admit that they constitute nothing but a purely formalized axiomatic system. However, when one speaks of axiomatization to a mathematician, one almost has to warn him of the danger of overestimating it, while, when one speaks to a physicist, one has to make a real effort to overcome his mistrust (this holds true even for those physicists who refuse to adhere to any type of "model" whatsoever). The reason for the mistrust lies in the fact that the statements of an axiomatic system have no "physical meaning"; but then this mistrust means that, somehow, we attribute to the words "wave" and "particle" a richer function than that of simple shorthand, viz., the function of signifying, and thus of denoting specific aspects of physical reality (in contrast to the explicit statement that they do not express any interpretation).

Let us assume that a consistent physicist, faced with this objection, answers that indeed quantum theory is nothing but a formal system. In this case there are two possibilities: Either he believes that the theory nevertheless "constitutes" the physical entities, in the sense in which Hilbertian mathematicians maintained that the axioms of a mathematical theory constitute, i.e., give substance to mathematical entities (in our case, one should thus say that the elementary particles are the equations of quantum mechanics--even Heisenberg at times declared himself in favour of this interpretation). Alternatively, he believes that it is only a type of formal algorithm, almost a computing machine which, without saying anything determinable "content-wise," nevertheless allows the correlation of experimental data.

The first alternative is contradicted by the fact that, in reality, quantum theory is utilized and thrives as an attempt to explain facts which it itself does not produce but are given; and, characteristically, it is neither more nor less subject to experimental control than any

other physical theory. The second alternative is also beset with problems, due to the theoretical inadequacies of the so-called "phenomenological" theories.

The correct position seems to lie somewhere in between: On the one
hand, one must guarantee such a freedom from prejudicial interpretations
and such a versatility with respect to adaptation to the experimental
data as can only be supplied by the adoption of an explicit formal system,
while, on the other hand, it is important that the system not be allowed
to become purely formal.

7.THE RECOURSE TO AXIOMATIZATION

The above exigence has been expressed merely as a need to overcome
the present difficulties of quantum physics by means of an adequate
axiomatization, which, in order actually to be adequate, must possess
its own semantics. Indeed, from what has been said above, it is clear
that the greatest effort in such an axiomatization must be directed
toward semantics, and this in two respects, both of which are very important. It must do so, in the first place, in its capacity as an accurate explanation of the primitive terms and a precise characterization
of the semantic components of many concepts, which are presently utilized
in a perhaps excessively "global" fashion, and, in the second place, in
its capacity as a search for a physical reference for the meanings of
certain terms.

One need not really start from scratch: In such an endeavour quantum
mechanics had been substantially axiomatized by Dirac as early as 1930,
though his work was not presented as an axiomatization, and therefore
requires a little reading between the lines to be recognized as such.
On the other hand, von Neumann is responsible for an explicit axiomatization (1932), which is quite noteworthy even from a formal point of
view. It is certainly not an exaggeration to say that, apart from purely
handbook-like and pedagogical expositions, the formal presentations of
quantum mechanics are to this day of the Dirac-von Neumann type. No
explicit reference is made in them to either waves or particles, but the
formalism is interpreted by utilizing such concepts as position, momentum,
linear superposition and so on.[9] Currently, other interesting attempts
at axiomatization are being carried out, although it is not possible to
discuss them in this article.[10]

We wish now to try to clarify why we believe that this very move
toward axiomatization can help in overcoming those radical difficulties
which the principle of complementarity recognizes without really trying
to solve. In order to introduce our considerations, it is appropriate
to discuss a well-known example.

As everyone knows, the rise of non-Euclidean geometries at first
created considerable difficulties. Geometry, in fact, seemed to demand
that *one and only one* line parallel to a given line could pass through
a given point, while on the other hand admitting *no* such line and *more*

than one of them as well. This at first led a few scientists to speak
of a crisis in logic, and others to speak of the fallacy of some of the
three different geometries. The advent of the quantum theory had similar
effects: There have been (and still are) scientists who have seen in it
the origin of a crisis in logic, and there have been (and still are)
those who believe it possible to obtain a "classical" explanation of
the new quantal events.

The analogies go further than that: It is well known that, when
considered within a broader frame, Euclidean geometry appears as the
limiting parabolic case both of non-Euclidean hyperbolic geometry and
of non-Euclidean elliptic geometry, just as classical physics appears to
be the limiting case of quantum physics. Not only that: Each non-Euclidean geometry admits its own Euclidean model, much in the same way as
given parts of the quantum formalism admit, separately, classical models
such as those of the wave and the particle. On the other hand, there is
also in geometry something similar to the "correspondence principle,"
since to every Euclidean theorem there corresponds a non-Euclidean theorem of analogous "structure," in which however the different "conditions
of parallelism" introduce essential modifications of the results, much
in the same way as when in physics one takes the "quantum conditions"
into account.

A scrutiny of similar analogies could go on, but what has already
been said is certainly enough to suggest a few useful considerations.
Let us start by recalling the path by which a substantial resolution of
the conceptual difficulties raised by non-Euclidean geometries was eventually reached. It was by the gradual coming into being of the notion of
geometry as a hypothetical-deductive system, that is, by its reduction
to a complex of axiomatic systems. It was then that people realized that
many purported contradictions did not exist, since the difference of the
axiomatic context assigned, necessarily, a different meaning to linguistically identical terms. We can say therefore that people realized that
it was not the same straight line which admitted one, more than one, or
no parallel lines and that, besides this difference in its relation to
other geometrical entities, such a straight line was endowed at times
with different properties (such as the property of being in some instances not liable to indefinite extension. In the same way, a few typical quantities intrinsically changed nature: Thus length, in non-Euclidean geometry, can be referred to a "natural" unit of length (such as
one has for angles in Euclidean geometry); well-defined Euclidean concepts, such as those of equality and similarity among polygons, are
fused into one in non-Euclidean geometries, and so on.

For the same reasons, we not only can but must say of the
wave or particle spoken of in classical mechanics and in quantum mechanics: "It is not the same particle," "it is not the same wave," since
the two contexts are different. Consequently, in the same way that a
parallel line does not have to be unique when it is not a Euclidean parallel line, so the particle does not have to be incompatible with given
wave-like characteristics when it is no longer the particle of classi-

cal mechanics and when the wave-like characteristics are again different from what they used to be in the classical context.

Finally, just as the "straight line" loses the most intuitive of its
properties, namely that of indefinite extension, when going from the
Euclidean to the non-Euclidean context, so the "particle" might lose
some essential intuitive property, such as its exact localizability or
velocity.

The analogy with non-Euclidean geometries helps us to understand
the true nature of a few other difficulties, which are apparently unavoidable because of their being purely logical.

Let us consider the famous example of the substantial difference
which sets the classical statistics of particles (Boltzmann type) apart
from quantum statistics (Fermi-Dirac and Bose-Einstein types). As is
well known, the difference lies in the fact that, given two particles
α and β and two cells A and B in the classical case one must treat the
situation in which α is contained in A and β in B as theoretically distinct from that in which they are reversed, while in the quantum case
these two events must be unified into one since they are indistinguishable.[11]

The difficulty, in this case, appears evident: Even though we cannot
conceive of putting a tag on the particles, it would nevertheless seem
unquestionable that these two practically indistinguishable cases are
clearly logically distinct from one another. The trouble is that, if we
do our statistics assuming that these two cases are distinct, the quantum experiments will prove us wrong, while they will be in complete
agreement with our calculations when we violate this "logically necessary" condition.

In order to understand the situation, we might return to the already mentioned example involving the similarity and equality of polygons.
Given two triangles, it would seem that three theoretically distinguishable states of affairs are possible: They are equal, they are different but have the same proportions (i.e., they are similar), or they are
neither equal nor similar. In hyperbolic non-Euclidean geometry these
three cases are reduced to two, since similarity and equality are there
identical. Does this perhaps constitute a scandal from the point of view
of logic? One feels one should say yes, since it seems unquestionable
that, logically, equality differs from similarity. On the other hand,
one must recognize that the coincidence of the two instances is not, in
hyperbolic geometry, a simple factual question, i.e., something that
just happens to be the case. No geometry has a place for "factual"
occurrences and, in fact, it is *logically necessary* that similar polygons (in this geometry) also be equal, since there is a theorem, which
does not hold in Euclidean geometry, according to which the equality of
homologous angles (condition of similarity) implies the equality of
homologous sides (condition of equality). In other words, one might say
that non-Euclidean polygons have as a logically necessary property that
of being equal if and only if they are also similar, just as Euclidean
polygons have as a logically necessary property that their being similar does not imply their being equal.

The question is by now very clear: Even the allegedly "purely logical" distinctions are always purely logical only in relation to a given
context, so that in the Euclidean context the semantic difference (at
the level of intension) between similarity and equality implies an ex-
tensional difference, while in the non-Euclidean context this implication does not hold.

In the same spirit, we may say that the "logical distinction"
between the possible cases of distribution of two particles in two cells
can only be maintained if our picture of particles located inside cells
can be applied correctly; but we do not know whether, in a quantum context, it can be. Actually, the fact that the "logical distinction"
between the cases discussed leads to results contradicted by experiment
at a quantum level can provide a proof of the fact that we *cannot* represent particles to ourselves according to that scheme. Likewise, the
need to use logically different statistics can be seen as proof of the
fact that we are faced with a new context, within which the statistics
that are confirmed by experiment provide results which are logically
justified[12] rather than being accepted *in spite of the logical difficulties*.

Summarizing the implications of the above considerations, we might
suggest that, once their import has been fully comprehended, the most
important step toward the resolution of the logical difficulties to
which the complementarity principle tries to give an answer has been
taken. The principle indeed informs us that the classical wave and particle concepts are incompatible: We have in fact seen that their being
incompatible in the classical context does not prevent them from being
compatible in some other context. Naturally, in order to be able to see
things in this way, it is necessary to go through a phase of purely
formal speculation, in which the classical concepts are considered as
pure "elements" of a new semantic combination. At this point, however,
one faces the problem of effectively reconstructing a physical (as
opposed to a purely contextual) meaning for these new terms; and until
a solution to this problem has been found, any step undertaken, no
matter how important, is not decisive.

8. IMPLICIT FORERUNNERS OF A CONTEXTUAL CONSIDERATION OF THE MEANINGS
     OF PHYSICAL CONCEPTS

It is worth stating explicitly that the greater part of the considerations of the preceding section are not only in keeping with many
fundamental theses of the Copenhagen school, but must actually be considered as a self-conscious explication of a few of its consequences
which have hitherto remained implicit. Its representatives have the
constant refrain that recourse to classical concepts is essential to
the presentation of the quantum theory; and they offer as a justification something which, from the outside, appears very much to lean on
practical considerations(namely, on the fact that the experimental

findings of the atomic level investigation are after all invariably
macroscopic).[13]   However, if we look at the situation more closely, this
justification appears to be of an epistemological nature, having to do
with the fact that, in order to construct a theory, it is necessary to
start by utilizing the concepts which are already available (waiting
to modify them until later, if such a need should arise), since it is
unthinkable that one should start from scratch.[14]

This can easily be allowed for, but at a certain point one cannot
help but ask oneself what it means to *modify* these concepts: Clearly
it can only mean one thing, that is, to change their *meanings*. But this
is equivalent to admitting that after a certain point these notions
acquire the status of purely linguistic entities, the intensions and
extensions of which are no longer exactly determined. The notions are
constructed little by little *ex novo* as each new context emerges into
which they are inserted, just as the traditional terms of Euclidean geometry took on new meanings upon finding themselves in a new context.

It is strange that Heisenberg, for instance, often approached this
view, but never became completely aware of it.

The fact that the scientists of the Copenhagen school did not take
this essential "linguistic" component of the problem and the semantic
function displayed by axiomatization into account, has prevented them
from finding the correct way out of the difficulties connected with
the use of the incompatible classical concepts.

9.PROPOSAL FOR THE SOLUTION OF THE DIFFICULTIES

Althougn the Copenhagen school had moved a considerable way in the
direction which, in our judgement, is capable of leading us out of the
logical difficulties inherent in the "dualism" of wave and particle,
it never quite took the decisive step of acknowledging that the classical concepts, thought of at a formal level, can appear as elements of
a *new* semantic combination. Within this combination the contradiction
disappears, since it is not *formally* connected with the concepts themselves, but rather with what the concepts classically *denote*, and
through which they are related to heterogeneous entities.

It is natural to ask why the Copenhagen school did not take this
step, and the answer is not difficult to find: Its representatives remained forever faithful to the belief that the classical concepts necessarily drag their original meanings along with them, even into new
contexts, and thus never become able to *denote* something new. This
belief would appear to be, at first sight at least, a necessary consequence of what we have presented as the "central thesis" of the Copenhagen school (we will discuss this shortly), i.e., the thesis according
to which we have no other choice, when constructing quantum physics,
than to utilize classical concepts. Indeed, we can investigate microscopic objects only by means of macroscopic instruments, so we find
ourselves in the position of "questioning" the microcosm with classical

questions and of being able to understand its "answers" which again are
of a classical nature. In other words, even if we do not want to use
the unpleasant words "wave" and "particle," we cannot investigate a
microscopic object without setting up experiments which deal with it as
if it were at times a particle and at times a wave.

Now, it seems that we can agree to asking questions of the microcosm
(and getting answers from it) by way of macroscopic instruments, but
this is not an indication of the hopelessness of our being able to
understand microscopic objects with concepts other than the classical
ones. For instance, we can calculate the distance of an inaccessible
point only by making use of measurements which involve "accessible"
operations such as the handling of rulers and protractors. Nevertheless,
such a distance is expressed and understood as something which cannot
be measured by moving rulers about. On the contrary, in order to express
it, we must use concepts which are inconceivable at the level of everyday measurements of length, such as light years. In the same way, we
can "question" the world of electrical phenomena with "mechanical" instruments and receive answers of a mechanical type. This does not prevent us from recognizing the particular character of electrical phenomena, nor from introducing new concepts such as charge, induction, and
so on, all of them non-mechanical, though all were undoubtedly established on the basis of their "mechanical effects" as recorded by
measuring devices.

We are now in the position to say that the fact that we utilize
classical instruments (so to speak) in examining microsystems does
not imply that we are compelled to utilize "classical concepts" in their
description: It is enough if we are able to explain how these non-classical entities can produce observable "classical effects". We can establish
that they indeed are nonclassical in a fashion similar to the one which
we utilized in establishing the specificity of electrical phenomena
versus mechanical phenomena: The former exhibited examples of repulsive
forces, of rate-dependent forces, and so on, which have no parallel in
mechanics. This entitled observers to say that here one was dealing
with nonmechanical facts, and to embark on a new conceptualization which,
despite the use of many mechanical concepts, could not be reduced to
them. Similarly, the anomalies one encounters when utilizing classical
concepts in quantum physics are an indication that here one is faced
with something new and has to begin a new conceptualization which,
despite the use of many classical concepts, is not reducible to them.

For the above reasons, in the final analysis, we do not believe
that we need adhere to the fundamental thesis of the Copenhagen school,
which is entrenched in the methodological approach of an extreme operationalism, according to which concepts simply denote single operations
(or single systems of operations).

We wish now to try to illustrate this possibility with an example:
Mechanics employs the concept of attractive force (typically, gravitational force) and of repulsive force (e.g., that of elastic recoil) and
habitually measures masses via the determination of forces which depend

on distance. If we now consider electrostatics, we find that within it
the concept of electrical charge denotes "something" which can also be
measured by determining forces that are distance-dependent, and we again
find the notion of attractive and repulsive force. In other words, the
present concepts, *taken individually*, are all mechanical, but what they
characterize is no longer mechanical, since they are differently connected: That "something" which in mechanics can be measured via the determination of distance-dependent forces concerns only attractive forces,
while in electrostatics the analogous "something" concerns both attractive and repulsive forces. It is this very difference (other things
being equal) which compels us to use a new term to denote this "something," calling it, for instance, "charge," instead of "mass." Furthermore, we realize that a given body can be shown to possess both mass
and charge, that is, it lends itself to being considered as an object
of both theories.

Let us now consider micro-objects: Let us admit that, *taken individually*, the available concepts are all classical: This is in a certain
sense unavoidable, not so much because of the measuring devices one is
obliged to use, but rather because there are no other concepts at one's
disposal when starting to construct this new branch of physics. We observe however that in the classical case no one concept simultaneously
contains within itself the concepts of position, momentum, and linear superposition (similarly, the mechanical concept of mass does not imply
that mass might give rise to both attractive and repulsive forces); in
the case of micro-objects, on the other hand, we are dealing with "something" in which such characteristics simultaneously do unite (similarly,
the notion of electrical charge includes the possibility of such a charge
interacting with other charges through both attractive and repulsive
forces).

It is this fact which obliges us to recognize that this "something"
must be denoted by a new concept; and this something, even if it is
a given microscopic object, e.g., an electron, can take part in processes for which a classical description is adequate and in others for
which the quantum description is necessary (similarly, an electrified
body in certain experimental contexts can manifest itself only as mass
and in other contexts behaves in a way understandable only if one takes
its charge into account).

Without attributing a greater value to this analogy than it deserves (it might certainly show weaknesses given close scrutiny, as all
analogies would when stretched beyond the limit within which they are,
indeed, analogies), it seems that it might nevertheless help us to
understand *how*, in fact, a different combination of old concepts might
give rise to authentically new concepts. In the case of electric
charge, indeed, we obtained a new concept by funnelling into one single
intension mechanical characteristics already known, but which in mechanics do not appear together; and we have called the concept thus born
from the fusion of the earlier notions and characterized by the new
intension "electric charge".

This kind of reasoning can be repeated in the case of microphysics:
The concepts of position, momentum, linear superposition, and so on are
undoubtedly classical, but if it is true (as it is) that in classical
physics they never concur in determining the intension of a single concept, it is by now obvious that fitting them together into one single
intension gives rise to a new, non-classical concept. We might lack a
word for it (or we could simply use that of "micro-object"), but the
concept, at any rate, is there. It remains to be seen whether it really
denotes something; but here the problem appears to coincide with the
problem of the truth of the theory, i.e., the problem of the testability
of the theory and of the degree of confirmation which it might gain from
supporting instances.

In these arguments the reader has seen the "contextual" view of
the meaning of physical concepts at work, which allows us to conceive
of concepts as authentically new when they are arrived at via the composition of already known intensions. This also puts us in a position
suitable for investigating the problem, left until now up in the air,
of clarifying what "type" of objects micro-objects are. One must remember that, earlier on, two antithetical considerations were encountered: On the one hand, the experimental behaviour of such entities
prompts us to say that the experiments "have to do with" particles *and*
that they "have to do with" waves. On the other hand, the uncertainty
relations compel us to say that these entities are neither waves nor
particles. What are they then? The answer is that all we can ever say
about a concept is what is expressed in its intension: Consequently, if
we possess, as we do, a list of the characteristics which constitute
its intension, we already have all we need to be able to say "what"
thing, if any, it denotes. To ask for more, means in effect to yield
to the "visualization" requirement--the inessential nature of which we
have already recognized--or rather to the gnoseological illusion which
pretends to find a hidden essence behind the "appearances." In fact,
if I ask: "What type of object is a micro-object?" and I receive as an
answer "It is an object which is characterized by this, this, ... and
this property", then I have obtained an adequate reply to my question.
If I want more, it is an indication that I am caught unawares in the
trap of "needing to find the essence" or that I am still nostalgic with
regard to the "classical model", when I should realize instead that I am
dealing witn another area of physical experience.

By now the answer to another question that was posed earlier and
left unanswered emerges clearly. The questions was: Is it possible to
understand *how* classically incompatible concepts can be reconciled?
Certainly it would have been expedient, if possible, to find this "how"
in an "interpretation" of atomic events within a new classical model:
This would also have greatly consoled the backers of the "central thesis" of the Copenhagen school, according to whom we can only understand
by utilizing classical concepts. Not only did this not happen, however,
but it is very unlikely to happen for reasons of principle, namely that,

since the reality one has to deal with is *new*, one ought not to expect that it can be completely understood using *old* concepts.

Our proposal consists in saying that the "how" emerges precisely from leaving the model aside: If we look for a classical model, we find ourselves in trouble, if on the other hand we focus on the formal structure, we can say that the various concepts at play are *formally* non-contradictory, and that, once *unified*, they give rise to a new intension. The problem of understanding what corresponds, as *denotation* to this intension is divided into two parts: One is void, since it consists in asking oneself what "type" of object is denoted by a given intension, as if this were not already stated by the intension itself; and the other is legitimate and consists in asking oneself whether this intension does not perhaps correspond to an empty extension; that this does not happen is guaranteed by the fact that the whole theory has been verified (albeit not in an absolutely unequivocal way).

By this we do not mean to deny that some difficulty remains at a psychological level and at the level of the "adequacy" of the prevailing situation. Undoubtedly, even if we admit that the concepts utilized are indeed "new" thanks to the contexts, we have somehow to "pay" for their origin, that is, for the fact that they invariably result from the "composition" of intensional elements taken from the classical context and in compatible therein.

There appears to be only one way out of this uneasy situation, a way which is connected with a very promising perspective. This perspective is that, in order to be meaningful, the effort directed toward rendering theorization about quanta more satisfactory must involve searching for a few concepts which are *new*. They ought not to be new only in that they result from the combination of classical concepts according to a different pattern, but also because they substitute, wholly or partly, such classical "components" with something truly novel.[15]

In other words, one would have to find new concepts capable of adequately mastering the microworld, which were such that classical concepts might result as a specialization of them (in keeping with the fact that classical physics is a limiting case of quantum physics).

Here, too, we can obtain an idea of what this might mean from a consideration of non-Euclidean geometry: As is well known, Lobachevskij did not construct his non-Euclidean geometry simply by modifying Euclid's parallel postulate, but rather built a *new* geometry, moving from *new* primitive concepts (those of body, of contact between bodies, and of rigid motion), from which he was able to *define* the Euclidean concepts of sphere, circle, and straight line, but within a wider context in which the Euclidean geometry appeared as one possible case alongside the non-Euclidean cases. Alternatively, if we want to give a more limited example, the introduction, suggested by Gauss, of the *new* notion of "parallel angle," gives rise to a geometry in which Euclidean geometry appears as a limiting case (when such an angle is $90°$) but which itself is not subject to this limitation. How it is possible, in the case of

physics, to make a similar choice of such *new* concepts is not easy to
say, but there are researchers moving in that direction.[16]

Naturally this is not the right place for a discussion of these
attempts, which are still at an embryonic stage and imply considerable
difficulties (even mathematical ones) for the non-specialist. On the
contrary, we find it useful to ask: From where will it be possible to
obtain these new concepts? Whoever has followed us thus far realizes
that, while we are aware of the importance of the "inventive" contribution of the intellect in such a task, we nevertheless believe that
the elements in the construction must somehow be given. Until now these
elements were supplied, more or less, by pre-existing physical theories;
but once this source is exhausted, where shall we obtain them? Which
"models," if any, can be used as starting points for the extraction of
new concepts?

This is a very delicate question, one which hides beneath it an
even more fundamental problem: whether the human intellect can understand something only by forming an intuitive image of it. We can say
that Western philosophy has always accepted this thesis in one form or
another. This has been so from the time when scholastic gnoseology
stated (along with the Aristotelism of old) that the intellect does not
know, *nisi convertendo se ad phantasmata*, other than by basing itself
on intuitive images, to the time when Kant said that not only is intuition necessary for knowledge, but that the pure forms of understanding
themselves can only be thought through a certain "schematism" of the
pure intuition of time. We will not treat such a general philosophical
problem. Limiting ourselves to the concrete case posed by the preceding
question, we think we can give the following answer: Microphysics itself
borrows these concepts from certain models; while it also seems that such
concepts are being offered, to a greater and greater extent, by the most
seemingly abstract of mathematical theories.

Though they are "abstract," these models are not purely "formal,"
as they may be comprehended as a kind of "content" of our intellectual
intuition. They can be constructed and grasped through the creative force of reason, which is the source of the invention in all fields of human activity, from literature, to the arts and philosophy, but also including science, as so many scientists (e.g., Einstein) have stressed.
However, in order for these structures and constructions to have a
"physical meaning," we have to relate them to those aspects of reality
which are investigated by means of the typical tools of physics; for
it is only under this condition that they are bound to physical *refe-
rents*. This is certainly possible and happens by means of a suitable
intensional semantics based upon *operational criteria.*[17] It would take
us too far to consider this further aspect which, on the other hand, is
not essential to the logical clarification of the issues discussed in
this paper, though it plays an important role in order to establish that
this logical analysis is fully compatible with a "realistic" conception
of science.

NOTES

1. Feyerabend, p. 193.

2. Bohr, p. 90.

3. Bohr, p. 40.

4. Pauli, p. 89.

5. Schrödinger, p. 198.

6. "However, to satisfy these wishes--Heisenberg says--one has to give up something important...Quantum theory indeed led to the result that an atom is no longer a figure accessible to our intuitive representation." (Heisenberg, p. 79.) And again, evidencing how this non-intuitive character is justified by the very principles of the new physics, he adds: "In reality modern physics is much less intuitive than the old investigators of nature might have hoped. But we must not feel unsatisfied, since we learned from nature that this non-intuitiveness is strictly and meaningfully connected with the existence of atoms. Approximately, one could argue that a structure that presented itself as perfectly intuitive could not be indivisible....The indivisibility and the unity of elementary particles, which is admitted in principle, make understandable the fact that the mathematical figures of the atomic theory are devoid of intuitiveness." (Ibid.pp.99-100.)

7. All this, naturally, is said having in mind more the intentions and the "points of view" expressed (mostly in writings of a philosophical outlook) than the effective realizations achieved. Thus a few people (such as Schrödinger) maintain that there is no such thing as particles, but rather charge and matter distributions, which are mostly continuous; however, they are eventually compelled to admit that the density of these distributions behaves in a way which is incompatible with the very idea of continuous distribution. Others (such as Born) state that only particles exist and that the waves represent physical properties of the particles, but in the end they have to admit, as was already mentioned, that these are not the same thing as the particles of classical mechanics. Neither the former nor the latter, then, explain everything in terms of sole and true waves, of sole and true particles.

8. This perspective also allows one correctly to find a positive evaluation of the point of view according to which the uncertainty principle is able to circumvent the contradiction between wave and corpuscle. The principle in fact compels us to say that microobjects are not "authentically" corpuscles (since they cannot have exact position and exact velocity at the same time), nor "authentically" waves (since they cannot have exact frequency and exact amplitude).

Therefore, this principle indicates, as a possible way out of the contradiction, the fact that a microobject is neither wave nor corpuscle, rather than the fact that it is both.

9. See for example: Mckey, Varadarajan, Jauch, Fano, and Piron.

10. Remarkable approaches in this direction are to be found, e.g., in: Beltrametti-Cassinelli, 1986; Beltrametti-Cassinelli, 1982; Beltrametti-van Fraassen.

11. It does not matter, in the spirit of the present discussion, whether or not the consideration of distributions within cells can be viewed only as a pedagogical artifact which illustrates the meaning of certain symmetry properties of the wave functions, with respect to exchange of coordinates of two identical particles. We think that the consideration of such a distribution is the most natural way (if not perhaps the only one) to give physical meaning to this mathematical operation, and we wonder what, in that case, one might think of the difficulties one encounters.

12. We do not wish to mean by this that, from a purely logical point of view, there is no other possibility for the solution of this difficulty; our knowledge of microphysical facts could have intrinsic limitations such as to prevent us from distinguishing the two cases mentioned above, and from detecting all the logically possible distinctions even when statistics over facts are considered. In this case, it could well happen that the two limitations would concur to produce an agreement with experimental measurements. We believe that this second alternative is quite artificial, even if logically non-contradictory.

13. "In this context--Bohr says--we must recognize above all that, even when the phenomena transcend the scope of classical physical theories, the account of the experimental arrangement and the recording of observations must be given in plain language, suitably supplemented by technical physical terminology." (Bohr, p. 72.)
    "A microphysical system," Jordan adds," is endowed with well-defined properties only when it interacts with macrophysical observation devices: By this statement it becomes clear that *even though* the laws of macrophysics are mathematically a consequence of those of microphysics, microphysics itself can only be formulated *after* macrophysics and *must take from that its lead*." (Jordan, p. 43.)

14. "It seems therefore that science can only go through one path: It must utilize, without reservations, the concepts as they are, for the description of what it observes, and procede to revision only when obliged to from the results of experience. To ask that the clarification of concepts is undertaken from the start amounts to asking that the entire development of science is pre-determined by a logical analysis." (Heisenberg, p. 37-38.)

15. A favourable stand toward this position was taken by Schrödinger, who believed, indeed, that quantum mechanics should go through the authentic elaboration of completely new concepts.

16. Only by inventing some new concept, that is, new in this more fundamental sense, could we possibly overcome in a definitive manner the present uneasy state of affairs, which is not related to the regret of losing the old concepts but to the lack of new concepts capable of adequately replacing them. As Weizsäcker once said: "What leaves us unsatisfied is not, essentially, that the old intuitions fail, but that nothing new which is immediately understandable takes their place." (Weizsäcker, p. 31.) A proposal in this direction, e.g., has been advanced by F. Selleri, through the introduction of the concept of quantum wave (Selleri, 1969, 1971; Diner et al.) The essential novelty of this concept is represented by the acceptance of the de Broglie realist interpretation of the wave-particle duality, but not of the symmetrical nature of this dualism. In Selleri's approach, both particles *and* waves are simultaneously real, but the latter can be characterized *only* through *relational properties* with the particles: The observable properties of producing interference and stimulated emission. Such a possibility would imply an ontological priority of particles over waves, which would therefore belong to a weaker level of physical reality, containing objects which are sensible carriers of exclusively relational predicates.

On the basis of this new concept, which requires a radical emancipation with respect to the classical as well as the quantum mechanical notion of physical object, some experimental proposals have recently been developed (Popper-Garuccio-Vigier; Garuccio-Rapisarda-Vigier; Tarozzi; Gozzini.) The interest of these experiments is due to the fact that they would allow one not only to test this new realist interpretation vs. the one of the Copenhagen school--i.e., to discriminate experimentally between two different philosophical interpretations of a physical theory--but also the well-known axiom of the reduction of the wave function, whose essentially *ad hoc* nature is at the origin of some serious difficulties of the quantum theory of measurement (Tarozzi, 1985.)

REFERENCES

E. Agazzi, "The concept of empirical data. Proposal for an intensional semantics of empirical theories, in M. Przelecki *et al.* eds., *Formal Methods in the Methodology of Empirical Sciences* (Reidel, Dordrecht, 1976), pp. 143-157.
E. Beltrametti and G. Cassinelli, "Logical and mathematical structure of quantum mechanics," *Riv. Nuovo Cimento, 6 (3)*, 321-404 (1976).
E. Beltrametti and B. van Frassen, eds., *Current Issues in Quantum Logic* (Plenum, New York, 1981).
E. Beltrametti and G. Cassinelli, "The logic of quantum mechanics," in

Encyclopedia of Mathematics (Addison-Wesley, Reading, 1982).

N. Bohr, *Atomic Physics and Human Knowledge* (Wiley, New York, 1958).

G. Fano, *Mathematical Methods of Quantum Mechanics* (McGraw-Hill, New York, 1971).

P.K. Feyerabend, "Problems in microphysics," in *Frontiers of Science and Philosophy* (Pittsburgh, University, Press, Pittsburgh, 1962).

A. Garuccio, V. Rapisarda and J.-P. Vigier, "New experimental set-up for the detection of de Broglie waves," *Phys. Lett. 90 A*, 17, (1982).

A. Gozzini, "On the possibility of realizing a low-intensity interference experiment with a determination of the particle trajectory," in *The Wave-particle Dualism*, S. Diner *et al.*, eds., (Reidel, Dordrecht, 1984).

W. Heisenberg, *Wandlungen in den Grundlagen der Naturwissenschaft* (Hirzel, Zürich, 1949).

J.M. Jauch, *Foundations of Quantum Mechanics* (Addison-Wesley, Reading, 1968).

P. Jordan, *Das Bild der modernen Physik* (Stromverlag, Hamburg, 1947).

G. Mackey, *Mathematical Foundations of Quantum Mechanics* (Benjamin, New York, 1963).

W. Pauli, "Die allgemeinen Prinzipien der Wellenmechnik," in *Handbuch der Physik* (Springer, Berlin, 1933), vol. 24/1.

C. Piron, *Foundations of Quantum Physics* (Benjamin, New York, 1976).

K.R. Popper, A. Garuccio, and J.-P. Vigier, "An experiment to interpret EPR action-at-a-distance: the possible detection of real de Broglie waves," *Epistem. Lett. 30*, 21 (1981).

E. Schrödinger, "L'image actuelle de la matière", in S. Bachelard *et al.*, *L'homme devant la science* (Editions de la Baconnière, Neuchâtel, 1953), pp. 31-54 and 195-212.

F. Selleri, "On the wave function of quantum mechanics," *Lett. Nuovo Cimento 1*, 90 (1969).

F. Selleri, "Realism and the wave function of quantum mechanics", in *Foundations of Quantum Mechanics* (Il Corso SIF) (Academic, New York, 1971).

F. Selleri, "Gespensterfelder," in S. Diner, *et al.*, eds., *The Wave-particle Dualism* (Reidel, Dordrecht, 1984).

G. Tarozzi, "Two proposals for testing physical properties of quantum waves," *Lett. Nuovo Cimento 35*, 553 (1982).

G. Tarozzi, "From ghost to real waves: A proposed solution to the wave-particle dilemma," in *The Wave-particle Dualism*, S. Diner *et al.*, eds., (Reidel, Dordrecht, 1984).

G. Tarozzi, "A unified experiment for testing both the interpretation and the reduction postulate of the quantum mechanical wave function," in G. Tarozzi and A. van der Merwe, eds., *Open Questions in Quantum Physics* (Reidel, Dordrecht, 1985).

G. Tarozzi, "Experimental tests of the properties of the quantum mechanical wave function," *Lett. Nuovo Cimento 42*, 438 (1985).

V.S. Varadarajan, *Geometry of Quantum Mechanics* (Van Nostrand, Princeton, 1968).

G.F. von Weizsäcker, *Zum Weltbild der Physik* (Hirzel, Zürich, 1949).

CHANCE: CHAOS OR LOGIC?

M.Cini
Dipartimento di Fisica
Università "La Sapienza"
00185 Roma, Italy

ABSTRACT. The history of twentieth century physics is interpreted as
the result of a consistent undertaking by the scientific community to
retain for this discipline the characteristics of a science "of laws",
by excluding from its domain all those phenomena and theories which
could have introduced into physics some features of the sciences "of
processes". A brief survey of the roads abandoned as a consequence of
this choice is presented, together with a short review of recent attempts to formulate a theory equivalent to Quantum Mechanics by using
the theory of stochastic processes.

1. SCIENCES OF LAWS AND SCIENCES OF PROCESSES

It is common practice to divide the natural sciences into two
categories: on the one hand, the normative sciences, characterized by
the search for, and the statement of, necessary and universal laws of
nature and thus capable of making rigorous predictions, on the other,
the evolutionary sciences, viewed as incapable of achieving the status
of universality, since they are given over to the investigation of
nonrepeatable processes and thus are at most able to provide a hypothetical reconstruction of a succession of events within a context
that can no longer be modified. To this distinction corresponds a
substantial difference in epistemological status. In point of fact, it
is the disciplines of the first kind (the sciences of laws or of
"why") - based on the repeatability of the experiment, on the modifia-
bility of the initial external conditions, on the eliminatability of
factors considered to be of a secondary nature - which are regarded as
being fully entitled to be called sciences, since they are verifiable
or falsifiable (according to different points of view) with respect to
nature. Subjects of the second kind, on the other hand (the sciences
of processes or of "how"), are often maintained by epistemologists to
be second-class sciences, which owe to the "why" sciences those general laws that are held capable of providing the sole convincing explanations of the concatenation of events which the "how" sciences are

75

G. Tarozzi and A. van der Merwe (eds.), The Nature of Quantum Paradoxes, 75–87.

limited to describing as plausible. Only the complexity of the context, the multiplicity of the accompanying factors, and the incompleteness of the documentation relating to the events that actually
have happened would, according to the dominant conception, justify
recourse to these "surrogates" of science which are, it is said, essentially lacking in any autonomous epistomological status.

A number of questions then immediately spring to mind. What is it
that fixes the defining criteria of the different disciplines and
places them firmly in one category or other? Up to what point do these
criteria stem from the nature of the subject, that is, from the collection of phenomena and facts that constitute the (approximately
closed) ensemble to be interpreted and explained, and to what extent,
on the other hand, do they depend on the choice made by scientists who
select the data and the experiment to construct their science, thus
placing it on the one side or other of the dividing line? It is commonly maintained that it is the properties of the object investigated
that determine the nature of the discipline (the science of "why" or
the science of "how") that studies it, the scientist being limited to
simply taking note of what there is arranged before him or her. But,
in actual fact, this demarcation is much less objective than one might
think. The most recent developments in the field of biology provides
an obvious demonstration of this fact:

Up to the middle of the seventies, this "science with a schizo-
phrenic personality" - as M.Ageno defined it recently[1] - was present
in two completely different guises. On the one hand, functional biology, essentially a reductionist discipline, was denoted to the analytical study of organisms, determining their structures and internal
processes right down to the minutest details. It thereby tended to
interpret all biological phenomena in terms of events taking place at
this latter level and therefore, in point of fact, to reduce biology
to the physics and chemistry of the molecules involved. By contrast,
evolutionary biology considered living organisms as indivisible entities whose particular characteristics emerge only at the level of the
totality and are only partially deducible from an analysis of the
constituent subunits.

We are faced, then, with two biologies: the first of these is to
be assigned to the group of the sciences of laws, or of "why", the second to the sciences processes, or of "how". But the revolution that
has taken place over the last few years, following the discovery of
unexpected properties of eukaryotic DNA, has changed this picture
quite radically. Ageno[1] writes:

"One used to think of it (DNA) as some sort of archive of
the heritage of the organism, carefully protected from all
those external agents which could be the cause of altera-
tion.

Such a straightforward and clear-cut view of affairs
now seems to have gone completely by the board. Far from
being something invariant, which is in essence conserved
within the framework of the dynamics of the organism, DNA

now appears to be involved in a dynamic all of its own, one that is incessant and to a great extent dominated by random events... Faced with the enormous variety of organizational, regulatory and adaptational solutions which are seen to result (from this dynamic), molecular biologists now see themselves forced, a little at a time, to change the type of questions which, within the framework of their research into the functionality of the organism, they were wont to ask.

Now, face to face with the multiplicity of a priori equivalent solutions, the search for causes, the quest of the why, the reasons, shows itself to be surprisingly indecisive and unimportant, and molecular biologists are ever more led to ask themselves how each solution has come into being, through what chain of events, and in what general environmental conditions.

Thus natural science and functional biology are, in actual fact, finding their common root in the theory of biological evolution. For biological phenomena, there are no possible explanations other than evolutionary ones."

What has been happening in biology thus indicates that the traditional demarcation, in this case at least, seems to have been created more by a choice of the scientists themselves than by the properties of the objects under study. It appears, in point of fact, that the molecular biology community, aiming at retaining for its own discipline the normative science character that distinguished it from the start, has, as far as is possible, eliminated from this disciplinary field all those phenomena which would have compromised its image by the impossibility of fitting them into the framework of general and a-temporal laws. Similarly, it appears that evolutionary biology, by accepting this division of spheres of competence, continued - as long as this division did not experience a crisis - to consolidate its own original nature as a science of irreversible and nonrepeatable processes.

At this point, then, the doubt arises as to whether this kind of procedure, consisting in the choosing of the objectives of investigation and the corresponding interpretative categories so as to ensure consistency between the development of the discipline and the epistemological status attributed to it by definition, is typical not only of biology, but might be a common practice adopted by the different scientific communities to define their own identities. That this sort of thing happens in the humanities probably does not come as a surprise. But that it also constitutes the rule in physics, which has always been considered the science of laws by definition, may seem a difficult assertion to maintain. I am, however, here proposing to show that in this case, too, at least in recent times, affairs have in fact developed exactly in this way.

As we shall see, the history of twentieth century physics acquires a new interpretative dimension if one sees it as the result of a consistent undertaking by the scientific community to retain for this discipline the characteristics of a science "of laws" by, as far as possible, excluding from its domain all those phenomena, together with their interpretations, which could have introduced into physics some characteristics of the sciences "of processes".

From this point of view, one can, for example, understand why the sensational developments that have lately taken place in the field of the dynamics of complex systems[2] have been trailing by fifty years late the appearance of the pioneering work of Poincaré (who in essence laid the basis for them), despite the fact that this work has been available since the end of the last century. And one can also understand how these developments have come about in disciplinary sectors that have now become autonomous with respect to the fields that physicists define as physics.

The victory of quantum mechanics at the end of the twenties represents, if seen in the same light, the success of an effort to reassert physics as the science of laws - albeit at the price of giving up determinism in the strict sense - by thrusting back, and even outside its boundaries, everything concerned with the unpredictability, irreversibility, and randomness that characterize phenomena such as turbulence, dynamic instability, stochastic processes, the thermodynamics of irreversible processes, and so forth. This point of view also explains why physics has maintained and protected its image as a science of the simple, excluding complexity as much as possible, as one category that characterizes the reality selected as the subject of its investigation. Thus it is that we witness an unchallenged dominance of reductionist ideology, with priority being assigned to the search for the "elementary" constituents of particles, which still today consistutes the essence of the most prestigious branch of physics.

This reconstruction of the conflict between different strategic choices seems to me more satisfying than the traditional one, which focus, above all, on the opposition (at the 1927 Solvay Congress) between the submissive supporters of a classical deterministic vision of physics (Schrödinger and Einstein: "God does not play at dice") and the victorious upholders of a conception based on indeterminism.[3] Such a contrapositioning certainly took place in the stages that preceded the victory of quantum mechanics. But the very lightning nature of the victory won shows how the conflict was more a rearguard action than a real debate between perspectives that potentially could have given rise to alternative developments. Paul Forman's reconstruction[4] of the cultural and ideological climate of Weimar Germany that saw the establishment of quantum mechanics, however accurate and acute, seems to carry little weight if viewed as the principal explanation of the ready conversion to the new theory of the German physics community, given that a hypothetical faithfulness to a deterministic conception could not have been translated into a scientifically valid alternative. The reconstructed climate, on the other hand, becomes much more convincing if seen as the background to a compromise that saved the fundamental nature of the "scientificity" of the discipline (i.e., its being the source of an unchallenged logical and empirical legality), rather than as justification for an opportunistic surrender

to the dominant irrationalism. It may be said in passing that it is in
this way that the apparent contradiction-for which Forman provides no
convincing explanation - between the adherence to the theses of the
Vienna Circle on the part of many of the founders of the new physics
and the tendency, which certainly was present, towards adaptation by
the community to pressure exerted by the dominant ideological-cultural
environment, is resolved. At the same time, it also explains why it
was that Dirac, who, contrary to Heisenberg, saw quantum mechanics
more as the logical outcome of, rather than a radical break with,
classical mechanics,[5] immediately found his place among the fathers
of the new theory alongside the German physicists.

2. THE ABANDONED ROADS

I will start to argue for my thesis by showing that the choice
made at the end of the twenties to restore the character, essential to
physics, of it being a science of general and immutable laws, had as a
consequence the expulsion, from its accepted province, of a wide range
of problems raised by the crisis in classical physics, which then had
to wait decades for their subsequent development, often by enterpris-
ing members of other disciplines.

The main element of this reassertion consisted in bringing the
probabilistic aspects of the new mechanics back under the rules of a
logico-abstract algorithm through the elimination of their temporal
character. From this point of view, the new mechanics emerges more as
the legitimate heir of Newtonian mechanics than as its implacable
opponent. For, as we see, the equations of motion in both theories are
deterministic and reversible, and the temporal evolution of the quantities that represent the state of a system is no other than the de-
ployment of a succession of changes that contain nothing new, inasmuch
as they are potentially included in any one of the past or future
states of the succession itself chosen arbitrarily.

Let us therefore examine the consequences of this choice.

The first fact that should make one reflect is the disappearance
from physics of the concept of irreversibility from the end of the
nineteenth century right up to the present times. It is well-known
that Boltzmann, first, and then Planck maintained that the second
principle of thermodynamics was an absolute law of nature. And as
such, it played a central role for them in physical science. I cannot
here dwell on the details, which are moreover well-known, of the H
theorem and the debate between Boltzmann, Loschmidt, and Zermelo. I
recall only that Planck's program, which then unexpectedly let to the
black-body radiation law and to its interpretation in terms of quanta,
started off with the intention of demonstrating that the electromagnetic field contained within a cavity would irreversibly reach the equilibrium state at a given temperature by virtue of no more than the
classical equations of motion.[6] Only when this objective proved
unreachable (because of the reversibility of the equations themselves)
did Planck adandon the description in terms of a temporal evolution

and embrace the statistical one in terms of the probability of states, which Boltzmann had earlier adopted, after having himself run into the same difficulties. From then on the question of irreversibility remained a topic of concern primarily of physical chemists. Indeed, from the Onsager relations of 1931,[7] right up to the work of Prigogine's school,[8] this line of research has occupied the sidelines of physics, so much so that Prigogine was awarded the Nobel Prize for chemistry.

Time thus disappeared from physics in the sense that the evolution of a statistical ensemble towards equilibrium ceased to be a problem worthy of interest: Statistical mechanics was reduced to the calculation of partition functions at equilibrium. And it is exactly this point on which was based the introduction of discontinuity: a discontinuity that was to become the seed from which quantum mechanics sprouted. As Kuhn has demonstrated, it was only a few years after the introduction of Planck's constant in 1900, when Lorentz had convinced himself of the inevitability of the discontinuity hypothesis for the energy of oscillators in equilibrium with radiation, that the physics community accepted the necessity for quanta. But this inevitability is a direct consequence of the tacit and general acceptance of the hypothesis that every "reasonable" system is ergodic. Even Poincaré, as Ciccotti and Ferrari have recently shown,[9] in an extreme - and unsuccessful - attempt to find a way of saving classical mechanics, refused to consider the possibility that there might exist nonergodic dynamic systems as an alternative to the introduction of the quantum. One has to wait until 1954 for the Kolmogorov theorem[10] to discover that the property of ergodicity is far from obvious for complex systems. The problem was then opened up again, but history had by now run its course.

Another expulsion happened, for a different subject originating within classical mechanics, at the end of the nineteenth century with Poincaré's famous note on the three-body problem: that of dynamic instability. This was a theme which in its turn opened up an unexpected breach in Laplacian determinism, which seemed a necessary consequence of Newtonian mechanics. Developed in particular by the astronomers in the '60s[11] and taken up independently in numerous other contiguous and allied disciplines, the subject here is based on the fact that completely deterministic nonlinear dynamic systems can have a "wildly chaotic" behavior. Only in 1977 a conference was held at Como on the stochastic behavior of Hamiltonian systems which, for the first time, brought together astronomers, biologists, economists, physicists and mathematicians who were all working in this area. It is significant that the conference was called thirty years after the famous one at Como that gave QM its official baptism, thereby making the explicit claim, at least in the intentions of the organizers, that the 1977 conference represented an historic turning point analogous to the previous one.[12]

A third field of research which was delegated to the sidelines by the physics community, and which only recently received a great impetus in the opposite direction - coming back into physics via the back door after having been pushed out through the front door, as one might say - is that of stochastic processes. The pioneering work of Wiener at the start of the twenties was in essence ignored by physicists who, right up to the seventies, never took into serious consideration the possibility of utilizing the mathematical tools developed at that time to deal with problems of interest for their discipline. It is of interest to underline the fact that as Battimelli has shown[13] in his work on the birth of the discipline that grew up around the research into turbulence, these tools were adopted in the 1940s with notable success by workers in this field through the direct links that existed between Wiener and J.C.Taylor, the founder of the relevant theory. In the thirties, the very act, on the part of Kolmogorov and his Russian school, of having founded such a fruitful new discipline as the classical theory of probability has passed practically unobserved by physicists right up until the most recent times. The techniques of stochastic processes began, in point of fact, to become fashionable among theoretical physicists who dealt with statistical mechanics and field theory in the 1970s.[14]

It is too soon to say whether this change, which for the moment seems to consist essentially in the adoption of more efficient and flexible techniques, is the prelude to a conceptual change by one part of the physics community in respect of the attitude to be adopted towards problematics typical of the sciences of processes. Personally, I would maintain that the situation is evolving in a direction analogous to that depicted by Ageno in the biological field. Some signs of an evolution of this kind are visible. To mention only one of them, I wish to point out that in 1985 the number of symposia, meetings, and schools dedicated to the application of stochastic processes and probability theory to physical problems has increased an order of magnitude compared to three or four years ago.

3. WAS QUANTUM MECHANICS THE ONLY POSSIBLE SOLUTION?

The plausibility of my thesis would be strengthened by showing that indeed other physical theories might have been invented in order to provide a solution to the difficulties which Newtonian mechanics ànd Maxwellian electromagnetism had run into. This is of course difficult to prove without actually producing such a theory. I will however show that viable possibilities in other directions existed whose exploration was hindered by the prevailing ideological hostility towards any attempt to question the ultimate and definitive character of quantum mechanics. It would of course be foolish to deny that quantum mechanics had from the start advantages of simplicity and flexibility, particularly from the practical point of view, which could not but have attracted the interest and even the enthusiasm of the majority of active physicists. It is a fact, however, that only in the sixties did the first serious attempt get underway to formulate a theory equivalent to QM by using the theory of stochastic processes.

My purpose is now to briefly summarize some of the latest pro-

gress in this direction, in order to show that, had this line of research been given the opportunity for autonomous development - an
opportunity denied it during almost thirty years by the unanimous
acceptance of von Neumann's theorem as a decree outlawing any search
for an alternative theory - then the dominant paradigm in the field of
microphysics might have been challenged many years ago. The most elaborated attempt up to now is the theory known as stochastic mechanics
developed by Nelson.[15] This theory describes the behaviour of a
nonrelativistic system in configuration space under the influence of a
random disturbance of unspecified origin. The motion of the system is
therefore the result of the joint action of the classical and stochastic forces, leading to a continuous but non differentiable chaotic
trajectory, typical of a Markov diffusion process. From this point of
view, the wavelike behavior of a single particle predicted by quantum
mechanics is explained by showing that the time evolution of its position's probability density is the same as the one derived from the
corresponding Schrödinger equation. Substantial progress in the development of this theory has been made in the last decade, notably by
Guerra and his associates and by Jona-Lasinio and coworkers. Both
general and practical problems have been tackled. On the one hand, the
theory has been generalized to Riemann spaces[16] and the basic equations have been derived by means of a variational principle on a stochastic action.[17] On the other hand, the list of achievements is
remarkable; ways of treating particles with spin have been worked
out.[18,19] The typical quantum tunneling effect has been interpreted
as a consequence of diffusion, and the results of classical probability theory have been applied to obtain substantial progress in the
computation of the motion of particles in complicated potentials.[20]

The question of defining momentum in a theory where trajectories
in configuration space have no instantaneous velocity has also been
dealt with, and the validity of the uncertainty principle has been
demonstrated.[21] Quite recently, the usual quantum mechanical definition of probability in momentum space has been shown to follow from
the transition probability in configuration space in the infinite time
limit.[22,23]

Some difficulties are however still not easy to overcome. These
difficulties are in fact of the same kind as those faced by Schrödinger in the early days of wave mechanics, when he tried unsuccessfully
to give a physical meaning in configuration space to the wave function
of an N-particle system by connecting it with the actual motions of
the particles in the physical three-dimensional space.

For the same reason, the formalism of stochastic mechanics, which
stresses so heavily the role of the particle trajectories, cannot be
readily extended to the treatment of the quantum properties of fields.
In fact, attempts in this direction,[24] realized by replacing the
classical oscillators of the field's normal modes by Nelson's oscillators, are unable to explain the corpuscular properties of the field,
because in this theory oscillators do not have a discrete energy spectrum. It is therefore no longer possible to interpret, as one does in
quantum field theory, the n-th excited state of a given oscillator as

a state of the field with n quanta occupying the corresponding mode. In this framework it therefore seems impossible to describe the most central (and indeed the first one discovered historically) quantum effect: the emission and the absorption of light quanta. This obstacle shows, in my opinion, that the program of reformulating quantum theory in terms of stochastic processes should take as its point of departure the established fact that the electromagnetic field, whose wave-like behavior is guaranteed by classical physics, shows also, under suitable conditions, particle-like properties. In other words, it might be that a stochastic theory should give priority to the task of reproducing the particle-like properties of fields predicted by quantum theory, rather than aiming in the first place at an explanation of the wave-like behavior of particles.

In a certain sense the question is not new. Historically, the two fathers of quantum electrodynamics, P.A.M.Dirac[25] and P.Jordan,[26] held widely different opinions about the nature of particles. Dirac believed that particles and fields are two essentially different entities, as shown by their respective classical limits. Correspondingly, for him the quantization procedure, which transforms into q numbers the c-number classical variables of any dynamical system, while maintaining their mutual relations (Poisson brackets and equations of motion), endows the (classical) particles with wave-like properties and the (classical) fields with corpuscular properties. Jordan, on the contrary, was convinced that the existence of particles is always a consequence of quantization, both in the case of bosons (photons) and of fermions (electron and protons, in his time). For him, only fields exist in the classical limit, and therefore Schrödinger's equation (as well as its relativistic Dirac generalization) has conceptually the same physical meaning as Maxwell's equations.

Of course, since these two ways of looking at physical reality led ultimately to the same theory (even if the Feynman picture can be seen as a direct offspring of the Dirac viewpoint and the Tomonaga-Schwinger formulation as a development of Jordan's), the choice between them can be considered a metaphysical question devoid of physical meaning. This is no longer true, however, when one is looking for a new theory. It may well be, in fact, that the monistic "metaphysical core" of Jordan's program is more fruitful as a guiding line for the formulation of a stochastic theory that the dualistic approach of Dirac.

Quite recently a formulation of stochastic mechanics has been advanced in which to the diffusion process of Nelson has been added a discrete stochastic process.[27] This work treats the spin component as a discrete random variable and gives a probabilistic version of the nonrelativistic Pauli equation. Subsequently,[28] a stochastic version of quantum mechanics in terms of Markov random processes taking values on the discrete set of eigenvalues of a complete set of orthonormal functions in the system's Hilbert space has been proposed. I conclude this paper by rapidly surveying a recent attempt by M.Serva and myself[29] to explore the possibility of constructing a stochastic field theory based on the assumption that for any normal mode of a given classical field with amplitude $q_k(t)$,[1] the corresponding action variable J_k becomes a stochastic variable assuming only positive integer values in units of h. This of course means that the energy E_k of any normal mode is also a stochastic variable that can assume only positive integer values in units $\hbar\omega_k$. In other words, we assume as a starting point of our theory what is usually considered to be a consequence of quantization, i.e., of defining the amplitude q_k and its conjugate momentum p_k to be operators satisfying the commutator rule

$$[q_k, p_{k'}] = i\hbar\delta_{kk'}. \tag{1}$$

This does not imply, however, that our theory has a weaker explanatory power than quantum theory. It simply means that the mathematical entities introduced in the two theories at the outset are different: the assumption (1) is in fact equivalent to the assumption that $n_k = J_k/h$ can take only integer values.

The dynamical evolution of the field is described in our theory by means of the infinite set of discrete stochastic processes $\bar{n}_k(t)$. In principle, the state of the field at time t will therefore be defined by the probability density $\varrho(n_1, n_2, \ldots n_k \ldots; t)$ that $\bar{n}_1(t) = n_1$, $\bar{n}_2(t) = n_2, \ldots, \bar{n}_k(t) = n_k$. To obtain the time evolution of ϱ, the transition probabilities per unit time are needed in addition. It turns out, however, that these quantities, for simple states, are indeed very simple and have straightforward physical interpretations. Of course, in the case of free fields without sources, the number of quanta in each mode, if initially given, remains constant.

When the field interacts with a source without recoil, its normal modes are decoupled and the equations can be explicitly solved. I will briefly summarize the main results.

The probability (n,t) that the discrete stochastic variable $\bar{n}(t)$ of any given mode has the value n at time t satisfies a forward Kolmogorov equation of the form

$$\frac{d\varrho(n,t)}{dt} = \left[- p_+(n,t) + p_-(n,t)\right] \varrho(n,t) + p_-(n+1,t)\varrho(n+1,t) + p_+(n-1,t)\varrho(n-1,t), \tag{2}$$

where $p_\pm(n,t)$ are the transition probabilities per unit time from n to $n \pm 1$, respectively. It turns out that, for the ground state, these quantities are amazingly simple:

$$p_+(n) = g^2\omega, \tag{3}$$

$$p_-(n) = n\omega, \tag{4}$$

where g is dimensionless coupling constant. This means that, in this way of looking at things, emission of one quantum by the source (jump from n to n + 1) is a Poisson process with probability independent of n, while absorption (jump from n to n - 1) is a decay process with probability proportional to n.

This result shows that a reformulation of quantum theory exists (at least for this very simple field model) in terms of classical probability theory if one makes use of simple and well-known discrete stochastic processes. This means that, at least in this case, there is no need to introduce, in order to describe physical reality, the usual mathematical machinery of Hilbert space, with state vectors and operators, along with their often forgotten but well-known problems of interpretation.

Up to now the theory has been extended to deal with time-dependent sources. I am well aware of the fact that it is not a straight-forward matter to extend it to more realistic cases of interacting fields. It is however striking that the simple idea of introducing the old Plank-Bohr-Sommerfeld law $J = nh$ in to the frame work of the theory of stochastic processes had to wait several decades before getting a chance of being developed.

4. FINAL REMARKS

It seems to me difficult to deny, even apart from the preceding considerations, that scientific communities evaluate the contributions to the development of their discipline relying heavily on criteria of a metascientific nature. There are in fact many ways of selecting those aspects of reality to be considered as primary and many ways of establishing connections between them. Furthermore, there is a vast amount of arbitrariness in the definition of what consitutes a satisfactory explanation of the empirical relations in terms of theoretical constructs. It is therefore necessary to appeal to the forementioned criteria in order to chose among these different possibilities. This is indeed what scientific communities are doing all the time, adopting criteria of choice which may be different depending on the issue at hand, both when they reject or neglect a new proposal and when they accept or endorse another one.

The specific case discussed here, however, shows, in my opinion, that the most important question at stake for the practitioners of a discipline is the preservation of the traditional image of their discipline, which is shared by themselves as well as by members of other communities within science at large. Undoubtedly the image of physics as the science of laws by definition, could not at the beginning of this century tolerate being endangered. This single criterion of selection was sufficient, as I have tried to argue, for the acceptance of a theory considered satisfactory from this point of view, but at the expense of banning from physics a whole range of interesting problems.

However, times have changed recently. Biology, and indeed evolutionary biology, as we have seen, is replacing physics as a model of science. The sciences of complex systems are becoming more fashionable than those of elementary ones. The _Zeitgeist_ is now favorable for a change also in the physical sciences. We can only wait and see.

NOTE

1. The one-dimensional field u(x,t) is expanded in its normal modes as usual in the form:

$$u(x,t) = \sum_{k=1}^{\infty} q_k(t) \sin \frac{k\pi}{L} x, \quad (0 < x < L)$$

REFERENCES

1. M.Ageno, 'Importanza della concezione Darwiniana nella biologia odierna' ('Importance of the Darwinian conception in present day biology'), unpublished manuscript.
2. See for example the collections: Dynamical Systems, Theory and Applications, Springer Verlag, Berlin 1975 and Dynamical Systems and Chaos, Springer Verlag, Berlin 1983.
3. See for example: Max Jammer, The Conceptual Development of Quantum Mechanics, McGraw Hill, New York 1966.
4. P.Forman, 'Weimar Culture, Causality and Quantum Theory', Historical Studies Physical Sciences 3, 1 (1971).
5. M.De Maria, F.La Teana, Fundamenta Scientiae 3, 129 (1982).
6. T.S.Kuhn, Black Body Theory and the Quantum Discontinuity, Oxford University Press, London 1978.
7. L.Onsager, Phys.Rev. 37, 405 (1931).
8. I.Prigogine and L.Nicolis, Self Organization in Non-Equilibrium System, Wiley, New York 1977.
9. G.Ciccotti and G.Ferrari, Eur.J.Phys. 4, 110 (1983).
10. A.N.Kolmgorov, Dokl.Akad.Nauk.SSSR 98, 527 (1954).
11. G.Contopoulos, Astron.Journ. 68, 3 (1963); M.Henon and C.Herles, Astron.Journ. 69, 73 (1964).
12. Stochastic Behaviour in Classical and Quantum Hamiltonian Systems, Volta Memorial Conference, Como 1977, Springer Verlag, Berlin 1979.
13. G.Battimelli, unpublished manuscript.
14. C.De Witt Morette and K.Elsworthy, 'New Stochastic Methods in Physics', Physics Reports 77 (1982).
15. E.Nelson, Phys.Rev. 180, 1079 (1966).
16. D.Dohrn, F.Guerra, Lett.Nuovo Cim. 22, 121 (1978).
17. F.Guerra, L.Morato, Phys.Rev. D27, 1774 (1983).
18. D.Dohrn, F.Guerra, P.Ruggiero in: Feynman Path Integral, eds. S.Albeverio et al., Springer Verlag, Berlin 1979.
19. F.Guerra, Phys.Reports 77, 263 (1981).
20. G.Jona-Lasinio, F.Martinelli, E.Scoppola, Phys.Reports 77, 313 (1981).
21. D.De Falco, S.De Martino, S.De Siena, Phys.Rev.Lett. 49, 181 (1982).
22. D.Schucker, J.Functional Anal. 38, 146 (1980).
23. M.Serva, Lett. Nuovo Cim. 41 (1984).
24. F.Guerra, M.I.Loffredo, Lett. Nuovo Cim. 27, 41 (1980).

25. P.A.M.Dirac, Proc.Roy.Soc. A**114**, 243 (1927).
26. P.Jordan, Zts.Phys. **44**, 473 (1927).
27. G.F.De Angelis, G.Jona-Lasinio, J.Phys. A**15**, 2053 (1982).
28. F.Guerra, R.Marra, Phys.Rev. D**29**, 1647 (1984).
29. M.Cini, M.Serva, J.Phys. A, in press.

SOME CRITICAL CONSIDERATIONS ON THE PRESENT EPISTEMOLOGICAL AND
SCIENTIFIC DEBATE ON QUANTUM MECHANICS

GianCarlo Ghirardi

Department of Theoretical Physics, University of Trieste
and International Centre for Theoretical Physics,
Miramare, Trieste, Italy.

ABSTRACT. Some general methodological considerations aimed to guarantee
the necessary logical rigor to the present debate on quantum mechanics
are presented. In particular some misunderstandings about the
implications of the critical analysis put forward by Einstein Podolsky
and Rosen (EPR) which can be found in the literature, are discussed.
These misunderstanding are shown to arise from possible underestimates,
overestimates and misinterpretations of the EPR argument. It is argued
that the difficulties pointed out by EPR are, in a sense that will be
defined precisely, unavoidable. A model which tries to solve the
difficulties arising from quantum non separability effects when
macroscopic systems are involved, is briefly sketched.

1.   INTRODUCTORY CONSIDERATIONS

In view of the fact that the debate on the epistemological and
conceptual implications of quantum theory is often vitiated, in our
opinion, by gross misunderstandings, we consider it appropriate to
focus on some methodological principles which must be kept in mind in
order to guarantee the necessary level of rigor of such a discussion.
The occurrence of misunderstandings on this matter is not surprising if
one takes into account the conceptual difficulties that the formalism
presents, on which the critical debate is still alive, and which have
not yet been fully clarified. One can mention that, already in the
early stages of the development of the theory, even eminent scientists
have been unable to fully appreciate[1] the crucial points and the subtle
implications[2] of the deep analysis of Einstein-Podolsky and Rosen[2] (EPR).
We want to stress strongly the necessity of logical rigor and clarity
in carrying on the debate. This can be guaranteed only if the terms

89

G. Tarozzi and A. van der Merwe (eds.), The Nature of Quantum Paradoxes, 89–103.
© 1988 by Kluwer Academic Publishers.

that are used and the conceptual framework in which each critical
investigation finds its place, are very precisely defined. Obscurities,
lack of precision, and rash extrapolations, seriously damage this kind
of research, and, in particular, they reinforce the wrong and naive,
but very dangerous conviction, shared by several members of the
scientific community, that to investigate the conceptual and
epistemological implications of the theory is a more or less useless
task. This, in turn, increases the division between the so called "two
cultures," with all its harmful consequences.

To avoid the above misunderstandings, it is first of all necessary
that the object of any critical analysis be defined precisely, in
particular by making absolutely clear whether the analysis involves the
direct consequences of the formal scheme or only one of its possible
interpretations. The meaning of the terms that are used must be made
very precise; e.g., a source of misunderstandings is often the use of
terms such as "realism," *tout court*, to describe both the very
precise requirements of reality put forward by EPR and the much more
stringent requirements coming from various forms of philosophical
realism. The same applies to the term "locality," which has very
different conceptual connotations in the critical investigations of EPR
and those of J.Bell. Analogous difficulties arise when it is claimed
that a "new" interpretation of the formalism resolves all its
conceptual difficulties, without confronting the proposed interpretation
with all the known crucial points of the theory. As an example, we can
mention that it is sometimes stated that alternative interpretations,
such as the "propensity interpretation" or the introduction of non-
Kolmogoroffian models of probability, solve the EPR paradox, without
facing the problem of identifying which "elements of physical reality"
can be attributed to the physical system under investigation within the
proposed interpretation, a point which is central for discussing the
EPR reasoning.

Let us now specify some general methodological points.

2.   GENERAL METHODOLOGICAL CONSIDERATIONS

It is useful to recall[3] here some possible approaches to a critical
analysis of an established theoretical scheme.

A. One can check the compatibility of the theory with general
ideas about physical reality and the level of knowledge we can get of
it. This kind of analysis is, in our opinion, not only legitimate, but
necessary. As a typical example of this line of approach, we can refer
to the EPR paper.

B. One can check the compatibility of the theory with other
physical principles, which are viewed as "true" by the scientific
community, even though they are not included among the axioms of the
theory considered. Into such a context one can fit all recent attempts[4]
to prove that the quantum formalism would allow, through wave packet
reduction, faster-than-light communication between different observers.
As we will show, this line cannot lead to any significant result.

C. One can check the internal consistency of the theory. Within this approach fall all investigations direct to clarify whether one can describe the functioning of the measuring apparata, as postulated in quantum theory, in terms of the dynamics of the microconstituents of these objects. This crucial point of the quantum theory of measurement has in our opinion not yet been completely clarified.

3. THE EPR PARADOX

In spite of the many investigations on the EPR paradox which can be found in the literature, we consider it appropriate to reconsider it in a coincise way, and to focus on some points which will be useful in illustrating some misunderstandings about this subject.

Let us recall, first of all, some basic points of the formalism. In quantum mechanics an ensemble is described by the statistical operator ρ, which will be briefly called "a state" and is a trace–class operator with trace 1. We will assume that no superselection rule is present for the systems of the ensemble.

Rule 1. The physical predictions of the theory are obtained according to the following scheme. Let A be an observable, and P_k the spectral family of projection operators associated with the eigenvalues a_k of A. If the ensemble is in the state ρ, the probability $P(A=a_k|\rho)$ of finding the result a_k in a measurement of A, is given by

$$P(A = a_k|\rho) = \text{Tr } P_k \, \rho \tag{1}$$

from which it follows that the mean value of the results obtained in a measurement of A is

$$<A> = \text{Tr } A\rho \tag{2}$$

Rule 2. Reduction of the wave packet: For an ensemble in the state ρ_b , a non selective measurement of an observable A induces the following change of the statistical operator

$$\rho_b \rightarrow \rho_a = \sum_k P_k \, \rho_b \, P_k \tag{3}$$

ρ_a and ρ_b representing the state before and after the measurement, respectively.

We shall make use of two simple theorems implied by the quantum formalism.

Theorem 1. Given two different states ρ and ρ' there exist at least one observable X such that

$$\text{Tr } X\rho \neq \text{Tr } X\rho' \tag{4}$$

This theorem implies that different (as operators) states involve different physical predictions and are thus, in principle, experimentally distinguishable. This in turn implies that one cannot describe the same physical situation with different statistical operators. I shall not enter the debate on this assumption (which could be questioned in some

cases) since I will bear in mind, in discussing the EPR paradox, such
simple systems as the spin states of a two-particle system, in which
case the assumption is certainly correct.

There is a simple theorem[5] of quantum mechanics which is extremely
important for an appropriate understanding of the requirement of
completeness of the theory, even though it is not often quoted in the
discussions on the EPR paradox.

Theorem 2. For an ensemble of composite systems $S = S_1 + S_2$, which
is a pure state, the necessary and sufficient condition that there will
exist a complete set of commuting observables belonging to one of the
two subsystems S_1 and S_2 , for which the probability of getting a
given set of results equals 1, is that the state of the system be the
projection operator on a factorized state vector

$$\rho = |\psi><\psi| \ , \quad |\psi> = |\phi_k^1> \otimes |\chi^2> \tag{5}$$

Remarks on Theorem 2: We shall call states of type (5)
"factorized states" and denote them by ρ_F . In view of this theorem, it
is legitimate to think of an ensemble as composed of subensembles, such
that for each subensemble the result of a measurement of an appropriate
observable of subsystems S_1 or S_2 can be predicted with certainty if
and only if ρ is a statistical mixture of factorized states

$$\rho = \sum_i p_i \, \rho_{Fi}$$

Note also that, in the case of a spin - ½ particle, any spin component
is by itself a complete set of commuting observables.

We can now discuss in detail the EPR argument. We will use a
version of it which is slightly modified with respect to the original
one, but is now currently used. We will summarize the argument in a
sequence of steps denoted by capital letters from A to E.

A. Einstein's reality requirement: If, without in any way
disturbing a system, we can predict with certainty the result of a
measurement of one of its observables, then there exists an element of
physical reality associated with this prediction.

Comments on A. On the basis of Theorem 2, we can state
(a) If

$$\rho = \sum_i p_i \, \rho_{Fi} \ ,$$

then we can attribute elements of physical reality to the individual
subsystems of the composite systems.
(b) In the quantum formalism, the formal counterpart of the existence
of elements of physical reality (possibly different from subsystem to
subsystem) for all individual subsystems of the composite systems is
that the ensemble can be described by a statistical mixture of
factorized states.

B. Choice of the initial state and of the experimental set up.
Let us consider an ensemble of composite systems $S_1 + S_2$ that is a
pure case, the corresponding state being non factorized:

$$\rho = |\psi> < \psi|, \quad |\psi> = \sum_j | \phi_j^1 > \otimes | x_j^2 > \tag{6}$$

Here the states $| \phi_j^1>$ and $|x_j^2>$ are eigenstates of two observables F and G, of systems S_1 and S_2, respectively, belonging to different eigenvalues for different j values. Suppose, moreover, that all states $|\phi_j^1>$ are well localized in a spatial region far away from that where the states $|x_j^2>$ are localized. (One can refer, e.g., to the well known EPR-Bohm situation[6] involving the singlet state.) Let us measure on systems S_2 the observable G. Then, according to quantum rules (rules 1 and 2 above), we can predict with certainty the result of a subsequent measurement of the observable F for every individual subsystems S_1. According to the reality assumption A, we can conclude that, immediately after the measurement on the systems S_2, all systems S_1 possess elements of physical reality.

   Comments on B. If the measurement on S_2 is of the ideal type, then we see, from the postulate of wave packet reduction, that quantum theory describes the ensemble after the measurement by a statistical mixture of factorized states,

$$\rho = \sum_j p_j | \phi_j^1 > | x_j^2 > < x_j^2 | < \phi_j^1 |, \tag{7}$$

and therefore contains the formal counterpart of the elements of physical reality which have been recognized as present.

   We want to stress that we, neither here nor in what follows, shall resort to measurements of noncommuting observables. It is also worthwhile noticing that,up to this point, one could also have avoided using the quantum formalism to draw the conclusion that elements of physical reality can be attributed to the systems S_1 after the measurement. In fact, use of repeated experimental tests and of inductive inference alone allows one to assert that, after the measurement on S_2, the results of the measurements on S_1 can be predicted with certainty without disturbing S_1. Therefore, if we accept Einstein's requirement of reality, we can assert that all systems S_1 possess elements of physical reality. The recognition that quantum theory contains the formal counterpart of these elements can then follow by an analysis of the formal rules.

   C. Einstein's locality requirement. This requirement can be concisely rephrased as follows: The elements of physical reality which are possessed by a system cannot be instantaneously influenced at a distance. The requirement leads directly to admitting the existence of elements of physical reality for all systems S_1, even before the measurement on S_2, at least for an appropriate time interval.

   D. Completeness requirement. Any complete theory must contain the formal counterpart of all elements of physical reality that are present.

   E. Conclusions. The state before the measurement has been prepared in the pure nonfactorized state (6). By virtue of theorem 2, since ρ is not a mixture of factorized states, the quantum mechanical description does not contain the formal counterpart of the elements of physical reality for systems S_1 which, according to our previous discussion, must be recognized as present before the measurement. The conclusion follows: Quantum mechanics is not complete. Alternatively, one could argue in the following way: The recognitibn of the presence of elements of physical reality before the measurement, together with the assumption of completeness, leads one to conclude that before the measurement the state of the system must be a statistical mixture of factorized states. On the other hand, the state is given by Eq.(6). Since these two states are different, by virtue of theorem 1, we get a contradiction.

4.   MISUNDERSTANDINGS ABOUT THE EPR ANALYSIS

It is useful to discuss some misunderstanding about the EPR paradox which can be found in the literature.

4A.   Underestimates of the Paradox

It is sometimes stated that there is nothing paradoxical in the EPR situation and that those who find it disturbing do not take into account that it is perfectly possible, within standard probability theory, that an increase of information on a part of a system can induce instantaneously an analogous increase of information about distant parts of the system. The standard reference is to a classical experiment with a box which can be split into two parts, each part containing one ball of a color (black or white) that is different from that of the ball in the other part of the box. Such a box is then separated into two parts, and each part is carried far away from the other. Checking the color of the ball in one box, yields immediate knowledge of the color of the ball in the distant box. This reasoning, of which J. Bell has so clevery made a laughing stock with his example of Bertelmann's socks, shows that those who adhere to it have completely missed the essence of the EPR argument: The paradox, in Einstein's example, does not arise from the immediate acquisition of knowledge of far and previously unknown elements of physical reality, but from the fact that, contrary to what happens in the classical example of the partitioned box, in quantum mechanics it results contradictory to assume that the system S_1 possessed the elements of physical reality already before the measurement. For this reason, all statements that wave-packet reduction corresponds only to an increase of information on a part of a system, and that this can legitimately yield information on other parts, so that there is no problem at all, simply reveal a complete lack of understanding of the crucial points of the EPR analysis.

4B   Overestimates of the Paradox

In the literature one can find misunderstandings corresponding to an illegitimate overestimate of the paradox. Some authors claim[4] that EPR-type setups, with more or less refined technical improvements, would allow, through wave-packet reduction, faster-than-light communication between distant observers.
   That this conclusion is certainly false is a very simple consequence

of quantum rules and can be proved as follows.[7] For an ensemble of composite systems $S = S_1 + S_2$, associated with the state $\rho(1,2)$, all physical predictions concerning the ensemble of the subsystems S_2 are derived through the statistical operator $\rho^{(2)} = \text{Tr}^{(1)}\rho(1,2)$, where $\text{Tr}^{(1)}$ denotes partial tracing on the variables of system 1. The measurement of an observable on system S_1, changes the state from $\rho(1,2)$ to

$$\tilde{\rho}(1,2) = \sum_j P_j^1 \rho(1,2) P_j^1$$

where P_j^1 is the family of projection operators on the eigenmanifolds of the measured observable. If the measurement is performed, all physical predictions about subsequent measurements on S_2 can be derived from the statistical operator

$$\tilde{\rho}(2) = \text{Tr}^{(1)}\tilde{\rho}(1,2) = \text{Tr}^{(1)}\sum_j P_j^1 \rho(1,2) P_j^1$$

$$= \text{Tr}^{(1)} [\sum_j P_j^1] \rho(1,2) = \text{Tr}^{(1)}\rho(1,2) = \rho(2) \qquad (8)$$

where we have made use of the cyclic property of the trace and of the fact that

$$\sum_j P_j^{(1)} = 1 .$$

The above equation implies that the measurements on S_1 cannot alter any probability distribution about measurements on S_2. This conclusion, we stress again, follows from a simple, straightforward and correct use of the formal rules of the theoretical scheme.

Strictly related to the problem discussed above is an argument put forward by K. Popper in his Poscript to the *Logic of Scientific Discovery*.[8] In view of the large interest that this book has created, we think it is useful to discuss its reasoning in detail.

The author propose "a simple experiment which may be regarded as an extension of the EPR argument." A source S emits pairs of particles in opposite directions. The particles, crossing the slits at A and B, are detected by a belt of Geiger counters in coincidence (see Figure 1).

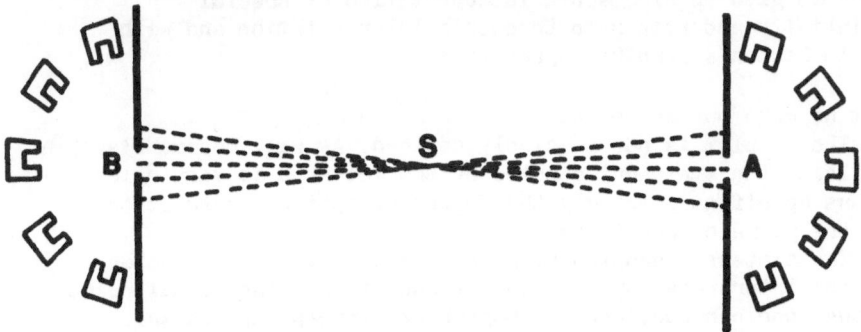

Figure 1. The experiment proposed by K. Popper in Ref. 8.

One then chooses to make the slit at A very narrow, keeping unaltered
the slit at B. Since the particle positions are correlated, the
localization at A induces an analogous localization for the partner
particles at B.

At this point it is appropriate to come back to the direct citation
of Popper's reasoning. In doing so, we label with Greek superscripts
the points which will be subsequently discussed in detail. According to
the author:

" We thus obtain fairly precise *knowledge* about the q_y
position of this particle (the one at B). (α) We have
measured its y position indirectly. And since it is,
according to the Copenhagen interpretation, our *knowledge*
which is described by the theory -- and especially by the
Heisenberg relations -- we should expect that the momentum
p_y of the beam that passes through B scatters as much as that
of the beam that passes through A, even though the slit at A
is much narrower than the widely opened slit at B. Now the
scatter can, in principle, be tested with the help of the
counters. If the Copenhagen interpretation is correct, then
such counters on the far side of B that are indicative of a
wider scatter (and of a narrow slit) should now count
coincidences: counters that did not count any particles
before the slit at A was narrowed. (ß) To sum up: if the
Copenhagen interpretation is correct then any increase in
the precision of our mere knowledge of the position q_y of the
particles going to the right should increase their scatter
and this prediction should be testable. (γ) If, as I am
inclined to predict, the test decides against the Copenhagen
interpretation... it would mean that Heisenberg's claim is
undermined.... (δ) What would be the position if our
experiment (against my personal expectation) supported the
Copenhagen interpretation -- that is, if the particles whose
y position has been indirectly measured at B show an increase
of scatter? (η) This *could* be interpreted as indicative of
an action at a distance, and, if so, it would mean that we
have to give up Einstein's interpretation of special
relativity and return to Lorentz's interpretation and with
it to Newton's absolute space-time."

Let us make our comments:

1. The problem is not precisely defined. As will become clear in
what follows, the statement denoted by the letter α is meaningless
unless one specifies precisely the degree of spatial correlation
between the pairs of particles.

2. The statement denoted by ß reveals a dangerous confusion
between the interpretation of a theory and its precise formal rules.
Even though one can consider the Copenhagen interpretation as
unsatisfactory or unacceptable, there is no doubt that the supporters
of this interpretation have never made use of predictions contradicting
those which are implied by the formalism. If one takes into account

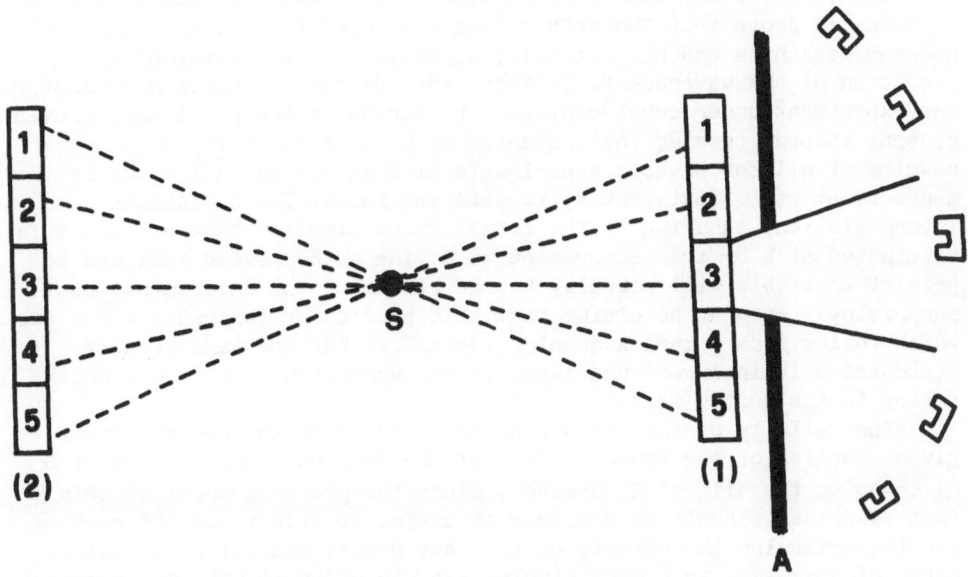

Figure 2. The experiment of Figure 1, analyzed with greater precision
to understand the mechanism of the reduction of the wave packet at A.
The state vector of the composed system is a linear superposition of
products of states localized in the regions denoted by the same number,
for the two decay products emerging from S.

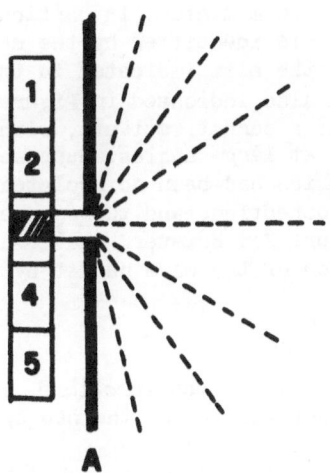

Figure 3. Effect of the reduction of the wave packet at the slit A of
Figure 2, when its opening is narrower than the transverse extension
of the state localized in region 3.

this fact, one easily sees that statement ß, as well as that denoted by
γ, actually prove that the author does not correctly use the rules of
quantum mechanics and has not fully appreciated the meaning of the
reduction of the wavepacket. In fact, what do quantum rules say about
the experiment under consideration ? By virtue of the previously proved
general theorem (see Eq.(8)), quantum mechanical rules *imply* that the
results of all conceivable experiments at B cannot be influenced by the
measurement at A. Statement δ, too, is the false: The Copenhagen
interpretation, adhering to the formal rules, implies that no effect is
originated at B by the measurement at A. The author seems inclined to
predict that this will actually be the result of the experiment, but,
surprisingly enough, he claims that this prediction contradicts the one
which follow from orthodox quantum mechanics. For the same reasons,
statement η is incorrect and leads to the surprising assertions which
follow in the quoted text.

The validity of the above statements rests on the general proof
given earlier of the impossibility of inducing physical changes at B
by changing the slit at A. However, since the previous proof of this
fact is rather formal, we consider it useful to illustrate the problem,
and in particular the effects of the wave packet reduction on distant
parts of a system, in a much simpler way, in order to make our argument
understandable also for those who are not fully familiar with the more
formal aspects of the theory. For this purpose, let us consider a
system $S = S_1 + S_2$ in the state

$$| \psi > = \sum_{i=1}^{5} | \psi_i^{(1)} > \otimes | \psi_i^{(2)} > \tag{9}$$

where $| \psi_i^{(1)} > (| \psi_i^{(2)} >)$ is a state describing particle 1 (2), well
localized in the spatial region denoted by the index i in Figure 2.
The transverse extension of the wave packet $| \psi_i >$ is obviously related
to the corresponding spread in momentum. Let us assume that the wave
packets correspond to small y scatters in momentum. In particular,
suppose that the wave packet 3, which is identified by the measurement
at A for the choice of the opening of the slit indicated in the figure,
corresponds to the angular dispersion also indicated in Figure 2. In
such a case, the selected particles at A cannot activate, with
significant probability, the counters at large angles. Suppose now that
the slit at A is narrowed. The beam which has been so isolated is
better localized in the transverse y direction, and the p_y scatters
after the slit is increased (see Figure 3). However, the more strict
localization at A implies the reduction of the wave packet by projection:

$$| \psi > \longrightarrow P_\Delta^{(1)} | \psi > \tag{10}$$

where $P_\Delta^{(1)}$ projects on the linear manifold of the functions $\psi^{(1)}$ which
are different from zero only in the inetrval Δ of the new opening of
the slit at A. From this we have

$$P_\Delta^{(1)} | \psi > = [P_\Delta^{(1)} | \psi_3^{(1)} >] \otimes | \psi_3^{(2)} > \tag{11}$$

showing that the component of the wave function referring to system 2 is localized exactly as before. This is the precise effect of wave packet reduction in the quantum formalism and also in the Copehagen interpretation of it. It is obvious that, if one chooses the wave packets $|\psi_i\rangle$ better localized than Δ from the beginning, then the measurement at A will induce the reduction to a better localized state in B; however, in such a case the large momentum scatters are already present also in the absence of any measurement.

It is hoped that this discussion will be useful in clarifying some serious misunderstandings that could be generated by Popper's argument.

4C. Misinterpretations of the Paradox

We want to comment here about another misleading argument concerning the EPR paradox. It has recently been claimed that the derivation of the paradoxical conclusions does not require the assumption of completeness of the quantum theory [9]. It is almost obvious that this cannot be true since, by the analysis of Sec.3, it should be evident that, if one does not require the completeness of the theory, one can consistently accept the existence of elements of physical reality both before and after the measurement (agreeing therefore with Einstein's realism and locality) without getting into any trouble. In such a case, the theory would contain the formal counterpart of the elements of physical reality which are present after the measurement, but would be unable to account for those which are admittedly present before (by locality). The misunderstanding originates from an incorrect use of the negation of the proposition defining Einstein's completeness requirement. In fact, to deny "the theory contains the formal counterpart of all elements of physical reality which are present," one should obviously assert "the theory does not contain the formal counterpart of all elements of physical reality which are present." As we have seen, this is actually the case for quantum mechanics if the Einstein's reality and locality requirements are accepted. Curiously enough, in Ref. 9 the negation of the previous statement is taken to be "the theory allows one to attribute simultaneously different elements of reality to the individual systems of the ensemble." From this assumption, a proof based on the use of Bell's inequality leads to the surprising conclusion that the paradox can be derived without using the completeness requirement.

The remarks of this section require a further logical analysis of the situation connected with EPR analysis. We discuss now what we shall call "the unavoidability of the EPR paradox."

5. THE UNAVOIDABILITY OF THE EPR PARADOX

We want to point out in this section that the conclusions which can be drawn from the EPR analysis are not dependent on a particular interpretation of the formal apparatus of the theory, but that they are, in a sense that we will define precisely, unavoidable.

From the discussion of Sec. 3, we see that the EPR reasoning can be summarized as follows: (i) If one accepts the reality and locality requirements of the EPR paper, and (ii) if the experimental correlations are appropriately described by the theory, then (iii) the theory does not contain the formal counterpart of elements of physical reality which must be admitted as present before any measurement takes place.

In what follow, we regard the first premise of the above statement as true, i.e., we accept the reality and locality requirements made in the EPR paper. Let us then see which logical possibilities remain open:

A. If the second premise of the statement is considered false, i.e., quantum mechanics would not predict correctly the experimentally observable correlations between subsystems of composite systems, then the conclusion (expressed in rough terms) would be: Quantum Mechanics is *false*. In such a case it would be useless to debate on its epistemological implications, and it would become a precise duty to start a systematic experimental investigation aimed at identifying precisely the disagreement with theoretical predictions and consequently trying to modify the theoretical scheme. After this, one should start again a critical analysis of the modified theory.

B. If one assume as valid also the second premise, i.e., that quantum predictions about correlations are correct, since the theory is not able to describe elements of physical reality which must be recognized as present, the logical possibilities that remain open are the following:

B1. There exists a theory which is predictively equivalent to quantum mechanics, but contains the formal counterpart of the elements of physical reality which are present before the measurement, besides the counterpart of those which are present after it.

We would like to remark that an example of a theory exhibiting these features, for the spin degrees of freedom of a system of two spin-½ particles, has recently been proposed by I. Pitowsky.[10] The model looks quite curious from a physical point of view; however it constitutes an example of a local deterministic completion of quantum mechanics. It has the same predictive content as quantum theory and, without having to introduce nonlocal effects to describe the EPR correlations, it meets Einstein's requirement of containing the formal counterpart of the elements of physical reality, both before and after the measurement. One of the model's merits, in my opinion, is that of making clear the necessity of distinguishing between the locality requirements made by Einstein and by Bell; in fact the model violates Bell's inequality, since it reproduces the correlations predicted by quantum theory, but it does not violate Einstein's locality.

Without discussing the delicate question of the possibility of extending such a model to other degrees of freedom, or taking a position about its internal consistency, which should be critically discussed in great detail, in the spirit of the logical alternatives we are discussing, we can state that, if alternative B1 occurs, then one can conclude that quantum mechanics is *incomplete* and Einstein was right.

B2. Another logical possibility is that a theoretical scheme equivalent to quantum mechanics could be formulated which does not contain the formal counterpart of the elements of physical reality, either before or after the measurement . Such a theory would avoid the unpleasant aspect of quantum theory that the measurement seems "to create" instantaneously elements of physical reality in systems far apart. This alternative can, however, simply be dropped, since the theory in question would imply a more drastic renunciation of the requirement that a description of physical reality be possible than does quantum theory itself. Actually, in such a case, quantum mechanics would be a completion (even though a partial one), in a realistic sense, of the theory under consideration.

B3. The only remaining logical alternative[2] is that any conceivable theory which is predictively equivalent to quantum theory would exhibit the same conceptual limitation, i.e., it would not contain the formal counterpart of all elements of physical reality which must be recognized as present. In such a case, quantum theory would be neither wrong nor incomplete; however the conceptual problems raised by EPR would still remain. In such a case, we would be compelled to accept either that there are definite limits to our ability of understanding and describing physical reality, or that even the absolutely natural (at least for us) requirements of reality and locality in the EPR sense must be reconsidered.

6. THE INTERNAL CONSISTENCY OF THE FORMALISM

The EPR paradox is strictly related to the socalled quantum nonseparability and, as we have seen, it engenders relevant epistemological problems. Quantum nonseparability, when macroscopic systems are involved, raises the basic problem of the internal consistency of the formalism. The typical example is given by the quantum theory of measurements. The difficulties arise from the fact that, if one assumes that also the system-apparatus interaction is governed by the basic linear laws of quantum mechanics, then the final state of the system plus apparatus is non factorized and involves linear superpositions of states for which the apparatus is in macroscopically distinguishable situations. The quantum theory of measurement asserts, that, in the process of measurement, the reduction of the wave packet takes place, leading to a statistical mixture in which the apparatuses are in definite macroscopic situations (e.g., linear superpositions of different pointer positions are eliminated).

The clash between these two descriptions constitutes the basic problem of the quantum theory of measurement. Various solutions have been proposed for this problem. We do not want to discuss them here in detail, but refer the interested reader to Ref. 11 and to the literature quoted there. We simply want to stress here that, in our opinion, to keep the standard quantum dynamics and to adandon or render ineffective for some systems the superposition principle (an assumption which is made in all proposed solutions), would amount to accepting, at least to a certain extent, a dualistic description of natural phenomena and to give up the program of a unified description of the behavior of all systems from the basic dynamics of microscopic systems.

We consider it appropriate to recall that recently we have

presented a model for such a unified description[11] which exhibits the
following features:
-- It makes predictions for microsystems that practically coincide with
standard   quantum predictions for extremely long times.
-- It allows a consistent derivation of macroscopic dynamics from the
basic dynamics of the microscopic constituents. The ensuing dynamical
behavior for macro-objects is largely compatible with classical dynamics
and leads to the suppression of linear superpositions of states which
are localized in spatially separated regions.
-- It solves the problem of the quantum theory of measurement by
leading, as a consequence of the system-apparatus interaction, to
statistical mixtures with fixed pointer positions for the apparatus.

The interested reader is referred to Refs. 11 and 12, for a
complete and systematic discussion of the model and of its implications.

REFERENCES

1. J.S. Bell, *J.Phys. (Paris)*, Colloque C2, Suppl. No.3, Tome 42
   (1981).
2. A. Einstein, B. Podolsky and N. Rosen, *Phys.Rev. 47*, 777 (1935).
3. G.C. Ghirardi and T. Weber, *Nuovo Cimento 78B*, 9 (1983).
4. N.H. Herbert, *Found. Phys. 12*, 1171 (1982); N. Cufaro-Petroni, A.
   Garuccio, F. Selleri and J.P. Vigier, *C.R. Acad. Sci. Ser. B 290*,
   111 (1980); F. Selleri, in *Dynamical Systems and Microphysics*,
   A. Blaquiere, F. Fer and A. Marzollo, eds.(Springer Verlag, New
   York, 1980).
5. G.C. Ghirardi, C. Omero, A. Rimini and T. Weber, *Nuovo Cimento 39*B,
   130 (1977); G.C. Ghirardi, in *Dynamical Systems and Microphysics*, A. Blaquiere, F. Fer and A. Marzollo, eds.(Springer-Verlag
   New York, 1980).
6. D.Bohm, *Quantum Theory*, (Prentice Hall, Englewood Cliffs, New
   Jersey, 1951).
7. G.C. Ghirardi, A. Rimini and T. Weber, *Lett. Nuovo Cimento 27*, 263
   (1980); A. Shimony, in *Proceedings of the International
   Symposium Foundations of Quantum Mechanics in the Light of
   New Technology*, S. Kamefuchi *et.al.*, eds. (Physical Society of
   Japan, Tokio, 1983).
8. K.R. Popper, *Quantum Theory and the Schism in Physics*, Vol.III
   of *Postscript to the Logic of Scientific Discovery*, W.W. Bartley
   ed. (Hutchinson, London, 1983).
9. F. Selleri, "L'altro labirinto", in *La Nuova Critica*, No. 62
   (Jouvence, Rome, 1982).
10. I. Pitowsky, *Phys. Rev. Lett. 48*, 1299 (1982); *Phys. Rev. 27* D,
    2316 (1983).
11. G.C. Ghirardi, A. Rimini and T. Weber, *IC/84/240 Trieste*, to
    appear in *Lecture Notes in Mathematics*, (Springer Verlag, New
    York, 1985).
12. G.C. Ghirardi, A. Rimini and T. Weber, in preparation.

NOTES

1. See the illuminating discussion of this point in the introduction of Ref. 1).
2. The possibility of a theory which contains the formal counterpart of the elements of physical reality that are present before the measurement, and not the counterpart of those which are present after words need not be considered, since it would encounter even more serious difficulties than those discussed under B2.

PART 2

EINSTEIN LOCALITY, SEPARABLE REALITY AND QUANTUM CORRELATIONS

PHYSICAL CONSEQUENCES OF REALITY CRITERIA FOR QUANTUM PHENOMENA

F. Selleri
Dipartimento di Fisica
INFN - Sezione di Bari
Via Amendola 173
I-70126 BARI

ABSTRACT. Physical realism has today a chance to free itself from two
old curses, thanks to the parallel developments of the Einstein, Podolsky and Rosen paradox and of the causal description of (single) atomic
systems.
First of all it can free itself from the accusation of being metaphysical, since it is capable of falsifiable predictions. Secondly it can
overcome Niels Bohr's interdiction against a description of two observables represented by non-commuting operators as simultaneously real.
These two lines of thought are here reviewed and developed.

1. DETERMINISTIC REALITY

"If, without in any way disturbing a system, we can predict with certainty (i.e., with probability equal to unity) the value of a physical quantity, then there exists an element of physical reality corresponding to
this physical quantity".
    This is the famous "criterion of physical reality" proposed by
Einstein, Podolosky and Rosen in 1935.[1] It is a very weak criterion of
reality, very carefully formulated and rather general. It appeared to
them (and it still appears today) "far from exhausting all possible
ways of recognizing a physical reality". This criterion can have many
applications to the macroscopic domain where it appears as
trivially true: For instance, if we can predict with certainty that the
length of this table is two meters (within an error of some millimeters)
then there exists an elements of physical reality corresponding to the
length of the table. Notice that EPR would not say that the length
itself is real; firstly, because the unit of length is conventional and
contains therefore subjective aspects; furthernore, because what appears
us immediately as extension in space might result from a physical situation rather different from our perception, which gives rise to a
final mental model built by the observer which is only a topological
distorsion of the true physical situation. Even in this complicated
case the EPR-criterion would apply, since it assumes <u>only the existence</u>
of an element of physical reality without specifying its nature and its

107

G. Tarozzi and A. van der Merwe (eds.), The Nature of Quantum Paradoxes, 107–129.
© 1988 by Kluwer Academic Publishers.

relationship to the measured value of the chosen physical quantity.
Einstein, Podolosky and Rosen regarded their criterion to be in agreement with classical as well as with quantum mechanical ideas of reality.

The EPR reality criterion thus rests on the idea that if it is
possible to predict with certainty the outcome of an act of measurement,
then there exists an element of physical reality associated with the
measured physical quantity. Some further qualifications of the EPR reality criterion are the following:

1) - The element of physical reality is thought of as existing
already before the act of measurement is concretely carried out. From
a philosophical point of view this means that a step into ontology has
been taken. Every positivist of strict conviction would have much to
say against this hypothesis, which seems indeed very natural if, e.g.,
our planet has to be considered as existing even before man appeared
on it (and this amounts to 99.9% of its history!) and the galaxies as
existing also when no telescope is observing them.

2) - The element of physical reality is regarded as associated with
the measured object and not with the measuring apparatus: The latter
has been cecked to work properly so that it could lead to different
results of the measured physical quantity. If we can predict with certainty that this apparatus will given an a priori known result by interacting with the given object, this means that the decisive factor producing the result of the measurement lies in the object itself. This does
not mean that an active role of the apparatus, so typical of quantum
measurements, is denied in general. The EPR reality criterion applies
in fact only to those very particular situations in which the outcome
of an act of measurement is predictable with certainty.

3) - The postulated element of reality is viewed as the cause of
the exactly predictable outcome of the measurement: realism and causality are therefore strictly tied together in the EPR reality criterion.

The element of reality is also assumed to be not necessarily coincident
with the result of the measurement. Therefore if we take for instance
an electron with the wave function

$$\psi(\vec{x}) = \psi_0 \exp\{i\vec{p}\cdot\vec{x}/\hbar\} \qquad (1)$$

we can predict with certainty that an apparatus measuring the electron's
momentum will give the result p . Then there exists an element of reality associated with the electron momentum and this statement is emphati-
cally not identical to the idea that the electron has a well defined
momentum before we measure it. The EPR reality criterion postulates only
the existence of a concrete (but in general unknown) cause of the exactly predictable result of a measurement of the particular (and perhaps
only conventional) physical quantity that we call momentum and that we
are at least able to define operationally.

4) - If past experience has shown that one has the capability to
predict with certainty the outcome of a measurement in a certain physi-

cal) situation, then we attribute an element of reality to the object even if no future measurement of the physical quantity with which it is associated will actually be carried out. With this natural assumption we exclude only that the future act of measurement may retroactively determine the present element of reality. In other words we assume that causal connections work from the past toward the future and not viceversa, or that there exists an "arrow of time".[2] Thus in the case of an electron with wave function $\psi(\vec{x})$ written above, we attribute an element of reality (corresponding to the momentum $\vec{p}$) to the electron. This element of reality is an objectively real property of unknown nature which is the cause of the result "momentum = $\vec{p}$", if and when a momentum measurement is carried out and which exists also if no measurement is carried out. Even if it might not look so at this point, this conclusion, coupled with the assumption of separability, leads to many extremely important consequences.

The idea of separability is the reasonable assumption that two arbitrary objects interact very little when their mutual distance is very large. Already in this first rough definition of separability we avoid assuming that the reciprocal interaction of the two objects is exactly zero, both because a very small (e.g. gravitational) interaction persists also at large distances and because it is not necessary for our tasks.

One can wonder about the definition of a "very large" distance. Experience tells us that there is no universal definition of this, but that such a distance, different from case to case, exists for all known situations. Thus if 10^{-10} cm is already a "very large" distance for nuclear interactions, much larger distances are needed before one can neglect electromagnetic and gravitational effects. For this argument of separability it is of course essential that all known interactions (gravitational, electromagnetic, weak and nuclear) tend to zero with increasing distance. Therefore, even if some go to zero faster (that is, exponentially) there must exist a distance beyond which they are all negligible.

The only important fact from the point of view of the EPR paradox is that such a distance, defined as "very large" in the above sense exists in every known case. Therefore we assume that: Given two physical systems, a minimal distance between them exists beyond which their mutual interactions become negligibly small.

Again one could wonder about the meaning of negligibly small interactions , but also here it is not difficult to specify that with this expression we imply: unable to create the elements of reality of either system. The latter statement could perhaps be considered too strong, but even a weaker versions of it suffices for our purposes: "capable of creating the elements of reality of either system at most in the 1 percent of the cases." The figure 1% could of course be lowered almost at will, but it is largely sufficient for our purposes since there exist 40% quantum mechanical violations of some predictions which can be deduced from the EPR reality and separability conditions.

The hypothesis of separability can be applied to quantum systems. These systems are known only through their wave functions, which can fill broad regions of space: There are however also situations in which the

spatial localization is very good: Take for instance two particles α
and β with wave functions $\psi_\alpha(x_1)$ and $\psi_\beta(x_2)$, respectively. The wave
function of the two particles considered as a single system is then

$$\psi(x_1,x_2) = \psi_\alpha(x_1)\,\psi_\beta(x_2).$$

Suppose furthermore that $\psi_\alpha(x_1)$ is a Gaussian function with modulus
appreciably different from zero only in a region centered around x_{10}
and of width Δ_1. Similarly, let also $\psi_\beta(x_2)$ be Gaussian but have
around x_{20} a width Δ_2 (See Fig. 1).

FIG. 1   Gaussian wave packets for a separable physical situation.

We shall call the two systems separable in the previously defined
sense if the distance $|x_{20} - x_{10}|$ is much larger than either Δ_1 or Δ_2.
If α and β propagate to the left and to the right, respectively, so
that the distance between the centers of the two packets increases linearly with time, it is not difficult to show that the Scrödinger equation
allows for situations in which the conditions for separability improve
with time.

It might be objected that a Gaussian fills the whole space and not
only a finite volume, so that there is after all a nonvanishing probability of finding α and β near to one another. This is of course
true, but the point is that this probability is extremely small, as small
as one might wish at some distance (say, less than 10^{-100}), given the
extremely rapid decline of gaussians with distance from their mid
points.

We see then that quantum situations exist in which very reasonable
conditions for a good separation in space are satisfied, thus leading
to the satisfaction of the above separability hypothesis. One makes
reference exclusively to such situations in developing the EPR paradox,
as for as the spatial part of the wave function is concerned.

The EPR reality criterion and the hypothesis of separability constitute together what is often called "Einstein locality". Many authors
treat reality and separability as two different ideas and defend reality
but not separability[3], or vice versa[4]. I am instead convinced that
these two ideas are different aspects of a conceptual unity which might
be called "modern physical realism". Defined in this way realism becomes
a falsifiable idea, since it leads to the validity od Bell's inequality[5]
which is instead violated by quantum mechanics. Therefore it is possible,
at least in principle, to choose experimentally between this brand of

realism (Einstein locality) and the predictions of quantum theory: both
cannot be right!

It should be stressed that Einstein locality leads <u>directly</u> to
empirical consequences that are incompatible with those of quantum mechanics without the need for any further assumption[6] (such as the
hypothesis that quantum mechanics is a complete theory - or an incomplete one). This result was not anticipated by Einstein and is full of
profound implications for physics. To give up the existence of separable
elements of reality means renouncing one of the most important strongholds of natural philosophy with immense consequances on our world outlook. To give up the correctness of quantum mechanics means that a
scientific revolution in fundamental physics has to be carried out. Even
if these outcomes might be seen totally unacceptable, logical thinking
leads pitilessly to the above alternative.

2. CRITICISMS OF EINSTEIN LOCALITY

We have defined "Einstein locality" as consisting of the EPR reality
criterion plus separability. Many arguments are usually put forward
against its validity, so that the majority of physicists today looks at
it rather negatively. In spite of this, we believe that a modern physical
realism based on Einstein locality can be defended and even could become
the basis of future theoretical physics. The conditions for this to
happen together with a discussion of the counter-arguments are given
below:

1. A first argument contrary to the EPR reality criterion is that
it defines as real <u>only</u> what in quantum mechanics is described as an
eigenstate of some Hermitean operator, since only in the case of eigenstates it is possible to predict with certainty the outcome of a future
measurement. Now, why should a physical system described by a state
vector which is a superposition of eigenstates be less real?[7] This
criticism is in a way certainly acceptable, but it should be stressed
that it objects only to the limited validity of the EPR reality criterion, without undermining its validity where it applies. Einstein, Podolsky, and Rosen were well aware of this limitation of their proposal and
wrote "... this criterion, while far from exhausting all possible ways
of recognizing a physical reality, at least provides us with one such
way." We can add that this criterion has in itself a tremendous predictive power and that it is in the interest of physics to exploit it fully, even if its limited applicability is recognized.

2. A weak point in the EPR reality criterion resides in the predictability <u>with certainty</u> of the physical quantity that is being considered[8]: How can one ever be absolutely certain of anything in a concrete experimental situation? The predictability with certainty appeared
in 1935 to be a consequence of quantum theory for certain types of
state vectors; the situation has however changed today, after the
Wigner-Araki-Yanase theorem [9], which proved the necessity of imperfect
measurement, even in principle, if quantum theory has to be considered
correct for all physical interactions, including measurements.

The resulting criticism can be overcome by giving a probabilistic generalization of Einstein locality, as will be shown in the next section.
It is however fully justified insofar as it is directed against the 1935
definition of Einstein locality.

3. The EPR idea of separability insists on the fact that two
quantum objects that are spatially separated should be unable to influence one another. Some people object that the quantum formalism does
not represent a reality in ordinary space time: The Schrödinger equation
for n particles represents a wave propagating in a 3n dimensional space
(configuration space). It is enough to take n>1 to have a nonphysical
space: It could then happen that separation in ordinary space does not
suffice to assure separation in configuration space. Related to this is
the idea that the state vector of two spin - 1/2 particles is an element
of an abstract spin space and that two objects separated in our space
could be contiguous in spin space. From this point of view, Einstein,
Podolsky, and Rosen look like naïve realists who do not understand the
subtleties of the quantum formalism.

The answer to all this is of course that the mathematical structure
of quantum mechanics is very useful and accurate in making empirical
predictions, but that it does not need to be true in nature. Those
who think so seem rather naïve Platonists unable to appreciate the
complexity of the relationship between the mathematical formalism and
the physical world. Mathematics is man-made and so are configuration
space and spin space. The four-dimensional space time exists on the
other hand in some sense independently of humans. Separability in ordinary space should therefore be the most important fact, particularly
because, as was stressed before, all known interactions decrease rapidly with distance.

4. Related to the previous point is Bohr's complementarity. Bohr's
answer to the original EPR paper does not question the correctness of
the EPR reasoning once all its implicit and explicit premises are accep-
ted$^{(10)}$. Rather, Bohr implies that the quantum mechanical formalism cannot be adapted to a philosophical point of view like that of Einstein.
Bohr argues in favour of "a final renunciation of the classical ideal of
causality and a radical revision of our attitude towards the problem of
physical reality", and stresses that his notion of complementarity is a
"new feature of natural philosophy" which implies a "radical revision of
our attitude as regards physical reality".

His line of reasoning is the following. All measurements performed
on atomic and subatomic systems must necessarily be prepared, carried
out, and, when results are obtained, expressed in classical terms. This
is so because the physicist lives in a macroscopic world where classical
laws and, even more important, classical conceptions (space,time causality, etc.) hold and have become the unavoidable means for expressing
all human experience. The experimental physicist will therefore naturally try to express his experimental results in classical terms, that is
he will try to express the regularities of the microscopic world as
causal processes in space and time. This is however impossible to do,
and the physicist finds himself suddenly exposed to an irrational ele-

ment because the existence of the quantum of action h implies a finite mutual interaction between the measured objects and the measuring instrument. This gives rise to effects on the observed system which cannot even be eliminated logically.

In other words, the existence of h implies a perturbation of the measured object which is completely unpredictable and thus, in a way, irrational. These facts not only set a limit to the extent of the information obtainable by measurements, but they also set a limit to the meaning which can be attributed to such information. Bohr's opinion is that an independent reality cannot be attributed to atomic phenomena: The word phenomenon should exclusively be used to refer to the observations obtained under specified circumstances, including an account of the whole experimental apparatus. From this point of view physics must

deal exclusively with acts of observation, all reference to an unobserved elementary reality being banished from a truly scientific reasoning. Obviously, then, no paradox exists when one considers two correlated systems.

From Bohr's point of view, the EPR assumption about the elements of reality appears immediately useless: One can postulate it, but it means only that there is an element of reality associated with an act of measurement concretely performed, for there is no other reality which can be considered. In particular, the EPR conclusion that position and momentum correspond to two simultaneously existing elements of reality appears totally unjustifield (Bohr writes that it contains "an essential ambiguity") because one can never perform simultaneous measurements of position and momentum; if there is no concrete measurement there is also nothing real to which an element of reality can be attributed.

If this is Bohr's point of view, it should be stressed that it attempts to forbid a causal completion of quantum theory (that is, a causal description in space and time of individual atomic events, a description necessarily more detailed than that provided by quantum mechanics, which discipline makes only statistical predictions). Therefore, Bohr's complementary has the same aim as von Neumann's theorem.[11] The existence of causal space time models which do precisely what von Neumann's theorem tried to forbid [12] not only makes this theorem obsolete but shows at the same time how arbitrary Bohr's conclusion was. This point is further discussed in our fourth section.

5. It is often stated that the EPR reality criterion, by associating elements of reality with observables that cannot be measured simultaneously, essentially gives to both observables well-defined values (counterfactuality). This point of view arises from a positivistic misunderstanding of the EPR realistic position. The point is perhaps best illustrated with a classical esample: We can predict with certainty that a bottle of water cooled below $0°C$ will solidify into ice; the EPR reality criterion allows us to associate an element of reality, say λ_1 , with this first property of water. A second property is that water becomes vapor above $100°C$, and also this can be predicted with certainty; the EPR reality criterion associates then a second element of reality, say λ_2, with this new property of water. The simultaneous existence of these elements of reality, λ_1 and λ_2, which reflect nothing else but

the interaction between water molecules, which is today well known, does not imply of course that we should be able <u>at the same time</u> to make a bottle of water freeze and boil. The situation with quantum measurements is analogous.

6. The most formidable challenge to Einstein locality is considered to come from experimental evidence. [13] Röhrlich expressed the opinion[14] that "Local hidden variables theory is dead. It received its <u>coup de grace</u> by two precision experiments carried out.... in Paris". As a matter of fact, all the experiments performed with atomic cascades in order to test Bell's inequality have been analyzed with the help of some additional hypotheses. [15] It is enough to negate logically these hypotheses in order to restore a full agreement between Einstein locality and quantum predictions as far as the existing empirical evidence is concerned. [16] Further experiments never performed up to now become however interesting. [17] The situation is therefore fully open, and opposite claims reflect more than anything else old-fashioned ideas biased by ideological choices which hamper logical thinking.

The incompatibility between Einstein locality and quantum theory remains however at the experimental level. It would be enough to carry out the usual photon correlation experiments with nearly ideal detectors in order to choose finally between these alternatives. But this remains unfortunately a remote possibility. Other promising avenues of investigation are those involving photon experiments with polarizers and $\lambda/2$ plates, [17], $K°\bar{K}°$- correlation experiments, [18] and coincident detection of correlated pairs of atoms. [19]

3. PROBABILISTIC REALISM

As we have seen in the previous sections, the original formulation of the EPR paradox, based on the idea of deterministic and separable elements of reality ("local hidden variables"), cannot be considered fully satisfactory today. The problem is that determinism very probably does not apply to the quantum domain, a fact brought to full light also by the clarification of the important distinction between causality and determinism obtained from modern epistemology. Determinism is a narrow particular form of causality. Therefore it is possible today to defend causality, but almost impossible to believe that determinism is universally valid: From this point of view it has finally been decided that laplacian determinism is dead, that Brownian motion is an <u>objectively casual</u> phenomenon, and that <u>objective casuality is part of causality</u>.

The original proof of Bell's inequality[5] was based on a deterministic approach. Given the fact that in the quantum domain we have every reason for believing that a deterministic description is not really applicable, people tried immediately to find probabilistic formulations of Einstein locality. The widely accepted approach of this type is the 1974 Clauser-Horne theory[20], in which joint probabilities are postulated to be factorable at some level (that is, after fixing a finite number of variables).

I have shown elsewere that the Clauser-Horne approach contains unsatisfactory features.[21] In the present section I propose therefore a new formulation of probabilistic Einstein locality, in which probabilities are associated with real physical properties of the ensembles for which they are predicted. Before discussing this point it will however be useful to make a brief digression into the different points of view concerning the nature of probability.

The classical definition of probability, due to Laplace, expresses the numerical value of a probability as the ratio between the number of favourable cases and the number of (equally) possible cases. Many objections have been raised against the classical theory. Frequently quoted is the fact that "equally possible" is nothing but another expression for "equally probable" so that the definition is circular. Attemted responses to the critical status of this notion of probability have led in our century to several new formulations of the theory of probability, which are usually classified as follows:

(1) - The subjective interpretation, in which a central role is played by the "degree of rational belief" which we may grant to a given statement about the future development of some given systems.

(2) The objective interpretation, which treats every numerical probability assignement exclusively as a statement about the relative frequency with which the event considered will occur within a sequence of similar events.

(3) - What can be called the agnostic position, based on the idea that mathematicians have clearly defined probability as a special case of measure theory, and that, as a consequence, the scientists need no longer bother about the problem of defining more physically what a probability really is.

There does not seem to exist any widely accepted definition of probability in objective terms and for a single object. Yet this has become a relevant problem in the discussion of the EPR paradox since the Clauser-Horne formulation[15] of local realism is really consistent only if probabilities are interpreted at the same time objectively and as refering to individuals!

Popper's propensity formulation of probabilities in quantum physics tries to bridge the gap between the objective properties of a single system and the objective probabilities in the statistical ensemble of many repetitions of the system under study.[22] It allows the single system to have objective tendencies or propensities of varying strengths and shows that, if these are interpreted as tendencies to realize themselves, they lead to corresponding statistical frequencies in a virtual (or in a real) statistical ensemble.

Coming back to the main argument under discussion, the probabilistic formulation of the EPR paradox, we repeat that recent research has shown that the Clauser-Horne formulation of probabilistic Einstein locality contains unsatisfactory features.

The main criticism is probably the following: Factorability means statistical independence and one does not see a-priori why separability should be _equivalent_ to statistical independence.

The problem is that one can hardly believe that, even at some deep level of reality (that is, after subtraction of some variables), two physical systems should __always__ be statistically independent: They could just happen to be correlated "all the way down". A reformulation of probabilistic Einstein locality is clearly needed.[23]

Our proposed formulation of Einstein locality rests on the idea that probabilities, which can be correctly predicted before measurements are performed, can be attributed as physically real to that statistical ensemble for which they are predicted. More precisely, let S be a given set of physical objects S_1, S_2, etc., of the same type (e.g., electrons):

$$S = \{S_1, S_2, \ldots S_N\}$$

Suppose a measuring instrument I(R) is given with which one is able to measure a physical quantity R on the system composing S. Suppose furthermore that the possible outcomes of the measurements of R are r_1, r_2, ..., r_n and that N is so much larger than n that even the least probable of these eventualities is to be found by I(R) a very large number of times. Assume finally that one can predict correctly that in some subensemble S' of S the results r_1, r_2, ...,r_n will be found with respective probabilities p_1, p_2, ...,p_n. Under these conditions it is natural to conclude that the foregoing probabilities belong to S' in some sense.

We can therefore state the criterion for probabilistic reality: If it is possible to predict the existence of a subset S' of S

$$S' = \{S_{i_1}, S_{i_2}, \ldots, S_{i_\ell}\}$$

without disturbing in any way the objects composing S' and S; if it is also possible to predict correctly that future measurements of R on S' will give the results r_1, r_2, ..., r_n with respective probabilities p_1, p_2, ..., p_n ; then it will be said that a physical property Λ' belongs to S' that fixes the values of the probabilities.

The previous statement can be called the "generalized reality criterion (GRC) as it constitutes an obvious generalization of the famous Einstein, Podolsky and Rosen (deterministic) reality criterion. Some comments perhaps useful for a deeper understanding of the GRC are the following:

(1) Prediction of the subset S'. This means that there is available a method not consisting of direct measurement or interaction for detecting the existence of a subset S' of S for which the above probabilities can be predicted. In practice, in an EPR-type situation one will detect S' by performing measurements on a set T of physical systems individual-

ly correlated with individual systems of S. Thus, if an observable $\bar{R}$ is found in a subset $T' \subset T$ to have, for instance, the fixed value $\bar{r}_0$, it is often possible to predict that another observable R will be found (in the subset S' of the systems individually correlated with the systems belonging to T') to have the values r_1, r_2, ..., r_n with frequencies p_1, p_2, ..., p_n.

(2) <u>Correctness of the previous prediction.</u> How does one know that a certain prediction is correct? Well, nothing more is needed than is usually done in science: If a broad empirical experience composed of repeated measurements shows that the probabilities p_1, p_2, ..., p_n invariably show up in S' for the results r_1, r_2, ..., r_n when S' contains the systems physically correlated to those composing T'^n(T' being defined as the set of T systems for which the experimental result $\bar{R} = \bar{r}_0$ has been found) then we have every reason for concluding that our predictions are correct and, therefore, that the GRC can be applied.

(3) <u>Generalized separability</u>. In practical applications, the GRC is used together with the idea that <u>measurements performed on the T systems do not generate the probabilities relative to any subset S' of S.</u> This can be viewed as a generalized separability principle (GSP). We can also say that GRC and GSP taken together are equivalent to the <u>probabilistic Einstein locality.</u>

We next show that the present approach leads to the conclusion that the ensembles S and S' in general cannot be homogeneous. (24)

Consider a large ensemble E of correlated pairs $S_i + T_i$. Suppose a double-valued observable A(a) = ±1 is measured on the set $S' = \{S_i\}$ and that either the observable B(b) = ±1 or the observable B(b') = ±1 are measured on the set $T = \{T_i\}$. The measurement of B(b) divides E in two parts: $E_+(b)$ is the subset for which B(b) = +1 has been found, while $E_-(b)$ is the subset for which instead B(b) = -1 obtains.

Suppose that:

$$
\begin{array}{llllllll}
p(a+| \ b+) & \text{is the probability of finding} & A(a) = +1 & \text{in} & E_+(b), \\
p(a-| \ b+) & " \ " & " & " & " & A(a) = -1 & \text{in} & E_+(b), \\
p(a+| \ b-) & " \ " & " & " & " & A(a) = +1 & \text{in} & E_+(b), \\
p(a-| \ b-) & " \ " & " & " & " & A(a) = -1 & \text{in} & E_-(b).
\end{array} \quad (2)
$$

If these probabilities can be predicted correctly, as we suppose, then one can attribute them to $E_+(b)$ and $E_-(b)$ by applying the GRC. They are of course real properties of some unknown subsystems also if no measurement of B(b) has been performed: In fact, if this were not the case, we should admit·that the measurement of B(b) on the T systems creates these probabilities for the S systems, in violation of the GSP.

Other probabilities can be introduced for a different splitting of E into $E_+(b')$ and $E_-(b')$ arising from the measurement of B(b') on the T systems. For example, one may define p(a+| b'+) as the probability of finding A(a) = +1 in $E_+(b')$.

But $E_+(b')$ is necessarily composed partly of elements of $E_+(b)$ and partly of elements of $E_-(b)$: If these two ensembles were homogeneous,

in the usual sense that every conceivable subset of them had a probability exactly equal to the one of the whole ensemble, then there should
obviously exist a real number α, such that $0 \leq \alpha \leq 1$, for which

$$p(a+|\ b'+) = \alpha p(a+|\ b+) + (1 - \alpha)p(a+|\ b-) \ . \tag{3}$$

The parameter α expresses the fraction of population of $E_+(b')$ which
belongs to $E_+(b)$ $\lfloor$and $(1 - \alpha)$ the fraction of $E_+(b')$ belonging to $E_-(b)\rfloor$.
Eq. (3) gives $p(a+|\ b'+)$ as a weighted average of $p(a+|\ b+)$ and $p(a+|\ b-)$.
But this simply means that $p(a-|\ b'+)$ must necessarily be internal to the
interval of positive numbers $p(a+|\ b+)$ and $p(a-|\ b-)$. This cannot however
be generally true: Even if it were true by chance, it would suffice to
repeat the reasoning with the arguments b and b' interchanged to find now
a condition which is not satisfied; the new condition will, of course,
be that $p(a+|\ b+)$ should lie inside the interval of positive numbers
$p(a+|\ b'+)$ and $p(a+|\ b'-)$. It can easily be shown that such conditions
can all be satisfied only in the very particular case in which all conditional probabilities are equal to each other. Since there is ample empirical and theoretical evidence that this is not generally true in the
quantum domain, the basis of the previous reasoning is undermined: The
ensembles under consideration cannot be homogeneous!
        The previous conclusion implies that if probabilities reflect ob-
jectively real properties of the ensembles, in the sense admitted by the
GRC, then they must be the averages of probabilities which are different
for different subsets of the considered ensembles.
        We have been led to the conclusion that e.g. $E_+(b)$ and $E_-(b)$ - and
therefore E - cannot be homogeneous ensembles as far as the probability
of finding $A(a) = +1$ is concerned. A symmetrical conclusion holds for E
in the case of the probability for $B(b) = +1$. The same is true obviously
also for the probabilities of finding the results -1. It is then useful
to introduce homogeneous subensembles of E, in the following way:

$$\sigma_1(a), \ \sigma_2(a), \ \dots, \ \sigma_m(a)$$

are homogeneous subsets for $A(a) = +1$ and for $A(a) = -1$, while

$$\tau_1(b), \ \tau_2(b), \ \dots, \ \tau_n(b)$$

are homogeneous subsets for $B(b) = +1$ and for $B(b) = -1$. Obviously

$$\sigma_1(a) \cup \sigma_2(a) \cup \dots \cup \sigma_m(a) \ = \ \tau_1(b) \cup \tau_2(b) \cup \dots \cup \tau_n(b) \ = \ E. \tag{4}$$

The notation reflects the fact that a subset that is homogeneous for the
probability of, say, $A(a) = +1$ is not in general expected to be homogeneous for the probability of $A(a') = +1$, if $a' \neq a$.
        One can next introduce subensembles which are homogeneous for two
or more measurable quantities. Considering $A(a)$ and $B(b)$, the homogeneous
subensembles are

$$E_k(a,b) \ = \ \sigma_i(a) \cap \tau_j(b) \ ,$$

where the single index k has been chosen, for simplicity, to correspond in a one-to-one way to the pair of indexes i,j.

The previous approach can be easily generalized to an arbitrary number of observables of the first and of the second object: Considering r dichotomic observables $A(a_1)$, $A(a_2)$, ..., $A(a_r)$ of the first object and s dichotomic observables $B(b_1)$, $B(b_2)$, ..., $B(b_s)$ of the second one, the following homogeneous subensembles of E can be introduced

$$\sigma_1(a_1), \ \sigma_2(a_1), \ ..., \ \sigma_{m_1}(a_1) \qquad \text{homogeneous for } A(a_1),$$

$$\sigma_1(a_2), \ \sigma_2(a_2), \ ..., \ \sigma_{m_2}(a_2) \qquad \text{homogeneous for } A(a_2),$$

$$...$$

$$\sigma_1(a_r), \ \sigma_2(a_r), \ ..., \ \sigma_{m_r}(a_r) \qquad \text{homogeneous for } A(a_r),$$

$$\tau_1(b_1), \ \tau_2(b_1), \ ..., \ \tau_{n_1}(b_1) \qquad \text{homogeneous for } B(b_1),$$

$$\tau_1(b_2), \ \tau_2(b_2), \ ..., \ \tau_{n_2}(b_2) \qquad \text{homogeneous for } B(b_2),$$

$$...$$

$$\tau_1(b_s), \ \tau_2(b_s), \ ..., \ \tau_{n_s}(b_s) \qquad \text{homogeneous for } B(b_s).$$

Obviously, the union of the ensembles of every line gives always E, just as in (4). By means of suitable intersections one can introduce smaller subensembles in which all the considered observables have constant probability. One can write:

$$E_k(a_1, a_2, ..., a_r, b_1, b_2, ..., b_s) = \qquad (5)$$

$$= \sigma_{i_1}(a_1) \cap \sigma_{i_2}(a_2) \cap ... \cap \sigma_{i_r}(a_r) \cap \tau_{j_1}(b_1) \cap \tau_{j_2}(b_2) \cap ... \cap \tau_{j_s}(b_s)$$

for a typical subensemble having homogeneous probabilities for all the considered dichotomic observables (in number of r + s). In the previous definition the single index k has been chosen, for simplicity, to correspond in a one-to-one way to the set of indexes $(i_1, i_2, ..., i_r, j_1, j_2, ..., j_s)$. The notation can be simplified if one introduces a "vector" V having r + s components in the following way:

$$V = (a_1, a_2, ..., a_r, b_1, b_2, ..., b_s), \qquad (6)$$

since the homogeneous subensemble and its population can respectively be written

$$E_k(V) = E_k(a_1, a_2, ..., a_r, b_1, b_2, ..., b_s), \qquad (7)$$

$$N_k(V) = N_k(a_1, a_2, ..., a_r, b_1, b_2, ..., b_s). \qquad (8)$$

The basic probabilities which are constant for all the pairs of a given subensemble $E_k(V)$ can be written

$p_k(a_1\pm)$ probabilities for $A(a_1) = \pm1$,

$p_k(a_2\pm)$ probabilities for $A(a_2) = \pm1$,

. . . .

$p_k(a_r\pm)$ probabilities for $A(a_r) = \pm1$,

$q_k(b_1\pm)$ probabilities for $B(b_1) = \pm1$,   (9)

$q_k(b_2\pm)$ probabilities for $B(b_2) = \pm1$,

. . . .

$q_k(b_s\pm)$ Probabilities for $B(b_s) = \pm1$.

Within every homogeneous subensemble the constant and local nature
of these probabilities insure that the joint probability for fixed values
of an A and a B observable can be obtained from statistical independence
as follows

$$P_k(a_i\pm, b_j\pm) = p_k(a_i\pm) q_k(b_j\pm) . \tag{10}$$

The same joint probability over the statistical ensemble E is then
given by

$$P(a_i\pm, b_j\pm) = \Sigma_k \{N_k(V)/N\} p_k(a_i\pm) q_k(b_j\pm) , \tag{11}$$

if N as before is the total number of pairs in E. If one writes

$$N_k(V)/N = \rho_k(V) , \tag{12}$$

one must obviously have

$$\Sigma_k \rho_k(V) = 1 , \tag{13}$$

so that $\rho_k(V)$ can be considered a probability density with constant value in every subensemble $E_k(V)$. The joint probability (11) can now be
written

$$P(a_i\pm, b_j\pm) = \Sigma_k \rho_k(V) p_k(a_i\pm) q_k(b_j\pm) , \tag{14}$$

which looks very much like a Clauser-Horne factorizability formula, the
role of the hidden variable λ being here played by the index k. There
is however a very important difference, because now the probability density (13) <u>depends on the values of the parameters of the considered observables.</u> In spite of this when the sum over k is performed all dependence on the arguments a_1, a_2, ..., a_r, b_1, b_2, ..., b_s disappears, with
the exception of a_i and b_j, as indicated in (14).

It is of course well known that from (14) one can easily obtain

Bell's inequality as well as all the other inequalities which together express the physical content of local realism. In the previous arguments only two assumptions were used, the probabilistic reality criterion and the generalized separability. Therefore a <u>universality claim</u> can be made: All theories satisfying these two assumptions must also satisfy the inequalities of the Bell type.

The importance of the previous proof of Bell's inequality is twofold: Firstly it puts that inequality on a more solid basis and, secondly, it provides a very natural theoretical background for some recent developments of the EPR paradox: If a physical mechanism of some type creates (S, T) pairs in which both objects have <u>high</u> (or both have <u>low</u>) detection probability, then the ensemble of detected pairs will not be a faithful sample of all emitted pairs. [25]

4. - BOHR AND VON NEUMANN

Bohr expressed his highly appreciative opinion about von Neumann's theorem for instance in 1938 when the conference on "<u>New Theories in Physics</u>" was held in Warsaw. [26] In the Proceedings of this conference, one can read how von Neumann gave an exposition of his theorem against hidden parameters [27] and how Bohr expressed his admiration while pointing out that one of his own papers, in more elementary ways, arrived at essentially the same conclusion.

The importance of this connection between Bohr and von Neumann is not stressed in the literature. In any opinion it is a very illuminating connection, especially because the paper Bohr has in mind is very clearly the one he had written three years before [28] with the title "Can Quantum - Mechanical Description of Physical Reality be Considered Complete? " This paper contains only two reference. The first one is the antagonistic reference to the Einstein, Podolsky, and Rosen paper with the same title, whose conclusions Bohr tries to undermine; the second one is to his book <u>Atomic Theory and Description of Nature</u>, [29] published by Cambridge University Press in 1934, which is regarded even today as the best exposition of Bohr's idea of complementarity (for instance it contains the 1927 contribution to the Como Congress). The reason why this important connection between complementarity and von Neumann's theorem has very rarely been noticed is probably that the people who were working on "impossibility proofs" of the causal completions of quantum theory did not quote one another. This is the case, for instance, with Heisenberg who gave a lecture at Vienna University in 1935 in which the "impossibility proof" was based on an analysis of the scattering of an α particle from a grating. [30]

It is nowadays commonly accepted that von Neumann's theorem, in all of its proposed versions, has failed in its purpose and is actually incapable of barring the causal completions of quantum theory. The best understanding of this fact has been provided by Bohm's hidden variable models, [31] which did in fact exactly what von Neumann's theorem was supposed to forbid: provide a causal formulation of the existing quantum theory. Further simpler models have been proposed which reproduce a limited but very meaningful part of quantum theory, the eigenvalues and the probabilities for all the spin observables of a single spin - 1/2 particle.[32] Few people seem to have realized that the existence of these models does not only provide a falsification of von Neumann's theorem but of Bohr's complementarity as well. Today we know how to give a causal description in space and time of most quantum phenomena (double-slit experiment, neutron interferometric results, and so on). There is even a movie, produced by Chris Dewdney,[33] in which deterministic particle trajectories reproduce the quantum mechanical interferences in several interesting instances. The main idea is here to associate a suitable "quantum potential" with the action of the wave on the particle. The result is that the wave and the particle, which together constitute a quantum object, both behave deterministically in space and time. Thus the arbitrariness of the conclusions which Bohr and other people thought they were able to deduce from the so-called "complementarity principle" is brought out in a most striking and convincing way. This is not the same, of course, as saying that one should subscribe to the validity of these models: Their value is mostly methodological in showing in a detailed rational way the obsolescence of all the "impossibility proofs", complementarity included. One such model is discussed later. Now we shall draw attention to the fundamental weakness of Bohr's conclusions by means of a simple macroscopic example:

A liter of water has the two following properties:

1) It boils at 100°C.
2) It freezes at 0°C.

This example is meaningful because the two processes (boiling and freezing) cannot be made to happen at the same time: They are mutually exclusive. The realist reasoning goes as follows. It can be predicted with certainty that our given liter of water will boil at 100°C. Other liquids do not have this property; therefore there must be something in the reality of the water which gives rise to its boiling when the temperature rises to 100°C. Einstein, Podolsky and Rosen would call this "something" an element of reality E_1, of water. It exists all the time, even at different temperatures, and is such that, when the right conditions are produced, it manifests itself by giving rise to the boiling process. In a completely similar way a second element of reality, E_2, can be introduced by starting from the freezing at 0°C. Thus it is concluded that water has simultaneously E_1 and E_2: This conclusion is fully upheld by modern physical chemistry, where E_1 and E_2 are nothing but different features of the interaction potential between water molecules.

The complementary idea, on the other hand, would start from the observation that one can never realize simultaneously the boiling and the freezing experiments on a given liter of water and would then go on to insist, for instance in Wheeler's words,[34] that "no ... phenomenon is a phenomenon until it is an observed phenomenon". In other words, there is no reality other than what is under observation. The reality of boiling water is there only when we actually make water boil and cannot be extended to different situations. Therefore, E_1 exists only when water boils, E_2 only when water freezes, and it is under no circumstance possible to attribute E_1 and E_2 to a given liter of water, since boiling and freezing are incompatible processes.

The refusal to accept an unobserved reality here blocks the way
of scientific discovery, exactly as it happens with the applications
of the complementarity idea to the microscopic domain. The adduced
example would perhaps not be convincing for Bohr himself, who always
refused to apply quantum theory to the macroscopic domain, but it
should be considered fair today, after the practically universal acceptance of quantum theory also for the description of macroscopic phenomena. It can obviously be assumed that wherever quantum mechanics
applies, complementarity applies as well.

Man lives in the macroscopic domain and for him conceptions such
as the description of events in space and time and the description of
events in causal terms have become unavoidable means of understanding
and of communication. Causality can mean just what in ordinary life is
intended with causal connections, but it admits also a more precise
and quantitative formulation: A process is causal if it takes place
according to well defined and identifiable rules. According to Bohr,
the most important of these rules is the law of conservation of energy
and momentum, and, in practice, in the atomic domain one can choose
almost indifferently between the expressions: "causal description" and
"description obeying the conservation of energy and momentum". Therefore, the physicist who studies the phenomena of the quantum domain
will naturally try to use his macroscopic preconceptions and thus
try to describe atomic processes as taking place both in space and time and in accordance with the rules of energy and momentum conservation.
However, he will find that it is not possible to do so, because quantum
observables are often described by noncommuting operators, cannot be
measured simultaneously, and are such that the measurement of one of
them in general destroys all previously accumulated knowledge of a
second one.

The roots of complementarity can best be exposed by discussing
the example of space localization (position measurement) and causality
implementation (momentum measurement). Spece localization can be obtained by measuring position with infinite precision ($\Delta x = 0$). After such
a measurement the wave function of the observed particle becomes the
δ function $\delta(x-x_0)$, if x_0 is the result obtained. But a δ function
can be written as a superposition with constant weight of all possible
plane waves, and this simply means that absolutely nothing is known
about momentum. All eventual knowledge about momentum, prior to the
position measurement, is in this way completely lost. No evidence can
therefore exist about momentum conservation, since no knowledge about
momentum is available. A concrete realization of localization in space
thus implies a necessary abandonment of the causal description. Symme-
trically, in a different experiment one could choose to implement the
causal description by measuring momentum with infinite precision: The
wave function would therefore become, as a result of the measurement,
a plane wave. But this would immediately imply that nothing would be
known about position, resulting in a complete loss of the description
of the atomic phenomenon in space. A concrete implementation of the
causal description would thus force the physicist to abandon any description in space. The two possibilities (space time and causality) are
thus seen to be mutually incompatible, to exclude each other. Bohr con-

cludes that, in the atomic world, it is in principle impossible to give a picture of quantum processes as developing causally in space and time.

The previous conclusion applies to any two noncommuting observables. Consider, for instance, a spin - 1/2 particle and its spin-component observables S_x, S_y, S_z.
It is well knón that ány two of them do not commute. This means that they cannot be assumed to be simultaneously measurable . Consider an electron in the initial state

$$|\psi_i> = \binom{1}{0} . \tag{16}$$

If the S_x component is measured, there can be only two results, <u>viz</u>

$$S_x = \pm \frac{\hbar}{2} . \tag{17}$$

Therefore the spin state <u>after</u> the measurement becomes

$$\text{either } \frac{1}{\sqrt{2}} \binom{1}{1} , \qquad \text{or } \frac{1}{\sqrt{2}} \binom{1}{-1} . \tag{18}$$

In either case, the S_y component becomes completely unknown, as can easily be cecked. Bohŕ would say that the implementation of the reality of S_x has made S_y completely undetermined. The opposite reasoning can obvĩously be maďe: S_y can become known, but then it is S_x which becomes necessarily compľetely unknown. One can thus say, with Bohr, that S_x and S_y are complementary aspects of reality: Either S_x is real, or S_y is real, but never the two of them at the same time.
The ḿost general observable in the spin space of a spin - 1/2 particle is given by

$$\Sigma = \alpha I + \vec{\beta} \cdot \vec{\sigma} , \tag{19}$$

where $\alpha, \beta_1, \beta_2, \beta_3$ are four real constants, I is the unit 2 x 2 matrix and $\sigma_1, \sigma_2, \sigma_3$ are the usual Pauli matrices. With the most general observable (19) we can associate the particular state vector

$$|\psi_i> = \binom{1}{0} \tag{20}$$

with no loss of generality, since every spinor is an eigenstate with eigenvalue +1 of some operator $\vec{\sigma} \cdot \hat{n}$, where $\hat{n}$ is a unit vector. In this sense, every spinor determines a direction $\hat{n}$, and it is always possible to rotate the axes in such a way that the z axis coincides with $\hat{n}$. The rotated spinor then becomes (20), while the rotated operators keep the form (19), which is the most general one.
As is well known, the eigenvalues of (19) are

$$\Sigma_{1,2} = \alpha \pm |\vec{\beta}| \tag{21}$$

and the corresponding probabilities for the state (20) are

$$P_{1,2} = \frac{1}{2} \left[1 \pm \frac{\beta_3}{|\vec{\beta}|} \right] . \tag{22}$$

We assume the previous predictions to be correct, that is, we assume
quantum mechanics to be a good theory. It is of course not necessary
that quantum mechanics be <u>true</u>, that is to attribute ontological existence to its theoretical structures: It is a good theory simply because
it makes correct predictions but the world can be considerably different
from what could naively be read into the theory itself.

It will next be shown that a model of the world (a model of spin
‑ 1/2 measurements) is possible without contradicting the empirical
predictions of the theory in any way. The model has the following features:

(i) It provides a causal description of quantum spin events.

(ii) It provides <u>at the same time</u> a description of these events in
space and time.

(iii) It describes all spin measurements as simultaneously predetermined by objective properties of the measured system, even though
two observables described within quantum mechanics by noncommuting matrices cannot be measured simultaneously.

Therefore, this model does twice what complementary tried to forbid.
The first achievement (causal description in space and time) is perhaps
not very relevant to a direct application of complementarity to spin
measurements, but the second one (simultaneous "reality" of all spin
components) definitely is.

The model contains assumptions both about the measured system and
the measuring apparatus: Of course, the role of the apparatus cannot
be eliminated in the quantum domain, given the existence of a minimal
action (Planck's constant, h).

Firstly, it is assumed that the system being studied is a spinning
sphere with angular momentum pointing in the direction $\hat{\lambda}$ which forms
an angle θ with respect to the z axis. Secondly, an apparatus built
for measuring the observable Σ (see (19)) is assumed to work as
follows:

(1) - It measures the sign of $\hat{\lambda} \div \vec{\beta}$ (<u>after</u> which it modifies the direction $\hat{\lambda}$ according to rules to be specified later).

(2) - It multiplies the result by $|\vec{\beta}|$.

(3) - It adds α to the new result.

The final result of these operations is

$$\alpha + | \vec{\beta} | \cdot \mu ,$$

where $\mu = \pm 1$ is the sign of $\hat{\lambda} \cdot \vec{\beta}$. It is in all cases coincident with
one of the two eigenvalues (21).

In order to reproduce also quantum probabilities, one needs to de-
fine the state $|\psi_i\rangle$, given in (20), within the model. One assumes

this state to represent a statistical ensemble of spinning spheres with $\hat{\lambda}$ vectors distributed according to the density function

$$\rho(\theta) = \begin{cases} 1/2 \cos\theta , & \text{if } -\pi/2 \le \theta \le \pi/2 \\ 0 , & \text{otherwise.} \end{cases} \qquad (23)$$

Obviously

$$\int_0^{2\pi} d\theta\, \rho(\theta) = 1.$$

in Fig. 2 one can see the (dashed) region in which the $\hat{\lambda}$ vectors can be found if the quantum state is as given by (20).

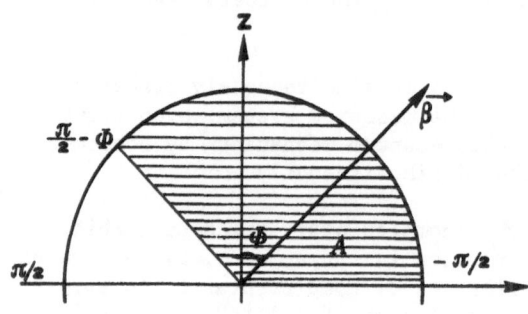

Fig. 2. A statistical ensemble of λ vectors included in the hatched area A_0 with density $\rho(\theta)$ (see Eq. (23)) represents the spin state $\binom{1}{0}$.

Given the rather mechanical nature of the steps 2) and 3) in the previously defined measurement process, the probability of measuring and finding the result $\Sigma_1 = \alpha + |\vec{\beta}|$ is obviously equal to the probability of finding $\hat{\lambda} \cdot \vec{\beta}$ positive. Therefore, one must only compute the integral of $\rho(\theta)$ over the hatched area A of Fig. 3, where $\hat{\lambda}$ and $\vec{\beta}$ form angles not exceeding $\pi/2$ (in modulus).

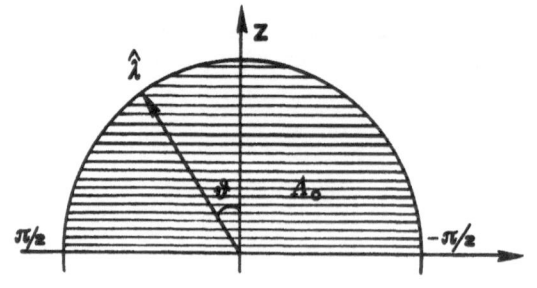

Fig. 3 λ vectors belonging to the area A form angles not exceeding $\pi/2$ (in modulus) with the vector $\vec{\beta}$.

The probability p_1 is therefore given by

$$p_1 \int_{-\pi/2}^{\pi/2 - \phi} d\theta\, \rho(\theta) = 1/2(1 + \cos\phi) = 1/2\, [1 + \beta_3/|\vec{\beta}|] , \qquad (24)$$

in full agreement with (22). Of course, also $p_2 = 1 - p_1$ comes out agreeing with (22).

The conclusion is that a deterministic model reproduces all the single-measurement predictions of quantum mechanics for spin - 1/2 particles (eigenvalues and probabilities). In order to reproduce subsequent spin measurements, one needs an assumption about the statistical re-distribution of λ's produced by the action of the apparatus on the electron: The idea is that every time a spin observable is measured, the λ vectors are redistributed according to a density similar to (23) but centered around the characteristic direction of the observable. In this way all the conceivable measurements turn out with quantitative features in full agreement with quantum mechanics.

The previous model is different from the one proposed by Bell[32] against von Neumann's theorem in one decisive respect: In Bell's case, fixing σ_1 meant always fixing also σ_2, and it was not possible to reproduce the physical properties of the quantum mechanical eigenstates of σ_1, for which σ_2 is not well defined. No such a drawback exists for the above model, which can fully account for all the quantum mechanical predictions for spin - 1/2 systems. The proposed model, could be concretely realized in a mechanical workshop by means of spinning spheres and with a suitably built "measuring apparatus". The resulting simulation would be perfect in all respects except for the numerical value of h, which is however a rather trivial multiplying factor of spin eigenvalues. Naturally the model is not "believable" in itself: Its existence shows merely that a door is open and that much "better" models should be possible. In the above model all the conceivable "spin components" are simultaneously real and summarized in the vector λ and in the working of the apparatus. In this way every conceivable measurement has a predetermined result. Notice that simultaneous measurements of different spin components are impossible because the λ's should come out to be redistributed according to two different (and then incompatible) distributions. It should also be stressed that Heisenberg's relations for spin components are valid as scatter relations over the statistical ensemble.

The interdictions coming from complementarity have been violated and the arbitrariness of Bohr's reasoning comes out very clearly to lie in the assumption that if two observables cannot be measured simultaneously, then they cannot both be real. It is all as simple as that!

NOTES AND REFERENCES

1. A. Einstein, B. Podolsky and N. Rosen, Phys. Rev. 47, 777 (1935).
2. The opposite idea that also retro-active causality holds in nature has been studied by:
   O. Costa de Beauregard, Nuovo Cimento B42, 41 (1977); B51, 267 (1979);
   C. W. Rietdijk, Found. Phys. 8, 815 (1978).
   J. A. Wheeler, in Mathematical Foundations of Quantum Mechanics,
   A. R. Marlow, ed. (Academic, New York, 1978).
3. D. Bohm and B. J. Hiley, Found. Phys. 14, 255 (1984);
   J. P. Vigier, Nuovo Cimento Lett. 29, 467 (1980).

4. H. P. Stapp, Found. Phys. 7, 313 (1977); 9, 1 (1979);
   P. Eberhard, Nuovo Cimento B46, 392 (1978).
5. J. S. Bell, Physics 1, 195 (1965).
6. F. Selleri, "Quantum Reality as an Empirical Problem", in The Concept
   of Physical Reality, E. Bitsakis, ed. (Zachoropoulos, Athens, 1983)
7. The same problem exists for mixtures.
8. F. Selleri, Found. Phys. 12, 645 (1982).
9. E. P. Wigner, Z. Phys. 133, 101 (1952);
   H. Araki and M. M. Yanase, Phys. Rev. 120, 622 (1960).
10. N. Bohr, Phys. Rev. 48, 696 (1935).
11. J. von Neumann, Die Matematische Grundlagen der Quantenmechanik
    (Springer – Verlag, Berlin, 1932).
12. D. Bohm, Phys. Rev. 85, 166, 180 (1952).
13. S. J. Freedman and J. F. Clauser, Phys. Rev. Lett. 28, 938 (1972);
    R. A. Holt and F. M. Pipkin, Harvard University preprint, 1974;
    J. F. Clauser, Phys. Rev. Lett. 37, 1223 (1976); Nuovo Cimento 33B,
    740 (1976);
    E. S. Fry and R. C. Thompson, Phys. Rev. Lett. 37, 465 (1976);
    A. Aspect, P. Grangier and G. Roger, Phys. Rev. Lett. 47, 460 (1981);
    49, 91 (1982);
    A. Aspect, J. Dalibard and G. Roger, Phys. Rev. Lett. 49, 1804 (1982);
    W. Perrie, A. J. Duncan, H. J. Beyer, and H. Kleinpoppen; Phys. Rev.
    Lett. 54, 1790 (1985).
14. F. Röhrlich, Science 221, 1251 (1983).
15. J. F. Clauser, M. A. Horne, A. Shimony and R. A. Holt, Phys. Rev. Lett.
    23, 880 (1969);
    J. F. Clauser and M. A. Horne, Phys. Rev. D10, 526 (1974).
16. T. W. Marshall, E. Santos and F. Selleri, Phys. Lett. 98A, 5 (1983).
17. A. Garuccio and F. Selleri, Phys. Lett.103A, 99 (1984);
    F. Selleri, Phys. Lett. 108A, 197 (1985).
18. F. Selleri, Lett. Nuovo Cimento, 36, 521 (1983).
19. T. K. Lo and A. Shimony, Phys. Rev. A23, 3003 (1981).
20. See Ref. 15 above.
21. F. Selleri, "On the notion of probability for pairs of correlated
    systems", in the Proceedings of the 2nd International Meeting on
    Epistemology, (Gutenberg, Athens, 1985).
22. K. Popper, Quantum Theory and the Schism in Physics; (Hutchinson,
    London, 1982), especially pp. 68-74.
23. F. Selleri and G. Tarozzi, Spec. Sc. Tech. 6, 55 (1983).
24. For a more detailed and general discussion of the following arguments
    see my contribution in: Microphysical Reality and Quantum Formalism,
    G. Tarozzi and A. van der Merwe, eds, (Reidel, Dordrecht, 1986).
25. The opposite statement is precisely the assumption made by A. Aspect
    in the analysis of Orsay's experiments. See references 16 and 17 above
    for a further discussion of the "additional assumptions" usually made
    in the analysis of EPR-type experiments.
26. Proceedings of the conference "New Theories in Physics", Warsaw (1938).
27. J. von Neumann; The Mathematical Foundations of Quantum Mechanics;
    (Princeton, University Press, Princeton, 1955).
28. See Ref. 10 above.
29. N. Bohr, Atomic Theory and Description of Nature, (University Press,

Cambridge, 1934).

30. W. Heisenberg, "Questions of Principle in Modern Physics", Lecture delivered at Vienna University on November 27, 1935; contained in Heisenberg's Philosophic Problems of Nuclear Science, (Fawcett, New York, 1952)

31. See Ref. 12 above.

32. J. S. Bell; Rev. Mod. Phys. 38, 447 (1966);
F. Selleri, Die Debatte um die Quantentheorie, (Vieweg, Braunscheweig, 1983).

33. C. Dewdney, in: "Microphysical Reality and Quantum formalism", G. Tarozzi and A. van der Merwe, eds. (Reidel, Dordrecht, 1986).

34. J. A. Wheeler; "This Incredible Quantum Business" lecture delivered in Beijing in October, 1981; contained in Wheeler's Physics and Austerity: Law without Law, (Center for Theoretical Physics, University of Texas, Austin, Texas, 1982).

POSSIBLE LINK BETWEEN EPR PARADOX AND WAVES

G. C. SCALERA
Istituto Nazionale di Geofisica
Via Di Villa Ricotti 42, Roma, Italy

ABSTRACT

It is shown that gedanken experiments using simple undulatory process are able to violate Bell's inequality perfectly in respect of locality and separability of the distinct measurement apparatuses. The possible failure of the corpuscle concept is proposed by arguments that lead to the unpleasant property of a wave-corpuscle link.

INTRODUCTION

Not much time has passed since I read in the concluding remarks of a recent book[1] on quarks that definitive results have been achieved with regard to matter and the universe. These are that the universe is one, whole and indivisible and that, because we are part of it, it is also conscious. The foregoing claim is only one of many ideas now on the market in a cultural context that propose a return to different forms of idealism and spiritualism. Since 1935, the battlefield, old as quantum mechanics (Q.M.), was the famous argument of Einstein, Podolsky, and Rosen,[2] in which the authors search for the theory's lack of completeness. Subsequently, von Neumann tried to demonstrate the impossibility of completing the Q.M. by hidden-variable theories. His work has build a sort of barrier for everyone who hopes to modify the theory by delving deeper into reality. Nearly thirty years ago D. Bohm[3] applied himself to the problem of the hidden-variable models in Q.M.. He obtained an important result that changed the state of the art regarding this subject. His paper was a criticism of the then-famous von Neumann theorem. Bohm demonstrated by a counter-example the serious limitations to this kind of self-sustaining proof. Thirteen years later, the result of a reasoning on the same subject carried out by J.S. Bell,[4] was an inequality, called Bell's inequality (B.I.), originally written as

$$1 + P(b,c) \geq | P(a,b) - P(a,c) | \tag{1}$$

where $P(a,b)$ is the value of the correlation function and a and b arbitrary directions of the polarizer axis. The inequality fixed an insurmountable limit for theories satisfying the principle of locality, i.e., for theories in which the impossibility of one measurement act to influence another measurement act in a spacelike separated region holds. Q.M. asserts that certain measurements carried out on particular states, such as the singlet state of a pair of photons in an atomic cascade, can violate the inequality (1) facing experimentally testable local hidden-variable theories. The theorem of Ref.(4) represented a great clarification of the Einstein-Podolsky-Rosen (EPR) argument [2] because it stated the actual configuration in the perfect coincidence experiments to be considered nonlocal. The situation described in EPR's paper

G. Tarozzi and A. van der Merwe (eds.), The Nature of Quantum Paradoxes, 131–140.
© *1988 by Kluwer Academic Publishers.*

was shown not to violate locality, and local models were developed to fit the value forecast for the correlation function by Q.M. for two parallel polarizers in the atomic cascade experiment.[5] EPR were only concerned with the completeness of the theory. In any case we refer in the following to the "EPR paradox".

Subsequently, Scalera,[6] starting from a roughly undulatory assumption, developed a simple separable model dealing also with the concept of contemporaneity of measurement which is able to achieve the maximum violation of another form of (1). The whole question is still open,[7] although someone has hastily claimed that the last "loophole" in Q.M. has been closed,[8] and the local hidden-variable theories are now quite dead.[9]

Several ways have been proposed to solve the EPR paradox. An incomplete list follows:

(1) The world is all in our minds. If this is so, no needs arises for locality because all "physical" processes occur in only one place: in your or the collective mind. We do not accept this Berkeleyesque solution, because of its reduction of physical matter to mere thinking and because it does not provide common ground in which all subsequent theories are rooted. It is such common ground that makes possible the progress of our knowledge, which otherwise would remain merely conventional.

(2) Reality is not unique. This is the position of the De Witt-Everett school, which claims that, if a measurement has more than one outcome, then all the different possibilities would become reality, thus producing somewhat ramifyng but not interacting universes. When the measurement apparatuses are two or more in number, as in an EPR's or Bell's inequality-testing experiment, the behavior of the two correlated particles is reproduced by setting ad hoc prohibition rules for the simultaneous outcome of two coincident uiverses to both of which we are sensitive. These meta rules, which have no limitation imposed by belonging to a real physical world but are placed behind these universes, can exactly reproduce the Q.M. correlation function. Also this position is difficult to accept because of its incompatibility with the laws of conservation, expecially that of energy. Every observation made can produce a sudden moltiplication of energy and matter as well as the multiplication of universes. This would be easy to carry out and to imagine if the worlds in question were not real but ideal. We are thus in the presence of a renewed version of idealism.

(3) Existence in a nonrelativistic context of signals with speeds larger than that of light, transmitting information from one istrument to another throughout the inevitable time lapse between the two distant measurement act. In a nonreltivistic context the presence of such signals is not a problem, because no limit is imposed on the velocities for which relativity is simple ignored. We do not belief that relativity is a final and exhaustive description of the nature. But we firmly belief that the effects described to us by the theory are surely a reflection of reality. This is something that cannot be ignored.

(4) The existence of signals traveling faster than light in a relativistic context. Such signals are called "tachions." It has been demonstrated that the introduction of these entities into the theory leads to inextricable paradoxes concerning the link between cause and effect. In particular inertial reference frames, the outcome of signal "B" before that of a signal "A," which is in fact the cause of "B," become possible. These difficulties, incapable of elimination, make the tachion hypothesis very improbable.

(5) Another way to violate Bell's inequality is through negative probabilities. The violation is effective, but the meaning and the conseguences of this radical modification of probability theory are not so clear.

(6) Lastly, I would like to mention Pitowsky's paper,[10] in which, with only slight modification of probability and without any negative probability, the author is able to violate Bell's relation. This work has inspired high hopes for a final solution to the EPR paradox by means of this sort of "Gaussian" revoluion in the theory of probability. But it is also true that until now no one has carried on thoroughgoing exploration of this hypothesis. Thus nobody is able to assure us that the paradox eliminated does not reappear at some other level in the modified theory.

These are all positions that claim the nonexistence of any EPR paradox because the world is different from the world we expect. My position, on the contrary, is that the paradox does not exist in the world as it is either, because of a lack of completness in our information on the phenomena under study; namely because of the very little importance given in our reasonings to the undulatory aspect of the objects involved.

LOCAL MODEL VIOLATING B.I.

The main ideas leading to the simple model of Ref.(6) are:

-1. Q.M. does not fail in its prediction in the vast majority of cases.

-2. The singlet state $S = (1/\sqrt{2}) \cdot (|+> \cdot -> + |-> |+>)$ is a very simple case indeed

-3. Q.M. is a wave mechanics theory governed by a Schrödinger equation, which is a wave equation valid also for macroscopic and locally behaving waves.

-4. Q.M. violates d.d.B.

Consequently, there must exist a separable macroscopic undulatory situation violating B.I..

The model was found, and it basically consists of a source S emitting two continuous and oppositely directed helix-shaped ribbons which themselves carry hidden variables in the form of unit vectors orthogonal to the propagation direction (Fig.1). A marked set of regularly spaced vectors is able to produce a tick of the photomultiplier every $t=L/c$. Passing through a "polarizer" with axis in the direction a, the preceding set of marked vectors are destroyed and replaced by another set in wich only direction a was marked on the ribbon. The polarizer must be a calcite crystal, because then no information would become lost, not even on the orthogonally polarized "extraordinary path." Absorption with the commonly used polaroid plates is not observable. We count the absence of the photon in the ordinary ray as its presence in the extraordinary ray.

A "photomultiplier" placed behind the "polarizer" integrates the energy contained in the ribbon, always starting from a marked direction and ending with the next marked direction. The counter ticks on only when two successive marked directions enter. We assume the counter is blind to the initial piece of ribbon arriving before the first marked direction. Therefore a "photomultiplier" placed behind a "polarizer" tickes regularly every $T=L/2c$, where L is the helix pitch. As stated above we explicitly assume that, without the "polarizer" the "photomultiplier" ticked every $T/2$ by virtue of the set of original double-density regularly spaced vectors that are destroyed by interaction with the polarizer and replaced by the new set of marked vectors. Such an assumption is necessary in order to achieve agreement with the calculations by coincidence rate as commonly performed in current experiments. With these simple hypotheses, together with the definition of a rectangular window of length T" for the coincidence time, we showed that B.I. is largely violated. We write the correlation functions in explicit form by coincidence rates:

$w(-,-) = 1$, (without the polarizers)

$P(a,b) = w(a+,b+) + w(a-,b-) - w(a-,b+) - w(a+,b-)$.

We easily verify that:

If $0° < \Theta_a - \Theta_b < 45°$, then $w(a+,b+) = w(a-,b-) = 1/2$, and $w(a+,b-) = w(a-,b+) = 0$;

if $45° < \Theta_a - \Theta_b < 90°$, then $w(a+,b+) = w(a-,b-) = 0$, and $w(a+,b-) = w(a-,b+) = 1/2$.

B.i. was posited in the form

$$\Delta = |P(a,b) + P(a,b')| + |P(a',b) - P(a',b')| \le 2 , \qquad (2)$$

and consequently $\Delta = 4$ was found for a particular selection of relative polarizer angles ($a=0°$, $a'=b=-30°$, $b'=30°$). Very short time delays are sufficient to produce great effects that may naively be judged as produced by nonseparability. The correlation function for this model is shown in fig.2. The time delay between measurements hitherto considered simultaneous, plays an essential role in defining the unusual property of this model and must compel us to consider Aspect's experiment in its true light, i.e., as a very limited conclusion about hidden variable theories. In an experiment of the Aspect type this logical model should give a violation of B.I. even if the instant of the direction change of the correlated beams is actually random and even if the first coincident or noncoincident couple pertaining to each short wave-train is included in the calculated averages. Only the difference between ensemble and run averages can be observed following Buonomano's prescriptions.[11] But none of these prescriptions is satisfied in Ref.(8) in which the choice of the instant of the direction change is made by sinusoidal function and the 15-20% photomultiplier efficiency does not allow the first impinging photon to be detected with certainty. Moreover, the coincidence time (18ns) is nearly twice the duration (10ns) of a single beam pulse.

Another amusing characteristic of this rough model is its consistency with two channel polarizer experiments (Fig.1). In fact if we find a tick on the channel A1, we never find a coincident tick on A2. This is consistent with the current interpretation in which the photon must appear only on one of the two possible paths. In the case of a crossed channel $(A_1 - B_2, B_1 - A_2)$ a total consistency holds but only for a particular value of the coincidence time, that is for $T'' = L /8c$. This means that coincidence rate calculated as in the Orsay experiment is meaningful only if a symmetrical correlation function holds in this simplified situation. Since this class of models has several concomitant similarities with the physical situation, it will be very unpleasant if they do not contain at least a small crumb of truth or some clues leading us toward it. Also in this version no flow of information runs from a photomultiplier to the other and the model wholly respects the separability.

We have demonstrated in this section that classical undulatory systems exist for which the Einstein-Podolski-Rosen paradox vanishes by time delay between independent measurements. No kind of corpuscular concept is needed for this aim. Bell's argument is strongly limited by the absence of this time delay hypothesis. It is customary in modern science to simplify phenomena and objects by not considering their inner time evolutions. A stone is considered always equal to itself in calculating its trajectory. This custom, which is useful for macroscopic systems, is less appropriate when applied on an atomic scale, conceiving of the microscopic entities as being akin to little stones whose states are not variable in time when undisturbed, but change abruptly during their reciprocal interaction. This is not true! A micro-entity, and its state, is an undulatory process, continuously and quickly varyng from a minimum to a maximum and so on. It is evolving also with respect to the shift in amplitude and direction of its main polarization axis. It is thus a material process in a general oscillating evolution on precisely the same time scale as that involved in the measurement acts. The true significance of the word "coincidence" must therefore be analized more fully before implementing a coincidence experiment.

We have to convince ourselves that the quantum processes are governed by waves, and no clear connection exists in the theory linking waves and the sudden apparition of the "particle" in measurement devices. Moreover the actual creator of the calculation methods, Schrödinger, guided by his idea, deliberately did not insert such a connection in the theory. I agree with him that the background of the quantal phenomena must be purely undulatory. The impossibility of a link between wave and corpuscle is, in my view, the major source of the several paradoxes afflicting Q.M..

CAN WAVE-PARTICLE ASSOCIATION HAVE ANY REALITY?

From the birth of Q.M., the corpuscle concept has made successive and completely different impressions in the minds of physicists. At the beginning there was no doubt as to the existence of microscopic partics surrounded by their own fields. The electron itself was described in this way, but this view could not long withstand the observed stability of Bohr's atom. An electron, rotating in a generally elliptical orbit, in a very short time loses all its energy by radiation. This contradiction with observed evidence marked the start of a succession of withdrawals of corpuscles with respect to the waves. They were the waves that were able to stabilize the orbitals of the atoms, closing in on themselves with perfect coincidence of phase. The corpuscle, the already ancient Newtonian entity, was no longer properly reconcilable with the new phenomena discovered from now on by physicists.

Nuance upon nuance, the meaning of the word "corpuscle" changed inexorably in the new world-view of the scientists. The only one who remained firmly adherent to the classical concepts was Landé,[12] who described light as a pure electromagnetic wave and the electron and other particles as corpuscles that are --if not neutral-- surrounded by their own electric field. Planck, the creator of the quanta, was convinced that the energy packet was not real but only a mechanism inherent to the radiation absorption by atoms. The Copenhagen School, in refining its interpretation of Q.M., remained very ambivalent about the matter. Wave and particle each became the ghost of the other. When a widespread wave impinges on a large screen, it instantaneously annuls itself on all the screen surface by an ad hoc process named "wave packet collapse", affixing the mark of its passage only in a single point. In that precise point --as we know from Q.M.-- the corpuscle mysteriously appears, and at the same time the wave disappears completely. Therefore the two entities are complementary: They are two aspects of the same entity which we cannot observe simultaneously; they cannot coexist.

Nowadays several other interpretations --each time with indefinable differences-- are often indifferently used by physicists and philosophers in their reasoning about Q.M. subjects. However, corpuscles are never denied. Bubble chamber tracks and double Compton effects keep the researchers away from this radical position. Only Schrödinger[13] had the intellectual courage to claim that the bubble chamber may be a mystification chamber and the wonderful tracks only unfortunate tricks played by nature on our prejudice. He was in an isolated position with respect to the other fathers of Q.M.. Indeed, there exist in nature several processes which, starting from a continuos and progressively increasing disturbance, produce a quantal result. An example is lightning, that is, a sudden discharge of slowly accumulated charge. Another is the sudden fracturing along a single plane of a rock sample that supports a critical load. Another amusing analogy can be found with a rectangular table covered with poker cards folded along their major axes and aligned in adjacent rows like paper-soldiers. When a fan is agitated in front of the first cards of each row, at a time depending on various factors, the first card of a row will fall and drag all the other cards in that row with it. A shortsighted observer will judge that an object has been moved along the table leaving a track of its trajectory.

In Einstein's conception, all the energy (or most of it) was concentrated in energy grains conceived as singularities of the field. Such singularities, even if not observed, obviously must have definite trajectories. Einstein did not develop his idea personally but was essentially in agreement with de Broglie[14] when he proposed the double solution theory (DST).

We are firmly convinced that the double solution theory, created in 1927 by Prince de Broglie, is a serious attempt, performed in a materialistic way, to provide an explanation of various strange microphysical phenomena. In particular, the DST can render the measurement process naturally understandable, thus avoiding the puzzling and unjustified wave packet collapse. But we are not convinced of the reality of its conclusion because of the unpleasant potential field deriving from it.

In very recent times solitons[15] have been proposed as corpuscular theories even in connection with the DST. Indeed great difficulty arises in these theories when a soliton impinges against a potential wall; the soliton become split along the trajectories of trasmission and reflection. It does not satisfy indivisibility.

We discuss now some simple situations based on realistic assumptions which lead us towards a logical and general exclusion of the concept of corpuscle.

Let us start with the simple reflection process. We know that the law of reflection is the identity of two angles, the incidence angle and the reflection angle. Those who say that this is also the law of the rebound of a rubber ball impinging on a rigid wall, hope that this corpuscular model is applicable to the micro-world. But this view, whilst valid if the objects are macroscopic, is entirely false in the microscopic domain. The corpuscle without a wave is incapable of fitting even this simple phenomenon. A "billiard-ball" photon, knocking against atoms and molecules, certainly does not meet a smooth and polished surface as it appears to us macroscopically. By contrast it impinges on objects (or wave structures), regularly or irregularly distributed, having bumps, as nuclei, single atoms or linked atoms.

A single particle without any field surrounding it would not be forced to rebound in only the one direction actually observed. In this case all the directions should be allowed with an angular distribution of probability in no way similar to a delta function. Particles alone, more often than not, generate more difficulties than advantages: Newtonian corpuscularists, indeed, are very few today.

To explain single photon reflection, we appeal to waves. The reflection law is what it is because the mirror surface, under the wave action, however weak, acts as a collective source, with the emission phases exactly as required to rebuild the wave front in the observed direction. Corpuscles alone cannot operate in this way. Only waves have the extension property that enables them to interact with a vast mirror area (or better mirror volume, because of some wave length deep interaction), which has to react in a "coherent" way to this interaction (with linearly distributed phase differences) on all of the affected area.

Similar processes are involved in the sort of tunnel effect produced when a beam splitter is used. The only difference is that the wave intensity is halved on both the allowed paths. We are, on the other hand, sure that the wave is present on both the paths by the proof of interference obtained by recombining the beams even at an intensity at which we are sure that only one photon at a time is present over a length of 1000 meters.[16]

The wave is present, and a sudden "click" on the photomultiplier, or a swift blackening of a single point on a photographic plate, tells us that a "corpuscle has arrived." Therefore both are real: this is briefly the reasoning leading up to Einstein's conceptions, Prince de Broglie's double solution theory, and general solitonic theories. We must now check whether these several corpuscle-wave associations endure against simple arguments starting from simple postulates and leading to unpleasant conseguences that are also present --even if not seen as contradictions-- in the DST.

(i) As far as corpuscles are concerned, we uniquely assume that, no matter how this entity is described (energy grain, singular solution, soliton, etc.), it is INDIVISIBLE. This means that it can run along only one of the possible paths, either reflected as a whole or transmitted as a whole. We will consider this assumption as true throughout the present argument.

(ii) The second realistic assumption is, in the light of the above-mentioned single photon interference, that the wave beam is roughly halved by the beam splitter.

(iii) The third is, considering the fact that the corpuscles appear denser in the interference maxima, that the particle is driven by a physical interaction with the wave.

(iv) The fourth is, in first approximation, that the corpuscle or the singularity is not able to regenerate the wave surrounding it.

These four assumptions lead us to an image of the semireflection process with two halved amplitude wave packets traveling in different directions, only one of which itself carries the singularity. If now we insert on one of the paths a screen with two pinholes (Fig.3), we can observe the interference pattern on a successive parallel screen. Can these halved amplitude waves guide the particle in the space with the same certainty? Could they produce the well known interference pattern with the same brightness? If the answer is "yes," we have to admit also the perfect guidance power of the wave at any level of its amplitude, even at the infinitesimal level. That is the wave, deprived of any realistic physical property, is reduced only to a sort of trace, similar to Tom Thomb's bread crumbs, that the particle decides to travel over.

The defenders of the localized grains of energy can still object that the corpuscle is "the source" of the linked wave and that it is able to "restore" any loss of wave intensity at the preceding level very quickly. This is, obviously, an ad hoc process, but it cannot be excluded in principle. Another problem could be the velocity of this hypothetical restoring process.

We have thus to search for a more general experiment able to achieve the same result if our fourth assumption is either TRUE or FALSE. We find a possible solution to our problems in a situation as represented in Fig.4. In this experiment we need only one photon to be present in the apparatus on the average. The requirement is a very-long-arm interferometer, like that used by Janossy and Naray in 1958.[17]

Let us first postulate that the corpuscle regenerates its wave field back to the last edge of its wave packet. We explicitly assume also that it cannot regenerate behind the semitransparent mirror if the two split waves, reflected and transmitted, are now, after a sufficient time, completely separated without any common bond on the first beam splitter surface. We can now operate successive half splittings by a succession of beam splitters on each of the two paths. With, shall we say, ten half splittings for each path, in the case when the particle crosses all the splitters in the proper direction, we obtain, in the recombined final part of the beam, a full amplitude wave with a full particle, superimposed on a 1/1024 amplitude wave. This 1/1024 amplitude wave is negligible with respect to the other: The result will be no apparent interference fringes.

The second alternative hypothesis is that no wave regeneration is taking place. In this case also the particle sees its wave halved at each passage through a beam splitter. The final result is, in the interference zone, an integer corpuscle quite stripped of its wave interacting with another extremely reduced wave. Can this 1/1024 amplitude wave exercise the same influence and drive the particle with the same certainty along the sinuous path of Fig.5?[18] We believe that this would mean a violation of the interaction principle. There exist in nature at least four kinds of interactions, namely the nuclear, the weak, the electromagnetic, and the gravitational. All four decrease with distance or, more precisely, with the decrease of the intensity of the relative field. Can we now conceive that the effects of the wave on the corpuscle remain the same, even if the wave field is extremely weak? We do not think so, although the quantum potential $(\Delta^2 R)/R$ of the DST has just this property: It produces a constant intensity guidance force however low the undulatory field.

The reasoning of the people who believe in DST is that, with the irrenounceable assumption of the existence of corpuscles, the quantum potential and its strange properties is a pain to be accepted. The assumption of a real corpuscle is, on the contrary, a serious limitation to an observation without prejudices of nature. Nonrecognition of the quantum potential as an anomaly does not produce any progress in the search of the truth.

A large number of experiments is open to a purely undulatory interpretation, starting from the old two-slit interference and arriving at the recent neutron interferometry[19] and at the modification of fluorescence lifetime in the presence of mirrors.[20] The most serious classical objection to the pure-wave picture is the macroscopic evidence of "particle tracks" in bubble-chamber devices. The only and solitary voice raised against the current interpretation of bubble traces was that, now remote in time, of the above mentioned Erwin Schrödinger.[13] More recently some indications of a possible resolution of this troublesome problem have come from the work of Cini et al.,[21] who have investigated the problem of wave-packet reduction considering the measuring apparatus as a macroscopic object. The proof, valid only for a two level microscopic system S, shows that no discontinuity exists between the status before the measurement act

$$| s > = p | u+> + q | u->$$

and the final status I u+> or I u->. The wave-packet reduction occurs as a continuous phenomenon and can be accounted for by the Schrödinger wave equation together with statistical considerations. The microsystem S is considered as not localized and thus as interacting simultaneously with all the N particles of the measuring apparatus, as in wave behavior. No proofs are yet available for the case of continuous autovector spectra, e.g., the position in a Wilson chamber, but only mathematical difficulties stand in the way of this future generalization. Who will accept this challenge and measure himself against this long-awaited proof?

We are strongly persuaded that the Q.M. correlation function is not such an indomitable beast: It is not impossible, at least in principle, to bring it back into the corral of separability. In order to do this --if experimental tests of this wide class of models should be positive-- we could be forced to abandon the concept of corpuscle.

REFERENCES

1.   B. McCusher, *The Quest For Quarks* (Cambridge University Press, 1983).

2. A. Einstein, B. Podolsky, and N. Rosen, *Phys. Rev. 47* , 777 (1935).
3. D. Bohm, *Phys. Rev. 85* , 166 (1952); 180 (1952).
4. J. S. Bell, *Physics 1* , 195 (1964).
5. A. Garuccio, G. C. Scalera, and F. Selleri, *Lett. Nuovo Cimento 18* , 26 (1977).
6. G. C. Scalera, *Lett. Nuovo Cimento 38* , 16 (1983); *40* , 353 (1984).
7. A. Aspect, P. Grangier, and G. Roger, *Phys. Rev. Lett. 47*, 460 (1981); *49* , 91 (1982).
    A. Aspect, J. Dalibard, and G. Roger, *Phys. Rev. Lett. 49* , 1804 (1982).
8. A. L. Robinson, Science 219 , 40 (1983).
9. F. Rohrlich, *Science 221* , 1251 (1983).
10. I. Pitowsky, *Phys. rev. Lett. 48* , 1299 (1982).
11. V. Buonomano, *Nuovo Cimento 38* , 16 (1983); *Ann. Inst. Henri Poincaré XXIX* , 379 (1978).
12. A. Landé, *Found. Phys. 1* , 191 (1971).
13. E. Schrödinger, *Brit. J. Philos. Sci. III* (1952), reprinted in *Ann. Fond. L. de Broglie*; *Ann. Phys.* (4), 82 (1927); translated in E.Schrödinger, *Mecanique Ondulatoire* (Ed. Alcan, Paris,1933).
14. L. de Broglie, *Compt. Rend. Acad. Sc. (Paris) 183*, 447 (1926); *184*, 273 (1927); *J. Phys. 8*, 225 (1927).
15. A. C. Scott, F. Y. F. Chu, and D. W. McLaughlin, *Proceedings of the I.E.E.E.*, 51 (1973).
    T. A. Minelli and A. Pascolini, *Lett. Nuovo Cimento 27* , 413 (1980).
16. A. J. Dempster and H. F. Batho, *Phys. Rev. 30* , 644 (1927).
    A. Adam, L. Janossy, and P.Varga, *Acta Phys. Hung. 4* , 301 (1955).
    E. Brannen and H. I. S. Ferguson, *Nature 4531* , 481 (1956).
17. L. Janossy and Zs.Naray, *Acta Phys. Hung. 7* , 403 (1957); *Suppl. Nuovo Cimento*, No.2, 588 (1958).
18. R. D. Prosser, *Int. J. Theoret. Phys. 15* , 169 (1976).
    J. P. Wesley, *Found. Phys. 14* , 155 (1984).
19. H. Rauch, W. Treimer, and U. Bonse, *Phys. Lett. 47A* , 369 (1974).
20. H. Kuhn, *J. Chem. Phys. 53* , 101 (1970).
21. M. Cini, M. De Maria, G. Mattioli, and F. Nicolò, *Found. Phys. 9* , 479 (1979).
    M. Cini, *Nuovo Cimento 38* , 16 (1983).

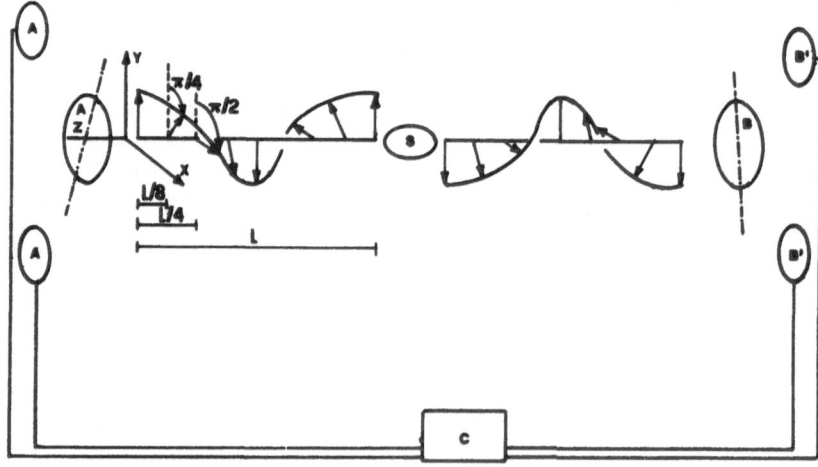

FIG.1. The helix model: A and B = polarizers, A_1, A_2, B_1, B_2 = photomultipliers, S = source, C = coincidence counter, L = helix step. The little arrows represent the hidden variables of the process.

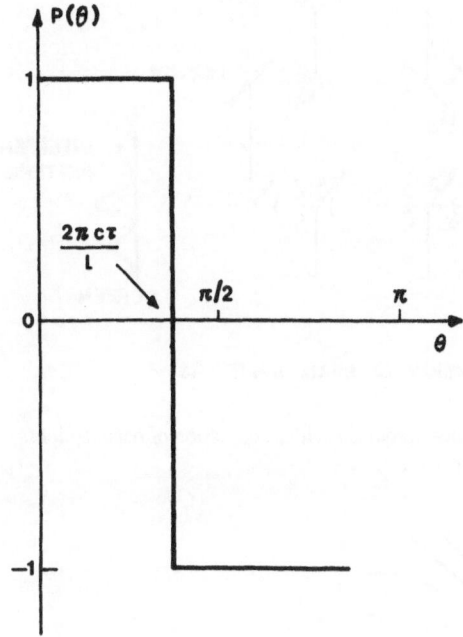

FIG.2. Correlation function for the original model by run averages.

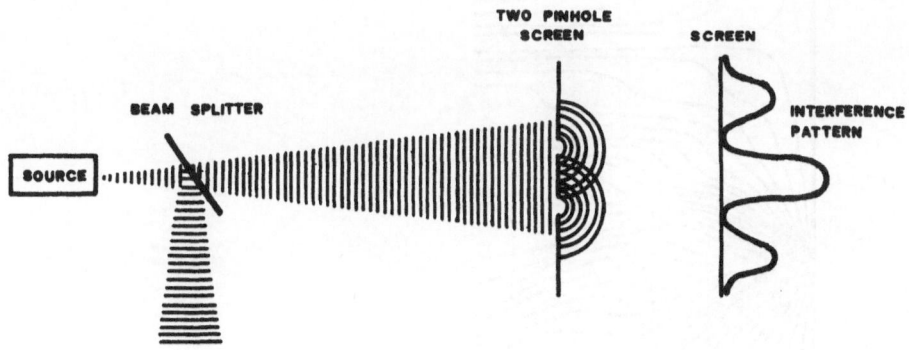

FIG.3. First experiment: interference of a halved wave by two pinholes.

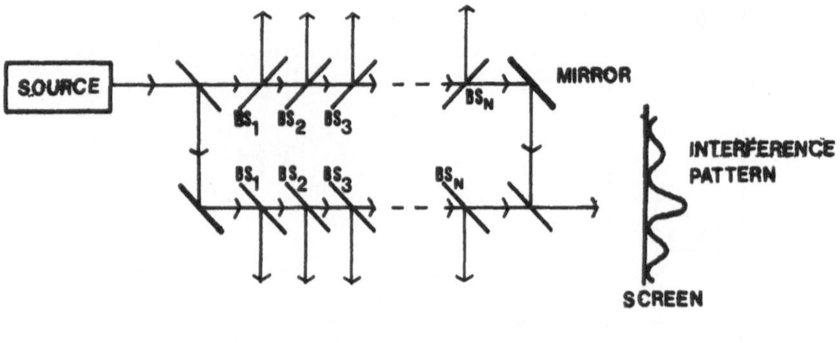

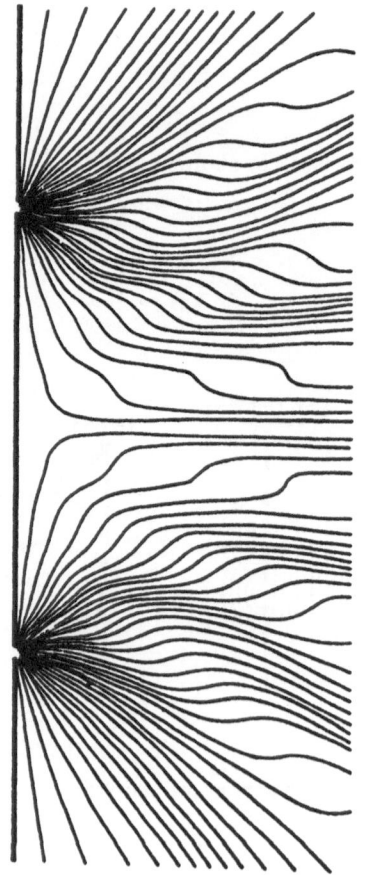

$BS_1 \cdots\cdots BS_N =$ SUCCESSION OF BEAM-SPLITTERS

FIG.4. Second experiment: interferometer with a succession of beam splitters.

FIG.5. Typical particle trajectories behind a double pinhole screen (Wesley, Ref.(18)).

A CRITICISM OF SOME RECENTLY PROPOSED MODELS THAT VIOLATE THE BELL
INEQUALITY

Saverio Pascazio [1]

Theoretische Natuurkunde, Vrije Universiteit Brussel
B 1050 Brussels, Belgium

ABSTRACT

Two models recently studied by G.S. Scalera and S. Notarrigo
are analyzed. These models can violate the Bell inequalities because
"facilitated" by the coincidence counting electronics. It is shown that
a hypothesis by Clauser and Horne in their well-known paper of 1974
explicitly forbids such a possibility for atomic cascade experiments.
Therefore, a local realistic solution to EPR problem, if it exists,
should be looked for in a different direction.

1. INTRODUCTION

Einstein, Podolsky and Rosen's paper[1] of 1935 has been a profitable
source for an immense quantity of work, papers, discussions and experiments. In 1965, Bell[2], sharpening EPR's argument, was able to deduce
an inequality which pointed out some nonlocal features of quantum
mechanical correlations. This inequality must in fact be satisfied
by every local realistic model, but it is violated in certain cases   by
quantum mechanical correlated systems. A few years later, Clauser,
Horne, Shimony, and Holt (CHSH)[3] could find a Bell-type inequality that
is susceptible to experimental inspection, and since then   many experiments have been performed. A complete survey of these experiments
up to 1978 can be found in Ref.4, along with the reasons why the atomic
cascade tests[5] have to be considered the best ones. The present experimental panorama strongly supports quantum mechanical predictions
against local hidden-variable theories; nevertheless it is worthwhile
to remark that the CHSH inequality, which is effective for atomic
cascade cascade experiments, can be obtained only at the price of

G. Tarozzi and A. van der Merwe (eds.), The Nature of Quantum Paradoxes, 141–154.
© 1988 by Kluwer Academic Publishers.

an untestable extra assumption, the so-called no-enhancement hypothesis. At present, the only way out of physical reality and/or locality-shattering conclusions seems to be represented by some "enhanced photon" models, investigated by Garuccio, Marshall, Santos, and Selleri[6]; such models can well fit the present experimental data (and so can violate the CHSH inequality) by denying the CHSH no-enhancement hypothesis. It is noteworthy that Clauser and Horne themselves wrote a paper in 1974[7] in which they could find a enhanced-photon model able to violate the experimentally tested inequalities, but they considered their own example to be highly implausible because of a dissymmetry between the two photon polarization analizers. This dissymmetry is not present anymore in the models quoted in Ref.6, but a recent paper by Caser[8] shows that enhanced-photon theories, in order to be quite equivalent to quantum mechanics, must exhibit dissymmetries between the photon polarization analyzers. There is certainly more to come on this subject, because Caser's results show that the way out of EPR problem is to be looked for among three possibilities:

(a) quantum mechanical predictions for correlated systems are all right;

(b) enhanced photons exist in nature and they are experimentally testable against orthodox quantum mechanical ones;

(c) quantum mechanics and enhanced-photon models are not experimentally distinguishable because of a space anisotropy (which would be responsible for the above-mentioned analyzers' dissymmetry).

In this note, we study another solution of the Bell inequality puzzle, which has been recently proposed by G. Scalera (GS)[9] and, independently, by S. Notarrigo (SN).[10] GS and SN claim that the validity of the Bell-type inequalities is strongly limited by the absence of time delay in the very definition of coincidence. We believe that GS and SN's arguments, even if conceptually sound, do not allow one to infer the radical conclusions drawn by the authors. It is certainly true that time delay is absent in Bell inequalities, but a Clauser and Horne's note, in Ref.7, justifies this lack by introducing a hypothesis on the coincidence window τ; as we shall see, this hypothesis does not hold in GS's and SN's models. Our attempt to show this falls into two parts. First we summarize the main characteristics of the GS and SN models. Then we discuss another system with the same peculiarities and we emphasize the common limits shared by models of this kind.

2. SCALERA'S AND NOTARRIGO'S MODELS

In GS's model,[9] a source emits two elix-shaped continuous ribbons in opposite directions along the z axis. A "polarizer" oriented in the a-direction (in the xy plane) simply marks the directions +a on the ribbon and does not modify it in any other way. A "photomultiplier" placed behind the "polarizer" integrates the energy contained in the ribbon, always starting from a marked direction and ending with the following one. A counter ticks on when two successive marked directions have entered the apparatus. The two ribbons are assumed to leave the source with vectors emitted at the same time lying on the same axis in the xy plane, and to screw in opposite senses with respect to their propagation directions.

If L is the helix pitch and c the ribbon velocity, then each counter ticks regularly every time

$$T = L/2c \tag{1}$$

Let a (b) denote the first (second) polarizer setting. If a = b, the counters tick at identical times and one does not get into troubles by defining a "coincidence"; but if a-b = θ , the second counter will always tick a time

$$t(\theta) = \theta L/2\pi c \tag{2}$$

after the first one. Therefore, if τ is the so-called coincidence window, one gets

$$P(\theta) = \begin{cases} +1, & \text{if } \tau > t(\theta), \\ -1, & \text{if } \tau < t(\theta) \end{cases}$$

where $P(\theta)$ is the correlation function. That is to say, a coincidence or an anticoincidence is observed, depending on the very definition of τ. Observe that, whatever the "inversion angle" for the correlation function,

$$\theta_0 = 2\pi c\tau/L, \tag{3}$$

i.e., whatever the coincidence window τ, it is possible to find polarizers' settings a,a',b,b', such that the Bell inequality

$$\Delta = |P(a,b)-P(a,b')|+|P(a',b)+P(a',b')| \leq 2 \tag{4}$$

is violated. Choosing, for instance,

$$|a-b| = |b-a'| = |a'-b'| = \frac{1}{3}|a-b'| = \frac{\theta_0}{2} \tag{5}$$

one gets $\Delta = 4$.

It is even possible to deal with two-channel polarizers (as in the second Aspect's experiment[5]) and with counting rates instead of correlation functions; to perform ensemble averages instead of time averages (according to Buonomano[11]); and even to randomize the model, breaking the rigid tick time regularity expressed by formula (1), but this does not add anything new to the ontology of the problem: The very causes of the violation of (4) are in fact contained in formulas (3) and (5).

Let us summarize the main steps leading to the violation of (4) in GS's system:
  (a) find a "suitable" model,

      (b)   settle the coincidence window τ,

      (c)   find angles $\theta_i(\tau)$ $(i=1,\ldots,4)$ such that (4) is violated.

There is a first remark which can be made: Note that in every experiment performed, the step (c) precedes the step (b): Angles θ_i that violate Eq.(4) are in fact obtained by maximizing the quantum mechanical prediction for the left-hand side of (4), which we may write $\Delta_{QM}(\theta_i)$, with respect to θ_i; later on, τ is fixed and the experiment performed. Therefore, the θ_i's are not and cannot be functions of τ. By contrast, in GS's model the Bell-inequality violating θ_i's (formula (5)) are functions of τ, through formula (3); if several experiments are performed, to test (4), and different τ's are chosen, it is quite possible that (4) would be violated for different θ_i's. Moreover, by varying τ within a suitable range, Δ always gets "shrunk," and (4) gets satisfied; such an effect, in our opinion, would already have emerged from atomic cascade experiments.[5] As we shall see, these first remarks are still valid in the model proposed by SN. We postpone others, different comments to the next section.[11]

The SN model[11] consists of two identical linear systems, each one made up of n+2 classical particles with harmonic interactions between adjacent particles only. The particles' equilibrium positions lie along the z axis, and vibrations can occur in the xy plane. The following assumptions are made:

      (a)   longitudinal elastic constants are stiff enough so that only transverse vibrations are allowed;

      (b)   transverse coupling and particle masses are identical;

      (c)   the particles with labels $\rho = 0$ and $\rho = n+1$ are constrained to their equilibrium positions.

In natural time units $T = 1/\omega$, the equations of motions are $(\rho = 1,\ldots,n)$:

$$\begin{cases} \ddot{x}_\rho = x_{\rho-1} - 2x_\rho + x_{\rho+1} \\ \ddot{y}_\rho = y_{\rho-1} - 2y_\rho + y_{\rho+1} \\ x_0 = x_{n+1} = y_0 = y_{n+1} = 0 \\ \dot{x}_0 = \dot{x}_{n+1} = \dot{y}_0 = \dot{y}_{n+1} = 0 \end{cases}$$

If the initial conditions,

$$\begin{cases} x_1(0) = a, \quad x_\rho(0) = 0, \quad \text{for } \rho \neq 1, \\ \dot{x}_\rho(0) = 0, \quad \text{for every } \rho, \\ y_\rho(0) = 0, \quad \text{for every } \rho, \\ \dot{y}_1(0) = b, \quad \dot{y}_\rho(0) = 0, \quad \text{for } \rho \neq 1, \end{cases} \tag{6}$$

are chosen for both systems, then the solutions of the equations of motions may be represented by the complex vector

$$|x_\rho(t)> = \begin{bmatrix} x_\rho(t) \\ -iy_\rho(t) \end{bmatrix} ,$$

with

$$x_\rho(t) = \frac{2a}{n+1} \Sigma_{k=1}^{n} \sin \frac{k\pi}{n+1} \sin \frac{\rho k\pi}{n+1} \exp(i\omega_k t),$$

$$y_\rho(t) = \frac{2b}{n+1} \Sigma_{k=1}^{n} \frac{1}{\omega_k} \sin \frac{k\pi}{n+1} \sin \frac{\rho k\pi}{n+1} \exp(i\omega_k t),$$

where the ω_k are the eigenfrequencies of the system[12],

$$\omega_k = 2 \sin \frac{k\pi}{2(n+1)} , \quad k = 1,2,\ldots,n.$$

Introducing the unit vector

$$|\theta> = \begin{bmatrix} \cos\theta \\ \sin\theta \end{bmatrix},$$

the ρth particle vibration amplitude along any direction θ of the xy plane will be given by

$$A_\rho(\theta,t) = \text{Re} < \theta | x_\rho(t)>, \quad \rho = 1,\ldots,n. \tag{7}$$

The SN "polarizers" are simply constraint on the particle $\rho = \sigma$ ($1<\sigma<n$) which allow vibration only along the direction θ of the xy plane. In other words, a polarizer is a constraint of the type $x_\sigma \sin\theta = y_\sigma \cos\theta$ ($1<\sigma<n$) on the system. Of course, the vibration component of every particle along the θ direction will not be affected by the constraint. A state of linear polarization is thus accomplished because, given the initial condition (6), labelled particles $\rho \geq \sigma$ will oscillate only along the θ direction. "Detectors" are devices which tick when the square modulus of the nth particle vibration amplitude $A_n(\theta,t)$ (formula (7)) crosses a given threshold with positive slope.

We shall not go into details describing the several computer experiments carried out by SN; he varies the number n of particles, the initial constants a and b (formula (6)), *the coincidence time* , the threshold of the "detectors," the duration of the counting time interval and the epoch t of entering the process. The main point is that, for certain values of the parameters, his model violates Bell inequality and that, in certain cases, it is even possible to obtain a correlation function quite close to the quantum mechanical one.

Why does SN's model violate Bell inequality? The explanation is suggested by SN himself in the last pages of his work; a drastic simplification of the model is there proposed: A single vibrating particle replaces the linear (n+2)-particle chain just described. In this case, from the vibration amplitude (formula (7)), we get:

$$|A(\theta,t)|^2 = A^2\cos^2(\omega t-\theta).\tag{7'}$$

Therefore, if T_h is the value of the detector threshold, the crossing times will be

$$t_r(\theta) = c(T_h)+ \frac{\theta+r\pi}{\omega}\;,\tag{8}$$

with $c(T_h)$ independent of θ and r an integral number. Let us now consider the Bell inequality in the form

$$\frac{n(\theta_1,\theta_3)}{n(\theta_3)} + \frac{n(\theta_2,\theta_3)}{n(\theta_3)} \leq 1+ \frac{n(\theta_1,\theta_2)}{n(\theta_3)}\tag{9}$$

which is the one investigated by SN ($n(\theta)$ is the single counting intensity with polarizer axis in the θ direction and $n(\theta_i,\theta_j)$ the coincidence counting intensity with polarizer settings θ_i and θ_j). If $\theta_1 = \theta$, $\theta_2 = 3\theta$, and $\theta_3 = 2\theta$, one gets, from (8),

$$t_r(\theta_2)-t_r(\theta_3) = t_r(\theta_3)-t_r(\theta_1) = (t_r(\theta_2)-t_r(\theta_1))/2 = \frac{\theta}{\omega},\tag{10}$$

so that, if the coincidence time τ is chosen to be

$$\tau > \frac{2\theta}{\omega} \quad\text{or}\quad \tau < \frac{\theta}{\omega},\tag{11}$$

then (9) is trivially satisfied.

"In fact, in the first case all correlated events in the three channels will stay inside the resolving time, while, in the second case, only one for each channel will be inside the resolving time."[11]

But if

$$\frac{\theta}{\omega} < \tau < \frac{2\theta}{\omega}\tag{11'}$$

then (9) gets grossly violated ($2 \leq 1$),

"because we have a coincidence in channel pairs (θ_1,θ_3) and (θ_2,θ_3), but not in channel pair (θ_1,θ_2), in spite of the fact that the three events in different channels are totally correlated. By opportunely weighting these two extreme cases, as, in practice, it was achieved by our model, we can obtain also the QM value."[11]

i.e. $\dfrac{n(\theta_i,\theta_j)}{n(\theta_k)} = \cos^2(\theta_i-\theta_j).$

It is quite evident that the criticism directed at the GS model holds also for the SN's: the Bell inequality is violated only when a certain coincidence window τ is utilized. It is even more clear, in this case, that, if τ is varied, the Bell inequality gets satisfied. Moreover, as in the other model, Bell violating angles strongly depend on τ, by Eq.(11').

3. A NEW MODEL VIOLATING THE BELL INEQUALITY

It is necessary now to carefully analyze the characteristics of Bell-type inequalities in order to try and understand the reasons why perfectly local models, like the ones discussed in the previous section, can violate them. Since the causes of the violations have to be ascribed to the coincidence counting electronics, it is not useful to consider the original Bell inequality[2] because this one applies only to ideal systems; we shall deal, therefore, with the CHSH inequality,[3,4,7] which is the one always tested in atomic cascade experiments.[5]

Defining $p_{12}(\alpha,\beta)$ as the probability of a coincidence count with polarizers settings α and β, a necessary constraint for any realistic local theory (that does not admit enhancement) is

$$-p_{12}(\infty,\infty) \leq p_{12}(a,b)-p_{12}(a,b')+p_{12}(a',b)+p_{12}(a',b')$$
$$-p_{12}(a',\infty)-p_{12}(\infty,b) \leq 0, \tag{12}$$

where ∞ denotes, as usual, the absence of the polarizer. Inequality (12) is derived in Ref.(7), by utilizing the well kwown no-enhancement hypothesis; in the same paper, indeed, a lot of work is done to show the necessity of such an assumption and the consequences derived from it. But also another, much less showy hypothesis is done by Clauser and Horne (CH) in a footnote to their paper: By defining the coincidence counting rate $N_{12}(a,b)$, they write (Ref.7, footnote 9):

"The practical criterion for a 'coincidence count' always involves a coincidence time window τ: pairs of counts separated in time by less than τ are defined to be coincident. This procedure may appear to make the definition of $N_{12}(a,b)$ ambiguous, since in general it will depend upon the experimenter's choise for τ. However, this dependence is usually insensitive to variations in τ which satisfy $s \ll \tau \ll 1/r$, where r is the average count rate at either detector, and s is the typical time separation of "true" coincidence pairs. *Thus, we tacitly require the experimental arrangement to be such that this condition obtains.*"

Of course, CH, being interested in a practically testable inequality, had to face the practical problem of defining τ without getting into trouble; this means that τ must be a compromise between the ideal, unattainable value $\tau = 0$ and a practical, experimentally meaningful value $\tau > 0$, that must take into account the running time of an experiment (which involves photomultiplier stabilities, source strenght and stability, etc.), the signal to noise ratio, the time separations

of pairs, and so on. In our opinion the above mentioned CH note has
not been carefully analized by GS and SN.

Let us consider indeed CH's condition

$$s \ll \tau \ll 1/r \tag{13}$$

which must be satisfied if one wants inequality (12) to make sense
from an experimental point of view. Of course, (13) cannot be directly
applied to GS's and SN's models, because it explicitly refers to
atomic cascade experiments. But it is quite evident that, in the
systems shown in the last paragraph, the logical analogy of (13) does
not absolutely hold. In GS's system, for instance, the average count
rate at either detector r and the typical time separation of coincidences (or anticoincidences) s are such that

$$s = \frac{\theta}{\pi} \frac{L}{2c} \lesssim \frac{1}{r} = \frac{L}{2c} \tag{14}$$

because of formulas (1) and (2). Note that s is a function of θ.
And in fact, GS violates inequality (4) just settling a "critical
value" for τ (formulas (2) and (3)) and choosing the right angles
(τ-dependent!) in formula (5). In this way, from formula (5), for
the angles $|a-b|=|b-a'|=|a'-b'|=\theta_0/2$, we get

$$s_1 = \frac{\theta_0}{2\pi} \frac{L}{2c} \lesssim \tau = \frac{\theta_0}{\pi} \frac{L}{2c} \lesssim \frac{1}{r} = \frac{L}{2c} \tag{15}$$

while, for the angle $|a-b'|=3\theta_0/2$, we obtain

$$\tau = \frac{\theta_0}{\pi} \frac{L}{2c} \lesssim s_2 = \frac{3\theta_0}{2\pi} \frac{L}{2c} \lesssim \frac{1}{r} = \frac{L}{2c} . \tag{16}$$

Both formulas (15) and (16) are in manifest disagreement with (13).
Even if the rigid tick time regularity of formulas (1) and (2) is broken
by introduction of some stochastic variant of the model (see second
paper in Ref.9), formulas (14), (15) and (16) would hold on the average,
and the same reasoning applies again.

It is not possible to analize SN's model in the same way because
of the necessary computer analysis, but we can deal with the simplified
one-particle version (formulas (7')-(11')). In this case, by the
same reasoning, from formulas (8), (10) and (11'), we get

$$s_1 = \frac{\theta}{\omega} \lesssim \tau \lesssim s_2 = \frac{2\theta}{\omega} \lesssim \frac{1}{r} = \frac{\pi}{\omega}, \tag{17}$$

where s_1 (s_2) is the typical time separation of coincidences if measures
at angles (θ_2,θ_3) or (θ_1,θ_3) ((θ_1,θ_2)) are perfomed.(Note that in
(17), $\theta < \pi/2$ because of the square modulus in (7')). Again, (17)
and (13) are in clear disagreement.

Let us summarize: CH's physically and experimentally sound condition (13) has to be satisfied if an atomic cascade experiment is
to be performed. In other words, if one wants to perform a good experiment to test the CHSH inequality (12), one has to choose τ such that
(13) holds. A comparison of formula (13) with formulas (15), (16), and

(17) shows that in GS's and SN's models bad values for τ have been chosen to test the inequalities (4) and (9), respectively. Anyway, even if we disagree with the two above-mentioned authors about their conclusions, we want to stress that they have pointed out a quite new feature of Bell-type inequalities, namely, that it is possible for a local realistic model to violate Bell inequality if "tricky" electronics is utilized. This result is absolutely not trivial, because it shows that, in a certain sense, Bell-type inequalities do not fully take into account the pitfalls that may result from the fact that the experimental definition of coincidence cannot be equivalent to the theoretical one.

We end this section by examining a very simple system which exhibits the same features as GS's and SN's ones, hoping that this model will clarify the ideas expressed in this note.

Our system is made up of three cubes of side lenght L. A spring of elastic constant k separates two of them (mutually connected by a light rod of lenght $l < L$ and each having the mass m) from the third one (Fig.1a). Now, let us suppose that the mass of the latter cube is

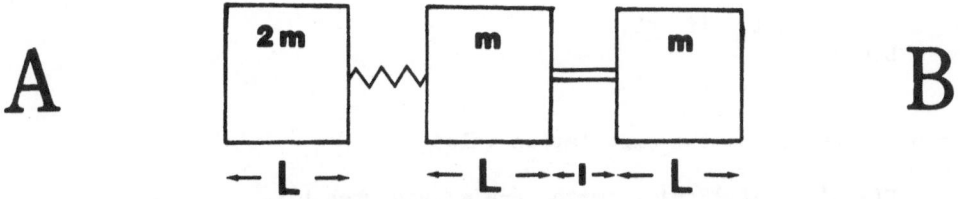

Fig. 1a. The proposed system: m and 2m are masses, L is the cube edge lenght, and l is the rod lenght. A and B denote the regions of space in which measurement are performed.

2m, i.e., equal to the combined mass of the other two, so that, when the spring is released, the two subsystems at its ends are projected into opposite directions with the same velocity (Fig.1b).

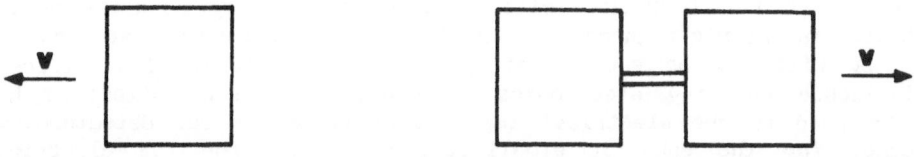

Fig. 1b. When the spring is released, the two subsystems are projected into opposite directions with the same speed v.

Imagine further that a source S, containing many identical systems of the type just described, shootes out "correlated subsystems" at a certain (average) rate r. We then want to carry out the following experiments side by side: First experiment: to count the number n of cubes detected; second experiment: to measure the global length x of cubes in the subsystem (x=L if one cube is detected, x=2L if a 2 cube-1 rod subsystem is detected).

We define the following dichotomic observables (A refers to measures carried out on the left and B on the right, in fig.1):

$$A(a) = \begin{cases} +1, & \text{if } n=2, \\ -1, & \text{if } n=1; \end{cases}$$

$$A(a') = \begin{cases} +1, & \text{if } x \leq L, \\ -1, & \text{if } x > L; \end{cases}$$

(18)

and, analogously,

$$B(b) = \begin{cases} +1, & \text{if } n=2, \\ -1, & \text{if } n=1; \end{cases}$$

$$B(b') = \begin{cases} +1, & \text{if } x \leq L, \\ -1, & \text{if } x > L. \end{cases}$$

(18')

As usual we define the expectation value

$$E(\alpha,\beta) = A(\alpha)B(\beta) \, , \quad \text{with } \alpha = a, a' \text{ and } \beta = b, b',$$

and we propose to test the Bell inequality in the usual form

$$\Delta = |E(a,b) - E(a',b) + E(a,b') + E(a',b')| \leq 2.$$

(19)

Our model is perfectly local and absolutely cannot violate (19) if the experiment is well done. Indeed,

$$E(a,b) = E(a',b') = -1, \quad E(a,b') = E(a',b) = +1$$

(20)

so that $\Delta = 2$ and (19) holds. In evaluating the correlation function, the coincidence counting electronics has not been taken into account. Actually it has been implicitly assumed that every coincidence was detected in the right manner. We shall return to this point later on.

It will now be shown that (19) can be violated if a "tricky" coincidence counting electronics is used. Of course, attention has to be paid to the electrical impulses delivered by our detectors. We assume, for the sake of simplicity, a δ shape for the electrical impulses: that is, if a detector clicks at the time t_c, a $\delta(t-t_c)$ signal is relayed to the counter. Also for simplicity, the two correlated subsystems are supposed to reach the respective detectors simultaneously (which means, strictly speaking, that the position of the source with respect to the two detectors is slightly asymmetric). Needless to say, the following reasoning holds even if we get rid of these two facilitating hypotheses. We assume that, if measurement of the first kind (counting the number n of cubes) are performed, the detector clicks as soon as the first face of the cube has entered it. If two-cube subsystems are detected, the two clicks will be $(L+1)/v$ seconds apart, where v is the (common) speed of the two subsystems. If, instead, measures of the second kind are carried out (measuring the lenght x of the subsystem), the detector needs some time because an "integration" has to be performed: If only one cube arrives, the detector will click the time L/v after the interaction (i.e., after the arrival of the first face of the cube); if the two-cube subsystem arrives, the detector will click twice, viz. at times L/v and $(2L+1)/v$, respectively, after the interaction. Let the time origin $t = 0$ be the instant at which the first detector (in A or in B, Fig.1) clicks. Let our coincidence counting electronics detect everything contained in the time interval $[0,\tau]$, where τ is the coincidence window. Define

$$\tau = \frac{L+1+\epsilon}{v} \tag{21}$$

with ϵ a positive constant chosen by the experimenter. If the bad choise

$$\epsilon < L \tag{22}$$

is made, (20) is violated. Indeed, in this case,

$$E(a,b) = E(a',b') = -1 \tag{23}$$

(first click in A and B at t=0, second click in B at $t=(L+1)/v < \tau$) and

$$E(a',b) = +1 \tag{23'}$$

(first click in B at t=0; clicks follow in A at $t_1=L/v < \tau$ and in B at $t=(L+1)/v<\tau$).Everything is hitherto identical to the results of formula (20). But

$$E(a,b') = -1, \tag{23''}$$

because A clicks at t=0 and B at $t_1=L/v<\tau$ and at $t_2=(2L+1)/v > \tau$ (!) if $\epsilon<L$. Therefore, the electronics will detect, as a coincidence, the arrival of one particle in A and of a L-lenght object in B. From (23), (23'), and (23''), we get $\Delta=4$, and thus (19) is badly violated.

We shall not digress here about the causes of this violation. The choice (22) is responsible for that. Every good experimenter would have chosen $\epsilon > L$, so as to detect coincidences in the right way, avoiding the situation outlined in formula (23''). Note that, if $\epsilon > L$, (20) holds again and (19) is not violated.

The model just shown can be improved; its symmetrization is straightforward: It suffices to deal with a mixture of 3-cube systems,

50% of which project the single cube to the right and 50% to the left. It is possible to deal with real electrical pulses instead of δ's and to take into account the possibility that the elastic constant k of the spring is not identical for every system. But we refrain from getting into entanglements because they do not change at all the ontological aspects of the problem. By contrast, it is worthwhile to remark that, in this model, the average count rate r at either detector can be made arbitrarily small, so that 1/r in formula (13) becomes arbitrarily large. But, once more, we can recognize that the typical time separation of coincidence pairs s depends on the measurements performed: For the correlation functions in formulas (23) and (23') we get $s_1 \lesssim \tau$; but, for the one in formula (23"), $s_2 \gtrsim \tau$, so that

$$s_1 \lesssim \tau \lesssim s_2 \ll 1/r. \tag{24}$$

Formula (24) represents certainly a small "step forward" with respect to formulas (15), (16), and (17) because the last, strong inequality moves us closer to CH's conditions (13). So, we can realize that Bell-inequality violations are due to situations of the type s~τ rather than τ~1/r. This is a quite interesting feature, because it allows us to infer that the condition $\tau \ll 1/r$ is generally connected to signal-to-noise ratio problems, while the condition $s \ll \tau$ rather deals with the right definition of coincidence. The above-mentioned CH footnote should now be completely clear: The definition of a coincidence count may apper ambiguous, because it depends upon the experimenter choice of τ. But this dependence is insensitive to variations in τ such that $s \ll \tau \ll 1/r$. In the three models shown in this paper, the definition of "coincidence" is strongly dependent on τ, because s~τ and CH's condition does not hold.

4. CONCLUSION

Many solutions have been proposed to solve the problems surrounding Bell-inequality violations. It has been claimed that our very conception of physical reality has been shattered by indisputable experimental results.[13] Many authors have proposed refuting Einstein's idea of locality. Actually, if one does not want to be forced to disown reality and locality, one is compelled to search hard for loopholes in CH's proof of their inequality; the research line followed in Ref.6 exemplifies this approach. But if every hypothesis made by CH in deducing their inequality is accepted, no way out is left: the CH inequality must hold, and no realistic explanation can support the quantum mechanical results. GS's and SN's approaches are certainly very useful, because they point out another, very hidden hypothesis that is needed in order to deduce the CH inequality. This hypothesis is contained in formula (13). Of course, if models are devised such that (13) does not hold, then Bell-type inequalities may be violated; but this is not a correct argument. In fact, in every atomic cascade experimental test of the Bell inequality Eq.(13) holds and, nevertheless, CH inequality gets violated. Accordingly, we think that the right way to go about proposing an alternative solution to the EPR puzzle would be to try and understand if the CH condition (13) is too restrictive with respect to hidden-variable theories and if its denial could lead one to criticize CH's "universality claim," according to which every enhancement-free hidden-variable theory is ruled out when (12) is found to be experimentally false. We hope that such an eventuality will be accurately checked by future research. In fact, it is our opinion that the concepts at stake (locality and realism) are far too important for us to leave even the smallest loophole unexplored.

The author gratefully acknoledge helpful discussion with Prof. J. Reignier. Thanks are also due to Instituts Internationaux de Physique et de Chimie Solvay (Brussels) for the financial support.

NOTE

(1) Work supported by Instituts Internationaux de Physique et de Chimie Solvay - Brussels.

REFERENCES

1.   A. Einstein, B. Podolsky, and N. Rosen, *Phys. Rev. 47*, 777 (1935).
2.   J.S. Bell, *Physics (N.Y.) 1*, 195 (1964).
3.   J.F. Clauser, M.A. Horne, A. Shimony, and R.A. Holt, *Phys. Rev. Lett. 23*, 880 (1969).
4.   J.F. Clauser and A. Shimony, *Rep. Prog. Phys. 41*, 1881 (1978).
     F.M. Pipkin, *Adv. At. Mol. Phys. 14*, 281 (1978).
5.   S.J. Freedman and J.F. Clauser, *Phys. Rev. Lett. 28*, 938 (1972).
     R.A. Holt and F.M. Pipkin, Harvard University preprint, 1974.
     J.F. Clauser, *Phys. Rev. Lett. 37*, 1223 (1976); *Nuovo Cimento 33B*, 740 (1976).
     E.S. Fry and R.C. Thompson, *Phys. Rev. Lett. 37*, 475 (1976).
     A. Aspect, P. Grangier, and P. Roger, *Phys. Rev. Lett. 47*, 460 (1981); *Phys. Rev. Lett. 49*, 91 (1982).
     A. Aspect, J. Dalibard, and G. Roger, *Phys. Rev. Lett. 49*, 1804 (1982).
6.   T.W. Marshall, E. Santos, and F. Selleri, *Phys. Lett. 98A*, 5 (1983).
     T.W. Marshall, *Phys. Lett. 99A*, 163 (1983); *Phys. Lett. 100A*, 225 (1984).
     A. Garuccio and F. Selleri, *Phys. Lett. 103A*, 99 (1984).
     T.W. Marshall and E. Santos, *Phys. Lett. 107A*, 164 (1985).
     F. Selleri, *Phys. Lett. 108A*, 197 (1985).
7.   J.F. Clauser and M.A. Horne, *Phys. Rev. D10*, 526 (1974).
8.   S. Caser, *Phys. Lett. 102A*, 152 (1984).
9.   G.C. Scalera, *Lett. Nuovo Cimento 38*, 16 (1983); *Lett. Nuovo Cimento 40*, 353 (1984).
10. S. Notarrigo, *Nuovo Cimento 83B*, 173 (1984).

11. V. Buonomano, *Nuovo Cimento* 57, 146 (1980).
12. H.C. Corben and P. Stehle, *Classical Mechanics* (New York, 1950), p.136.
13. B. d'Espagnat, *In Search of Reality* (Springer-Verlag, London,1983); *Phys. Rep;* 110, 201 (1984).

A PHENOMENOLOGICAL ANALYSIS OF THE EPR ARGUMENT

Vincenzo Fano
Department of Philosophy
University of Bologna, Bologna, Italy

ABSTRACT. A purely philosophical formulation of the EPR argument, in its original form, is presented. The philosophical concepts utilized to develop the reasoning are those of Husserl's semantics, as delineated in the first *Logical Investigation*:"Espression and Meaning." The analysis bring to light some semantical and ontologicaɪ premises implicit in the notions of physical reality and separability. Besides, the pragmatical character of the measurement is evidentiated. Measurement, which lies at the basis of the EPR argument and of quantum mechanics in general, is not a theoretical concept, but rather a meaning act. The conclusion is the following: The foundations of quantum mechanics imply not only a semantical problem, but also an ontological one.

1. INTRODUCTION

According to Wittgenstein, "philosophy aims at the logical clarification of thoughts. Philosophy is not a body of doctrine, but an activity." Furthermore, "philosophy sets limits to the much disputed spheres of natural science."[1]From this point of view, the scope of philosophy is reduced to that of merely legitimizing those scientific theories which are correct, i.e., logically consistent and empirically verified.

The foundations of quantum mechanics, and in particular the EPR argument, seem to reverse this conception.[2] The confirmation or the falsification of scientific theories reverberates upon philosophical thinking in general, hence science appears, in marked contradiction with Wittgenstein's idea, as a philosophy-clarifying activity. The interaction between philosophy and science is thus constituted as a reciprocal influence of the two disciplines, so that their mutual relationship is dialectic rather than logical.

While it is unthinkable to reduce philosophy to science or vice

155

G. Tarozzi and A. van der Merwe (eds.), The Nature of Quantum Paradoxes, 155–167.

versa, we need to construct a *scientific philosophy*, which, taking into
account the results of scientific endeavour, can also become the instrument of a *philosophical natural science*.

The present work, a phenomenological revisitation of the EPR argument, aims at moving along this direction.

2. HUSSERL'S SEMANTICS

A brief summary of the first logical investigation by Husserl will be
given, insofar as the structure of the relationship between *expression*
and *meaning* is concerned.[3]

A sign which expresses a meaning is called *expression*. While a sign
is simply something that stands for something else, an expression is a
meaningful sign. Expressions can belong in a dialog and thus have a communicative function, or they can belong to an isolated mental life. Presently, only the first case will be considered.

The expression in its communicative function is not a simple collection of sounds, but is invariably accompanied by *meaning-intending acts*
which make it comprehensible for the receiver. The subject of our interest is the "philosophy of language" of physical theories, where communication is not only verbal but also written. For the latter type of communication, Husserl's thinking can easily be extrapolated. Since in the
written expression, acts are, as it were, "imprisoned" on paper, it is
not necessary that he who formulates an expression repeats these acts
every time. The receiver of the written expression perceives also the
meaning-intending acts pertaining to the text that he is reading.

It is still   clarified how meaning-intending acts are comprehended by the expression receiver. The listener does not perceive only the
sounds, but the meaning-intending acts pertaining to the sounds as well.
It is the very perception of such acts which lends meaning to the expression perceived; it would seem, therefore, that the listener's comprehension is in turn constituted of meaning-intending acts relative to the
expression perceived as a collection of sounds. Husserl does not speak
in these terms; he only remarks that comprehension is not conceptual
knowledge, and that "there exists a mutual coordination between speaking
and listening, between manifesting certain lived experiences in talk and
receiving this information upon listening."[4]For our purposes,`we con-
sider comprehension as constituted by meaning-intending acts.

Expressions however, in their comunicative function, do not only
signify something to the listener, but express a certain referred object (*"Gegenstaendlichkeit"*), as well. While the meaning necessarily belongs to the expression, the referred object could eventually be missing. For instance, we might consider the expression "square circle,"

which certainly comunicates something to the listener, albeit something which does not exist. These considerations apply to the general case. On the other hand, for the expressions of theoretical physics, to which we want to apply Husserl's semantics, the referred object is essential if we want to avoid pure idealism. Accordingly the expression is accompanied not only by a series of meaning-intending acts, but also by a series of *meaning-fulfilling* (conferring, strenghtening, illustrating) *acts* "thereby actualizing the reference to the object."[5]Consequently, the comprehension of expressions involves the perception of both meaning-intending acts and meaning-fulfilling acts. And the comprehension of the referred object by the listener is constituted by fulfilling acts.

We can summerized Husserl's semantics with reference to Ogden and Richards semiological triangle.[6]The expression (sign) has a double reference: a mental one (meaning) and a bodily one (state of affairs). In Husserl's thinking, none of these three concepts exist statically for itself: The sign is an expression of a speaker (a writer); the meaning is constituted by meaning-intending acts which accompany the expression; the state of affairs is the referred object constituted by meaning-fulfilling acts.

This presentation of the concept of meaning is advantageous from the philosophical point of view, for the following reasons:

1. The pitfall of idealism is avoided due to the presence of referred object and of meaning-fulfilling acts. On the other hand, the referred object is not already given, as in naïve realism; indeed, the referred object is the fulfilment of a meaning, hence it has implicitly a subjective component.

2. The meaning and the referred object could appear as Kantian concepts. The former seems homologous to *empty* a priori forms of the intellect, and the latter seems homologous to *blind* sensitive intuitions. However there is a profound difference, since for Husserl the signification process and the relationship with the real reference are dinamically characterized as acts. As a consequence, in this context both intellect and reason are *practical*, and not only reason as in Kant's philosophy.

3. Finally, Husserl's semantics assigns a central role to communication. The usage of expressions, the treatment of meanings and the reference to reality are constituted in relation with the listener, therefore with communication of information.

It is possible to correlate these philosophical implications of Husserl's semantics with a few important aspects of present day epistemology.

1'. Popper maintains that physics is intrinsically realistic.[7] Schlick, even if considering realism as a non-verifiable hypothesis, believes that the existence of the external world is at least as reliable

as the best established law of science.[8] Carnap even tries to formulate
this hypothesis in an incompletely testable manner.[9] These istances correspond to the anti-idealistic connotation implicit in the concept of
meaning-fulfilling acts.

On the other hand, up-to-date epistemology (Kuhn, Hanson, and
Feyerabend) has shed light on the *theory-ladenness* of all observational
data. We may correlate the theory-ladenness with the meaning not yet
fulfilled, called by Husserl *meaning-intention*. Also, the term "referred
object" clarifies; in the first place one may speak of factual data only
insofar as they are comprised within a meaning-intention (referred) and
not for themselves; in the second place that one does not have to do
with objects in a strict sense ("*Gegenstaendlichkeit*" rather than "*Ob-
jektivitaet*").

2'. Kuhn[10] defines a *disciplinary matrix* as an ordered array of elements of various kinds which constitutes the common possession of those
who are engaged in research within a particular discipline. The elements
are: the *symbolic generalizations*, the *exemplars*, and the *models*. The
latter, inasmuch as they are the analogies preferred by the scientific
community, do not play a precise gnoseological role. Symbolic generalizations, on the contrary, are something of the type $(x)(y)(z)_\phi (x)(y)(z)$;
they are the formal part of the disciplinary matrix. Within the realm of
natural science they are not to be seen statistically as theorems of a
mathematical system, but as a set of applications of laws, which constitute the theoretical-physical sense. If laws are expressions, their application, namely symbolic generalizations insofar as belonging to a disciplinary matrix, are the meaning-intending acts pertaining to these expressions.

It must be noted but the assimilation of symbolic generalizations
to meaning-intending acts can be established since the definitions of
Kuhn's epistemological concepts are invariably related to the acting subject (i.e., the scientist or the scientific community), and therefore
have a pragmatic character similar to that of Husserl's semantics.

The disciplinary matrix is utilized by the scientist in order to
formulate predictions on experiments both for already-solved problems
(such as handbook exercises), and for unsolved problems (new application);
in both cases one has to do with exemplars; the former are already achieved,
the latter are yet to be achieved.

Kuhn uses a simple analogy to explain the learning process of exemplars. A child is in a garden with his father. He does not know ducks, geese
and swans, but only birds in general, hence he has a certain framing of
perceptive space as far as animals are concerned (meaning-intention). The
father accompanies the expressions "goose," "swan," and "duck" with the
suitable ostensions and the child apprehends the new concepts, thus mod-

ifying his framing of perceptive space (meaning fulfilment). Clearly, the various kind of birds are the homologous of exemplars.

It is not difficult to assimilate the acquisition, utilization, and creation of exemplars to meaning-fulfilling acts.

Another similar connection between Husserl's semantics and up-to-date epistemology emerges from the structuralist view of Sneed and Steg-mueller.[11] This most original approach to the structure and dinamics of theoretical physics identifies a theory of mathematical physics by means of an ordered pair (S,I), consisting of a formal structure S and a set of intended applications I. Stegmueller formulates also a precise definition of the term "a person *p* hold a theory (S,I)" and associates this term with the attitude of Kuhn's *normal* scientist. Consequently there are two manners for normal science to progress:

A. By the addition of a new intended application to set I.

B. By the addition of a new expansion of the core of the theory belonging to S, with new special laws and new special constraints.

The holding of the mathematical structure S of a theory and type (B) of progress of normal science are constituted by meaning-intending acts. On the other hand, the holding of a set of intended applications I and type (A) of progress of normal science are constituted by meaning-fulfilling acts.

These analogies between Husserl's semantics and Kuhn and Sneed's epistemology show the relevance of the phenomenological analysis for the foundations of modern physics.[12]

3'. As far as the third issue is concerned, we will just say that perhaps the relationship between information theory, or comunication theory and the foundations of natural science[13] has not yet been fully explored. The sociology of science is actually moving in this direction, but it is not homogeneous with the epistemology, and often philosephers look down up on it.

3.   THE EPR ARGUMENT

In our opinion the EPR argument, as it was originally proposed,[14] is an eminently philosophical reasoning concerning a physical theory.

What we want to do next is to evidence the general philosophical structure of the argument, applying Husserl's semantics to the work of Einstein and collaborators.

1. From the combination of Husserl's semantics and Kuhn and Sneed's epistemology, a physical theory is envisaged as a collection of expressions, of meaning-intending acts, and meaning-fulfilling acts. This is not a nominal definition since acts are neither language entities nor linguistically expressible. It is simply a useful classification for the

notion of completeness which we presently will define.

2. It is a necessary condition for the *completeness* of a physical theory that to every meaning-intending act pertaining to an expression of the theory there corresponds at least one meaning-fulfilling act.

This criterion is similar to that of EPR, insofar as it represents a necessary, and not a sufficient, condition, but it is different in another respect. Indeed, for EPR the incompleteness is determined by the lack od a theoretical counterpart corresponding to an element of reality, while in our case it is due to the over-extension of theory with respect to reality.

Furthermore, the EPR criterion calls for the possibility to define an element of reality outside any theory. Indeed EPR are compelled to give a "criterion for physical reality" for itself; this is a very questionable procedure from a philosophical point of view.[15]

In the following formulation of the argument, realism is not utilized from the logical point of view; from the philosophical point of view, however, realism is implicit in the distinction between meaning-intending acts and meaning-fulfilling acts, hence *a fortiori* in the completeness criterion.

When viewed through Husserl's semantics, the EPR completeness criterion is meaningless since the meaning-intending acts are essential to the expression, and only the meaning-fulfilling acts are accidental.

3. Let us consider a system S with one degree of freedom described by the wave function ψ. Then:

$$P \psi = -i\hbar \; \frac{\partial \psi}{\partial x}$$

is a meaning-intending act pertaining to the expression "momentum of system S," since the operator P expresses the quantum-mechanical sense of the observable momentum. Similarly,

$$Q \psi = x \psi$$

is a meaning-intending act pertaining to the expression "position of system S."

4. If the quantum mechanics is a complete theory, and it contains in itself meaning-intending acts pertaining to the expression "position of system S" and "momentum of system S," then meaning-fulfilling acts pertaining to the same expressions must also exist. However, the uncertainty principle prevents the expressions "position of a system S" and "momentum of a system S" from being simultaneously fulfilled. Therefore, either quantum mechanics is an incomplete theory, or simultaneous meaning-intending acts pertaining to the expression "position of a system S" and "momentum of a system S" are not possible.

The conclusion is similar to that of EPR, but the reasoning is altogether different. EPR utilize their criterion of reality to establish
the existence of reality elements corresponding to observables P and Q,
the uncertainty principle therefore prevents the simultaneous reality
of both in theory, hence the dilemma. By contrast, in our case P and Q
are meaning-intending acts for which the uncertainty principle prevents
the simultaneous fulfilment, hence the dilemma.

5. At the time $t = T$, S is *separated* into two subsystems S_1 and S_2.
Two physical systems are said to be separated when a meaning-fulfilling
act relative to the expression of an observable of the first system cannot fulfill the meaning of an expression of an observable of the second
system.

From a physical point of view this formulation of the separability
principle is identical to the one implicitly utilized by EPR and made
explicit in the following presentations of the argument.[16] From a philosophical point of view, on the other hand, the differece is considerable.
The separability of two physical systems is not conceived as the existence of a certain kind of object, but rather as a possible mode of fulfilment of the meaning-intention. In other words, the principle does not
express a modality of being, but a modality of the relationship between
thought and being. Hence its violation does not imply a denial of the
existence of a certain structure of reality but the abandonment of a
certain view of reality, which is implicit, for instance, in special
relativity and in other physical theories.

6. We can perform on the separate system S_1 both a meaning-fulfilling act pertaining to the expression "position of system S_1" and one
pertaining to the expression "momentum of system S_1," that is, the
measurement of the two observables, with the ensueing reduction of the
wave packet.

If u_1, u_2,... and v_1, v_2,... are respectively the eigenfunctions
of P and Q, and ψ_1, ψ_2,... and ϕ_1, ϕ_2,... the corresponding coefficients of the series development of ψ expressed as a function of the
variable characterizing the system S_1, i.e., x_1, then we obtain:

$$\psi(x_1, x_2) = \sum_{n=1}^{\infty} \psi_n(x_2)\, \mu_n(x_1)$$

$$\psi(x_1, x_2) = \sum_{r=1}^{\infty} \phi_r(x_2)\, v_r(x_1)$$

The reduction of these wave functions determines, for the first system,
the state $\mu_k(x_1)$ and $v_j(x_1)$ and, *consequently*, for the second system,
the states $\psi_k(x_2)$ and $\phi_j(x_2)$. In other words, the meaning-fulfilling
acts pertaining to the expression "position of system S_1" and "momentum
of system S_1" have constituted within quantum mechanics meaning-intend-

ing acts pertaining simultaneously to the expressions "position of system S_2" and "momentum of system S_2."

In this step the difference with respect to EPR is considerable. For them, indeed, the possibility to predict the values of P and Q for the second system guarantees the existence of two elements of reality which eventually will not be allowed to have a theoretical counterpart by virtue of the uncertainty principle. In our case, by contrast, predictions are a theoretical element of quantum mechanics which will not have a correspondence in reality in virtue of the uncertainty principle.

One might object that the possibility of performing on the first system meaning-fulfilling acts, pertaining to both expressions "momentum of system S_1" and "position of system S_1," implies the counterfactuality of returning to the first system after the first measurement. But the objection is not relevant, as EPR has already remarked; the first meaning-fulfilling act guarantees the existence of S_1, hence the possibility to perform the second fulfilling act. To deny this, implies a very poor concept of physical reality.

7. Let us assume *ab absurdo* that quantum mechanics is complete. Then, according to point (4), meaning cannot be conferred simultaneously on expressions "position of a system S" and "momentum of a system S," which contradicts the conclusion of step (6). Hence *quantum mechanics is not a complete physical theory*.

4. POSSIBLE SOLUTIONS

Let us analyse the possible solutions of the problem posed by EPR, presupposing that non-completeness of a theory is a defect.

1. The first solution is given by EPR in their conclusion of their article:

"While we have those shown that the wave function does not provide a complete description of the physical reality, we left open the question of whether or not such a description exists. We believe however, that such a theory is possible."

In other words quantum mechanics must be substituted or at least completed.

2. The second solution should be the orthodox one, according to which only measurements are meaning-fulfilling acts. But this is no solution, since in our reasoning exclusively meaning-fulfilling acts which are measurements are utilized.

The orthodox solution becomes the near caricature of itself, that is the impossibility of distinguishing between meaning-intending acts and meaning-fulfilling acts. It thus becomes impossible to define a criterion of completeness and the argument fails.

It is easy to find ambiguities in a sentence which claims to be a criterion of reality, whether formulated by EPR or by somebody else; by contrast it becomes a paradox to deny the possibility of existence of at least one criterion of reality, by suppressing the distinction between the two types of acts. A philosophical reasoning which involves a "constructive" utilization of this criterion is fraught with dogmatism, but, then, denying the existence of reality in absolute is just as dogmatic.

The flattening of Husserl's semantics into Berkeley's "esse est percepi" can follow two paths: the reduction of all meaning-fulfilling acts to meaning-intending acts (idealism), and vice versa, the reduction of all meaning-intending acts to meaning-fulfilling acts (empiricism). Even though the positions are gnoseologically distinct, they are attained via complementary paths, and the very thoughts of Berkeley is an example of the dialectics between solipsistic idealism and radical empiricism.

3. From our formulation there results a refutation of the argument which is not easily reducible to those usually found in the litterature: The validity of step (6) is refuted. According to this view, once the meaning-fulfilling act of the expression "position of system S_1" is performed, it is impossible to return in time and perform the meaning-fulfilling act of the expression "momentum of system S_1." This corresponds to denying that the first meaning-fulfilling act guarantees the existence of system S_1 indipendently of time. It actually only guarantees itself, namely the fulfilment of the first expression. Actually, the counterfactuality that we utilized introduces surreptitiously a criterion of reality for the first system, which can be denied, based on the principle that the meaning-fulfilling acts of one expression exclusively fill the part of meaning they have fulfilled.

The latter view can be likened to the one presented by Bohr in his reply to EPR. What he objects to is the possibility of performing measurements without perturbing the system:

> "Of course there is, in a case like that just considered, no question of a mechanical disturbance of the system under investigation during the last critical stage of the measuring procedure. But even at this stage there is essentially the question of an influence on the very conditions which define the possible types of predictions regarding the future behavior of the system."[17]

Bohr questions the value of EPR's principle of reality, therefore of the predictions concerning the second system. In our case, by contrast, in

order to fully relativize the meaning-intention to the measurement, one
needs to deny the very existence of the first system, even after the first
fulfilling act as been performed.

   This solution, even if consistent, leads to a·concept of physical reality at least as paradoxical as the one implicit in the second solution.
The relational conception of quantum states, when applied to our version
of EPR, is valid both for the second and for the first system. It thus refers to the concept of observation in general and to complementarity with
all its complex philosophical implications.

   4. The fourth and last solution consists in denying the separability,
hence the validity of step (5).

   In a sense, both denying the validity of the principle of reality of
EPR, the way Bohr does, and denying the reality of the first system, as
solution (3) does, are philosophical attitudes which *a fortiori* deny separability. In fact, in both cases, elements of reality ontologically independent are unthinkable, hence so are separated systems. However, from
the philosophical point of view, it is safer to deny separability, since
it is less ontologically committed. Furthermore, we have seen that separability is one of the possible fulfilments of meaning-intention which
does not prevent other modalities of fulfilment. The negation of separability leaves, like the other solutions, a vacuum of reality with respect to other physical theories such as, for instance, special relativity. What is left open, however, is the possibility to fill this vacuum
by reconstituting the referred object in a different way.

   The removal of separability implies a *holistic principle* which already was applied to the domain of physics by Aristotele. However, one
must carefully distinguish between this position, which might just coincide with that assumed by Bohm[18] in his more recent work and is tinged
with Aristotelian reminescences, and the holistic principle implicit in
Bohr's thinking, which is emphasized by Jammer.[19] The whole which differs from the sum of its component parts, in the first case is made up
of two microphysical systems, whereas in the second case it is made up
of a physical system and measurement apparatus.

   Furthermore, whenever the holostic principle is invoked one must
face the problem: Why is the sum of these parts lesser than the corresponding whole? Within Aristotelian physics, the answer is teleological,
i.e., the whole has a function which the single parts cannot have. In
Bohr's thought, on the other hand, it is the subject who, being an actor
and a spectator at one and the same time, is instrumental in creating the
separation between measuring device and system, which for themselves would
be united. By contrast, within the realm of contemporary physics, if we
deny separability, the question remains open. As we have already said,
the solution could be the very negation of the holistic principle, re-

sulting from a new subdivision into parts, that is, from a new modality of fulfilment of the meaning-intention.

We have thus analyzed the possible solutions which deny step (2) (completeness), (5) (separability), and the first part of step (6) (counterfactuality). Step (3) and the second part of step (6) classify some conceptual operations which belong within quantum mechanics as meaning-intending acts, and don't seem to pose any particular problem. Point (1) is instead a characterization of physical theories, which summarizes the philosophical assumptions of our argument, already presented in Sec.2. Finally, steps (4) and (7) are purely logical. There does not therefore appear to be any other possibility to refute the EPR argument in the present formulation.

5. THE PRIMACY OF SEMANTICS AND THE EPR ARGUMENT

According to Carnap,[20] in order to be able to talk about new entities one needs to introduce "new ways of speaking, subject to the new rules," an operation which he defines. construction of a new *framework*. One may thus distinguish two kinds of existence questions: (1) existence questions of new entities which could be placed inside the framework: internal questions; (2) questions concerning the existence of the framework itself: external questions. The former can be solved empirically or logically, while the latter are questions framed in a wrong way. In other words:"To be real in the scientific sense means to be an element of the framework."[21] Let us analyze now the solutions of the EPR argument previously outlined, on the basis of this "Carnap principle," which we could define as *the primacy of semantics over ontology* and which underlies the greatest part of modern philosophy of science.[22]

1'. The completion of quantum mechanics, which in our view coincides with the introduction of fulfilling acts not pertaining to theory, consists in stating the existence of entities not belonging to the framework of quantum mechanics (hidden variables). This is a manifest violation of the primacy of semantics. On the other hand, the replacement of quantum mechanics with another theory consists in the introduction of a new framework. It is an unlikely perspective, but it does not violate the primacy requirement.

2'. Solipsistic idealism and radical empiricism are philosophically paradoxical views, but are not in contrast with the Carnap principle.

3'. This perspective is perhaps the one must in line with the anti-metaphysical demand of Carnap's empiricism, but is not as tolerant, except perhaps in Bohr's thinking. Indeed, in the third solution it is not possible to state the existence of a physical system even after its mo-

mentum or its position have been measured, that is respecting the framework of quantum mechanics. The introduction of the complementarity concept nevertheless makes this conception more elastic, even though the
pitfalls of the inconclusive characteristics of every dialectical solution
cannot be avoided.

4'. The negation of separability is independent of the Carnap principle, since it is not a problem of language, but of the relationship between meaning-intention and its fulfilment, as is evident from the definition given in step (5) of the argument. This is not meant to imply that
separability is not also a problem of language, as the debate on phenom-
enalistic language and physicalistic language, initiated by Carnap himself[23] and continued by Goodman,[24] shows. We only mean that the problem
of constituting microphysical systems is not a purely semantic one.

From the preceding discussion on the interrelations between the solution of the EPR argument and the primacy of semantics, it appears that
the latter is badly shaken by contemporary physics. We may therefore say
that the foundations of quantum mechanics pose not only a semantic problem,
but an ontological one, as well, with all the related risks of dogmatism
and the associated profound philosophical themés.

REFERENCES

1. L. Wittgenstein, "Logish-philosophische Abhandlung", *Ann. Naturphilos.*
   14 (1921), Secs. 4.112, 4.113, and 4.111.
2. G. Tarozzi, "Réalisme d'Einstein et mécanique quantique: un cas de
   contradiction entre une théorie physique et une hypothèse philosophique clairement définie," *Rev. Synthèse 101-102,* 125, (1981).
3. E. Husserl, *Logische Untersuchungen* (Halle, 1900-1).
4. E. Husserl, *Logische Untersuchungen* (Halle, 1922-3), II, p.33.
5. *Ibid.*, p.38.
6. C. K. Ogden and I. A. Richards, *The Meaning of Meaning* (London,
   Routledge and Keegan, 1923).
7. K. R. Popper, "Realism in quantum mechanics and a new version of the
   EPR argument," in G. Tarozzi and A. van der Merwe, eds., *Open Questions in Quantum Physics*, (Reidel, Dordrecht, 1984), p.3.
8. M. Schlick, "Meaning and verification," *Philos. Rev. 45*, 367 (1936).
9. R. Carnap, "Testability and meaning," *Philos. Sc. IV*, 37 (1937).
10. T. S. Kuhn, "Second thoughts on paradigms," in F. Suppe, ed., *The
    Structure of Scientific Theories*, (University of Illinois Press,
    Urbana, Illinois, 1974).
11. J. D. Sneed, *The Logical Structure of Mathematical Physics* (D. Reidel,
    Dordrecht, 1971); W. Stegmueller, *The Structure and Dinamics of
    Theories* (Springer-Verlag, New York, 1976).

12. P. A. Heelan, "Horizon, objectivity and reality in the physical science," *Int. Philos. Quarterly* 7, 375–412, (1967).

13. J. F. Lyotard, *La condition postmoderne* (Les Editions de Minuit, Paris, 1979).

14. A. Einstein, B. Podolsky, and N. Rosen, "Can quantum-mechanical description of physical reality be considered complete?" *Phys. Rev. 47*, 777,(1935).

15. R. Carnap, "Empiricism, semantics, ontology," *Rev. Int. Philos. IV*, 20 (1950).

16. For example, F. Selleri and G. Tarozzi, "Quantum mechanics, reality and separability," *Riv. Nuovo Cimento, 4* (2), (1981).

17. N. Bohr, "Can quantum-mechanical description of physical reality be considered complete?" *Phys. Rev. 48*, 700, (1935).

18. For example, D. Bohm and B. J. Hiley, *Found. Phys. 5*, 93, (1978).

19. M. Jammer, *The Philosophy of Quantum Mechanics* (Wiley, New York, 1974), p.199.

20. Ref. 14.

21. *Ibid.*, p.22.

22. E. Melandri, *La linea e il circolo* (Il Mulino, Bologna, 1968), Secs.53 54.

23. R. Carnap, *Der logische Aufbau der Welt* (Benary, Berlin, 1928).

24. N. Goodman, *The Structure of Appearance* (Harvard University Press, Cambridge, Massachusetts, 1952).

PART 3

CAUSAL AND DETERMINISTIC INTERPRETATIONS
OF QUANTUM FORMALISM

CAUSAL VERSUS ORTHODOX INTERPRETATION OF QUANTUM MECHANICS

A. Pignedoli
Dipartimento di Matematica, Università di Bologna
Bologna, Italy

ABSTRACT

The present work constitutes first of all a discussion of the so called "orthodox" interpretation of the wave function Ψ of quantum mechanics and the so-called "unorthodox" related theories. The Einstein-Podolsky-Rosen argument, the theories of the hidden variables, and the inequalities of Bell's type are considered. Moreover the paper contains an exposition of the author's point of view concerning the validity of quantum mechanics. The meaning of probability (in the theory of knowledge) and of causality (within the ontological point of view) are discussed.

PART I. PRESENTATION AND DISCUSSION OF "UNORTHODOX" THEORIES AND COMPARISON WITH THE "ORTHODOX" INTERPRETATION

(1) The interest in the question dealt with in this article has been intensified recently in view of the investigations concerning the EPR argument as well as those concerning the opposite stand taken on the matter by Niels Bohr. There exist, undoubtedly, certain essential characteristics which endow a physical theory with a definite "credibility." Among these, the adherence to experimental facts comes first, followed by the suitability and versatility with respect to the prediction of experimental results. Also, one must consider how the doctrine can be formulated within a logically consistent framework and whether it can supply an interpretation endowed with clarity (an essential feature) when physically interpreting the whole of the mathematical formalism through which the theory is expressed. One must also not disregard the amplitude of the theory, i.e., the range over which it is valid. Finally, one usually throws in formal elegance as one of the essential characteristics of a physical theory. The latter, however, appears to us to be rather an unessential quality.

The body of the doctrine of quantum mechanics constitutes undoubtedly a far-reaching synthesis of phenomena. Its so called "most complete" form describes various systems containing different kinds of particles, which are allowed to interact with one another and can be

171

G. Tarozzi and A. van der Merwe (eds.), The Nature of Quantum Paradoxes, 171–195.
© 1988 by Kluwer Academic Publishers.

made to satisfy the corrections required by the special theory of relativity. What we mean by "most complete" form of quantum mechanics is actualy quantum field theory. In the case of an isolated single particle, the theory is essentially expressed as the relativistic wave equation in its various forms, while the nonrelativistic case, under neglect of the yields the classical De Broglie-Scroedinger wave equation. This, of course, concerns the analytical aspect of quantum mechanics: One encounters here the typical, interesting class of eigenvalue problems, which can be elegantly formulated in terms of those particular linear function spaces, namely the Hilbert spaces containing square-integrable functions. However, this body of doctrine, which deals with the internal structure of matter and energy and with the dynamics of single particle of systems of particles in their space-time evolution, must be expected to pose – or presuppose – fundamental questions of a general nature, pertaining to natural philosophy as well as to philosophy proper, in particular, gnoseology. The first philosophical attitude assumed toward the mathematical formalism was the "orthodox" interpretation of the Copenhagen school, headed by Niels Bohr. In order to "see" this interpretation in its true significance, we must digress a little over the axiomatic formulation of quantum mechanics (QM).

It is well known that the state of a given system in classical analytical mechanics is described by assigning 2n Hamiltonian conjugate variable, q_i, p_i, where n is the number of degrees of freedom of the system. The state of QM systems, i.e., micro-objects, is characterized a "wave function" Ψ (q,t), generally complex and a function of Lagrangian coordinates q and time t, but not of the conjugate momenta p. Considering first a case in which t is not contained in the wave function and the latter is known, for any physical quantity $\underline{a}$ of QM, it is possible to find the probability that the (real) quantity $\overline{a}$ satisfies the double inequality-equality $\alpha \leq a \leq \beta$, where α and β are any two real numbers. More precisely, if the system is in the state $\Psi(q)$ the probability that the q values are contained in a given set G is given by the ratio

$$P = \int_G |\Psi(q)|^2 dq \; : \; \int_R |\Psi(q)|^2 dq, \tag{1}$$

where dq is the volume element in the coordinate space R: The following condition must hold:

$$\int_R |\Psi(q)|^2 dq < +\infty, \tag{2}$$

that is, $\Psi(q)$ must belong to the Hilbert space of function L_2. If, with the usual symbols, we consider the scalar product

$$(u,v) = \int_R u(q)\overline{v}(q) \; dq \; , \tag{3}$$

we can state the normalization condition

$$\int_R |\Psi(q)|^2 \, dq = 1 \tag{4}$$

At the foundation of the theory one poses the following axioms:

I. With a quantity a there is uniquely associated a self-adjoint operator A within a space $Q \subseteq H$, where H indicates a Hilbert space, and this is symbolically represented as follows:

$a \longleftrightarrow A$ in Q.

II. If $a \longleftrightarrow A$ in Q, then the mathematical expectation $E_\Psi a$ of a in the state Ψ, with module $\|\Psi\|=1$, is given by the scalar product $E_\Psi a=(A\Psi,\Psi)$.

One further introduces the concept of "dispersion" of the quantity a from the real value α in the state Ψ by means of the relationship

$$D_\Psi a = E_\Psi (a-\alpha)^2 = ((A-\alpha I)^2 \, \Psi, \Psi)$$

$$= ((A-\alpha I)\Psi, \; (A-\alpha I)\Psi) = \|(A-\alpha I)\Psi\|^2, \tag{5}$$

where one has $M_A \subseteq Q$ for the domain M_A of the operator A, and where I is the operator unity. If the quantity a in the state Ψ with $\|\Psi\|=1$ takes on a "precise" value $a=\alpha$, its dispersion from the real value α is null, one has consequently $D_\Psi a=0$ and can write:

$$\|(A-\alpha I) \, \Psi\| = 0 \tag{6}$$

$$A\Psi = \alpha\Psi. \tag{6'}$$

The number α thus is found to be an eigenvalue of the operator A in Q, and the converse is true.
   As far as the dispersion of the quantity a from its expected value is concerned, one finds immediately:

$$D_\Psi a = \|(A-\alpha I) \, \Psi\|^2 = ((A-\alpha I) \, \Psi, \; (A-\alpha I) \, \Psi =$$

$$= (A\Psi, A\Psi) - \alpha(\Psi, A\Psi) - \alpha(A\Psi, \Psi) + \alpha^2 (\Psi, \Psi) =$$

$$= (A\Psi, A\Psi) - 2\alpha \, (A\Psi, \Psi) + \alpha^2 =$$

$$= \|A\Psi\|^2 - 2\alpha \, (A\Psi, \Psi) + (A\Psi, \Psi)^2, \; [\alpha = E_\Psi a = (A\Psi, \Psi)],$$

thus, finally

$$D_\Psi a = \|A\Psi\|^2 - (A\Psi, \Psi)^2. \tag{7}$$

One has therefore the following proposition:
   If $A \subset Q$ and $B \subset R$ are symmetric operators, with domains $M_A \subset R$ and $M_B \subset Q$, respectively, and if the permutation (8) is valid,

$$AB\Psi - BA\Psi = \frac{h}{2\pi i}\ \Psi, \quad \forall \Psi \in Qu, \text{ with } \|\Psi\| = 1,$$

then the following fundamental inequality holds:

$$\|Au\|\ \|Bu\| \geq \frac{h}{4\pi}. \tag{9}$$

From this, the next proposition, which essentially states Heisenberg's uncertainty principle, ensures, since $D_\psi a = \|(A-\alpha I)\ \Psi\|^2$ and $D_\psi b = \|(A-\beta I)\ \Psi\|^2$, with α and β real numbers, that one has:

$$D_\psi a \cdot D_\psi b = \|(A-\alpha I)\ \Psi\|^2 \cdot \|(A.\beta I)\ \Psi\|^2 \geq \frac{h^2}{4\pi^2}. \tag{10}$$

Let us now be more precise with regard to energy and let us consider a particular physical system consisting of a m with (nonrelativistic) mass m. If it is subject to a force due to a time-independent potential $U(x) = U(x,x_2,x_8)$, then, if E denotes the total energy and $p_r = mx_r$ ($r = 1,2,3$) the r component of momentum, the Hamiltonian function $s(x,p)$, $[p = (p_1,p_2,p_3)]$ will satisfy the relationship

$$s(x,p) = \frac{1}{2m}\ (p_1^2 + p_2^2 + p_3^2) + U(x) = E. \tag{11}$$

If one formally substitutes for the Hamiltonian the operator S -let's call it Hamilton-Schrödinger operator- in the space where the operator is defined, then, substituting formally for the components p, p_2, p_3 of momentum the operators

$$P_r = \frac{h}{2\pi i}\ \frac{\partial}{\partial x_n}, \quad r = 1,2,3, \tag{12}$$

one finally obtains:

$$Su = \frac{1}{2m}\ (P_1^2\Psi + P_2^2\Psi + P_3^2\Psi) + U(x)\ u =$$

$$= \frac{1}{2m}\ \left(\frac{h}{2\pi i}\right)^2 \left[\Psi_{x_1,x_1} + \Psi_{x_2,x_2} + \Psi_{x_3,x_3}\right] + U(x)\ u = \tag{13}$$

$$= -\frac{h^2}{8\pi^2 m}\ \Delta_2\Psi + U(x)u,$$

where Δ_2 is the two-dimensional Laplacian.

One now takes a subspace $\zeta' \subset H$ such that the operator S in ζ' is symmetric and self-adjoint, and one considers the problem of the determination of eigenvalue of the functional equation

$$S\Psi = E\Psi \text{ with } \Psi \in \zeta', \tag{14}$$

keeping in mind that the relevant Hilbert space H is

$$H = \{\Psi(x) \int |\Psi(x)|^2 dx < \infty \}. \tag{15}$$

If one lets

$$\frac{8\pi^2 m}{h^2} E = \lambda, \quad \frac{8\pi^2 m}{h^2} U(x) = q(x), \tag{16}$$

substitution in (14) yields

$$A\Psi = \lambda\Psi \quad \text{with} \quad A\Psi = -\Delta_2\Psi + q(x)\Psi. \tag{17}$$

In the particular case in which

$$q(x) = q(x_1, x_2, x_3) = q_1(x_1) + q_2(x_2) + q_3(x_3), \tag{18}$$

assuming

$$-\Delta_2\Psi + q(x) \Psi = \Psi_1''(x_1) \Psi_2''(x_2) \Psi_3''(x_3), \tag{19}$$

one obtains

$$-\Delta_2 \Psi + q(x) \Psi = -\Psi_1''\Psi_2\Psi_3 - \Psi_1\Psi_2''\Psi_3 - \Psi_1\Psi_2\Psi_3'' +$$
$$+ (q_1+q_2+q_3) \Psi_1\Psi_2\Psi_3, \tag{20}$$

that is,

$$-\Delta_2 \Psi + q(x) \Psi = \lambda\Psi_1\Psi_2\Psi_3. \tag{21}$$

Now, if $\Psi_1 \neq 0$, $\Psi_2 \neq 0$, and $\Psi_3 \neq 0$, one has

$$-\frac{\Psi_1''}{\Psi_1} - \frac{\Psi_2''}{\Psi_2} - \frac{\Psi_3''}{\Psi_3} + q_1+q_2+q_3 = \lambda \tag{22}$$

And, because the quantities

$$(-\frac{\Psi_1''}{\Psi_1} + q_1), \quad (-\frac{\Psi_2''}{\Psi_2} + q_2), \quad (-\frac{\Psi_3''}{\Psi_3} + q_3)$$

are only functions of x_1, x_2, x_3, respectively, they must be each one separately equal to a constant λ_1, λ_2, λ_3, such that $(\lambda_1+\lambda_2+\lambda_3)=\lambda$. We thus can write:

$$-\Psi_1'' + q_1(x_1) \Psi_1 = \lambda_1\Psi_1, \quad \text{for} -\infty < x_1 < \infty, \tag{23}$$

$$- \Psi_2'' + q_2(x_2) \ \Psi_2 = \lambda_2 \Psi_2, \quad \text{for } - \infty < x_2 < \infty, \tag{24}$$

$$- \Psi_3'' + q_3(x_3) \ \Psi_3 = \lambda_3 \Psi_3, \quad \text{for } - \infty < x_3 < \infty, \tag{25}$$

that is, our problem is reduced to a Weyl-Stone eigenvalue problem.

(2) We will now synthetically recall the fundamental equations of quantum mechanics and further on discuss them in so far as they have opposed meanings in terms of the "orthodox" (Copenhagen school) and "realistic" views.

We will in the following adopt the following symbols: $\bar{r}$ = position vector, $\bar{p}$ = momentum vector, $H(\bar{r},\bar{p})$ = Hamiltonian operator, $\Psi(\bar{r},t)$ = wave function.

The equation of motion for a particle in nonrelativistic quantum mechanics is, in its Hamiltonian form,

$$H(\bar{r},\bar{p}) \ \Psi(\bar{r},t) - \frac{ih}{2\pi} \frac{\partial \Psi(\bar{r},t)}{\partial t} = 0 \tag{26}$$

The probability of finding, at time t, the micro-object under study in x $\longleftrightarrow$ x + dx, y $\longleftrightarrow$ y + dy, z $\longleftrightarrow$ z + dz is

$$P(\bar{r},t)d^3\bar{r} = \Psi \Psi^* d^3\bar{r} . \tag{27}$$

The probability distribution of the momentum, $|\phi(\bar{p},t)|^2$, is defined via the following Fourier transform:

$$\phi(\bar{p},t) = \frac{1}{h\sqrt{h}} \int_{V_\infty} \Psi(\bar{r},t) \ \exp \ (\frac{2\pi i}{h} \ \bar{p} \cdot \bar{r}) d^3\bar{r}, \tag{28}$$

where h is Planck's constant.

The Heisenberg uncertainty principle for microbjects (particles and photons) can thus be written:

$$\Delta x \cdot \Delta p_x \geq \frac{h}{4\pi}, \ \Delta y \cdot \Delta p_y \geq \frac{h}{4\pi}, \ \Delta z \cdot \Delta p_z \geq \frac{h}{4\pi}, \ \Delta E \ \Delta t \geq \frac{h}{4\pi} \tag{29}$$

where E is the energy and where the uncertainties are given by

$$\Delta x = \sqrt{\langle (\delta x)^2 \rangle} , \quad \Delta p_x = \sqrt{\langle (\delta p_x)^2 \rangle} , \tag{30}$$

Here

$$\langle x \rangle = \int_{V_\infty} x \ \Psi \Psi^* d^3\bar{r} \tag{31}$$

is the average value of x and $\langle (\delta x)^2 \rangle$ is the mean square deviation of x defined by

$$\langle (\delta x)^2 \rangle = \int_{V\infty} (x - \langle x \rangle)^2 \, \Psi \, \Psi^* d^3\bar{r} \; ; \tag{32}$$

furthermore,

$$\langle p_x \rangle = \int_{V\infty} p_x \, \phi\phi^* d^3\bar{p} \tag{33}$$

is the average value of momentum along the x axis and

$$\langle (\delta p_x)^2 \rangle = \int_{V\infty} (p_x - \langle p_x \rangle)^2 \, \phi\phi^* d^3\bar{p} \tag{34}$$

is the mean square deviation of p_x.

The same holds for the other coordinates. Before we proceed to a discussion related to the axiomatic framework of quantum mechanics, we recall the relativistic extension of the Schrödinger equation. The starting point for such extension is to be found in the relativistic spin-free wave equation for a force-free electron (scalar potential V=0, vector potential $\bar{A}$=0), that is, in the celebrated equation generally known as the Klein-Gordon equation

$$\Delta_2 \Psi - \frac{1}{c^2} \frac{\partial^2 \Psi}{\partial t^2} - \frac{4\pi m_o c^2}{h^2} \, \Psi = 0 \, , \tag{35}$$

where m_o is the "rest mass" of the electron. However, the most satisfactory extension is Dirac's theory, where one has four wave functions, $\Psi_1, \Psi_2, \Psi_3, \Psi_4$, and where the existence of the spin is taken into account. For the case of an electron in an electric field with scalar potential V and vector potential $\bar{A}$, Dirac's equations are:

$$(-P_4 + m_o c)\Psi_1 + (P_1 - iP_2)\Psi_4 + P_3\Psi_3 = 0,$$

$$(-P_4 + m_o c)\Psi_2 + (P_1 + iP_2)\Psi_3 - P_3\Psi_4 = 0, \tag{36}$$

$$(-P_4 + m_o c)\Psi_3 + (P_1 - iP_2)\Psi_2 + P_3\Psi_1 = 0,$$

$$(-P_4 - m_o c)\Psi_4 + (P_1 + iP_2)\Psi_1 - P_3\Psi_3 = 0,$$

where

$$P_k = \frac{h}{2\pi} \frac{\partial}{\partial x_k} + \frac{e}{c} A_k, \; k=1,2,3; \quad P_4 = -\frac{h}{2\pi i c} \frac{\partial}{\partial t} + \frac{e}{c}$$

3. The discussion of the definitive system of axioms on which the formulation of QM is based and through which it is classified can now take place. The axioms considered are the following four:

Axiom I. There exists a wave function Ψ and, for any "observable" A, Ψ

yields the state probability and contains all the possible information on the physical system considered.

Axiom II. For any observable a, there exists a Hermitean linear operator A <——> a, for which $(\eta, A\xi) = (A_\eta, \xi)$, such that the eigenvalues and the corresponding eigenfunctions of such operators are connected with the wave equation.

Axiom III. At t=0, Ψ is completely determined by a "maximum observation," that is, by an observation that yields the greatest number of compatible and independent observables.

Axiom IV. The changes in the state of a system with time are given by the equation of motion.

We must not forget the following important proposition: "A necessary and sufficient condition for the compatibility of two observables a and b is the commutation of the corresponding operators, that is, the fact that A <——> a and B <——> b satisfy the commutation relation

$$[A,B] = AB - BA = 0 \tag{37}$$

As far as the QM of N-particle systems is concerned, this relation is introduced in two ways. The first consists in the Schrödinger and -equation method. With $m_1, m_2, \ldots, m_N$ denoting the masses of the single particles, $U(x_1, y_1, z, \ldots \ldots x_N, y_N, z_N)$ the potential acting on the system, one has the wave equation

$$- \sum_{r=1}^{N} \frac{h^2}{8\pi^2 m_r} \Delta_2^{(r)} \Psi + U\Psi = \frac{ih}{2\pi} \frac{\partial \Psi}{\partial t} , \tag{38}$$

where $r = 1, 2, \ldots, N$ and $\Delta_2^{(r)}$ denotes the Laplacian for the rth particle. The probability amplitude dP that results is

$$dP = \Psi \Psi^* dx_1 \, dx_2 \ldots dx_N. \tag{39}$$

The second way we will mention here consists in the "wave field quantization" method of Born-Heisenberg-Jordan, also called the "second quantization method." Here the operator A_k^+ (particle creator) and A_k (particle destruction operator) for the quantum state k are introduced, and they satisfy the following commutation rules:

$$[A_k, A_{k'}^+] = \delta_k^{k'}, \quad [A_k, A_{k'}] = [A_k^+, A_{k'}^+] = 0 , \tag{40}$$

for bosons, and

$$[A_k, A_{k'}^+]_+ = \delta_k^{k'}, \quad [A_k, A_{k'}]_+ = [A_k^+, A_{k'}^+] = 0 , \tag{41}$$

for fermions.

According to the orthodox view of QM, typical of the Copenhagen school (Bohr, Born, Heisenberg, Pauli, Dirac and von Neumann), only a probabilistic interpretation of the wave function Ψ is possible (due to the exclusion of every model) because of the uncertainty principle, and *it is not possible* – by way of *hidden variables* – to restore the *deterministic* characteristic of classical mechanics.

On the opposite side, one leds "realistic" views which can be summarized first with Einstein's statement, according to which there exists an objective reality of the physical world, irrespective of the knowldege deriving from our observations. In this connection, one has a sufficient condition and a necessary condition for a realistic vision of the physical world, and they are intimately connected with what is generally referred to as "EPR argument."

The sufficient condition can be stated as follows: "*If, without in any way perturbing the system, one can predict with certainty the value of a physical quantity referred to it, then there truly exists an element of physical reality corresponding to such quantity.*"

The necessary condition is also called "completeness condition" and is as follows: "*Every element of physical reality must have a correspondence in the physical theory in which such reality is viewed and described.*"

Here the attitude of Einstein, Podolsky, and Rosen before QM is clearly outlined. According to it, the description of the physical world given by the wave function is "incomplete"; or, it happens that when two operators cannot be permuted the two corresponding physical quantities cannot have at the same time physical reality. The attitude of the authors quoted is supported also by and ideal experiment which is named "EPR" paradox, and which can be formulated essentially in terms of the following observation: A particle P of spin $s_x=0$, situated on the x axis, disintegrates and give rise, in the two symmetric positions a and b,to two particles, of spin $s=+1/2$ and $s=-1/2$ respectively. *Such values, according to Einstein-Podolsky-Rosen, cannot be considered as due to an "instantaneous" action across a distance, but rather must be viewed as pre-existing.* However it is possible to discuss the experiment ..... that, if one repeats the experiment a number of times, it will be possible to have two sets of results, namely E^+ and E^-, characterized by $s_z^{(a)}=1/2$, $s_z^{(b)}=-1/2$ in the former and by $s_z^{(a)}=-1/2$, $s_z^{(b)}=1/2$ in the latter; it must also be remembered that the z direction, along which the spin is evaluated, does not depend on the system and that exactly identical results could be obtained choosing the y direction.

One thus concludes that the direction of the angular momentum has no objective existence. The EPR argument appears therefore as a "paradox."

On the other hand, the paradox can be discussed also on the basis of the three following statements:

(a) The difficulty originates in the "incompleteness" of Quantum Mechanics, which actually is an "incomplete" theory according to Einstein's point of view.

(b) There is no paradox, since the physical system considered is "not separable."

(c) On the other hand, one cannot state the non-separability of the system, if one thinks about the possible interactions which the system might have undergone and still undergoes in time.

In this reasoning, and in general, the problem of the existence of "hidden variables," if any, in view of which determinism might be resurrected, is still present. The question is the following: *Do such hidden variables actually exist?* Here the theoretical framework take over and leads to the formulation of von Neumann's theorem. The author starts from the hypotheses:

I. There exist physical observables a $\longleftrightarrow$ A which cannot be further analyzed.

II. Since $\langle a \rangle$ is the average value of a, the law of linearity holds:

$$\langle \alpha a + \beta b \rangle = \alpha \langle a \rangle + \beta \langle b \rangle \ , \tag{42}$$

where α and β are two constants.

Once these hypotheses are accepted, von Neumann demonstrates the following theorem: *"There exist "dispersionless" systems, that is systems for which the variables undergo no fluctuations."* This is equivalent to state, essentially, that *no hidden variables can be found* in terms of which the physical determinism can be restored.

Next to the adamant stand taken by von Neumann and against the "orthodox" interpretation of QM, one has the new theory of hidden variables due to Bohm and Bub. In the latter, one considers a Hilbert space H_1 of the non-hidden variables c_r of the system and a dual Hilbert space H_2 of the hidden variables of the same b_r; let such spaces be such that the two respective wave functions are defined in terms of the following series developments of eigenfunctions:

$$\psi \overset{\text{def}}{=} \sum_r c_r \chi_r, \quad \zeta \overset{\text{def}}{=} \sum_s b_s \zeta_s \ . \tag{43}$$

One has then the differential time development equation of Bohm and Bub:

$$\frac{\partial c_r}{\partial t} = \frac{h}{2\pi i} \sum_s H_{rs} c_s + \lambda \sum_s J_s \left(\frac{J_r}{K_r} - \frac{J_s}{K_s} \right) \ , \tag{44}$$

where H_{rs} is the matrix element of the usual Hamiltonian operator and (45) holds:

$$J_s = |c_s|^2 \ , \quad K_s = |b_s|^2 \ ; \tag{45}$$

λ is a numerical coefficient connected of course nonlinearly, with the c_r and b_s coefficients. When $\lambda=0$ the equation of motion of Bohm and Bub is reduced to a Schrödinger-like equation:

$$\frac{\partial c_r}{\partial t} = \frac{h}{2\pi i} \sum_s H_{rs} c_s \ . \tag{46}$$

The theory of Bohm and Bub has a bearing on the theory of "measurement," and in particular one has:

(a) For a given time interval $\Delta t = \tau$ one has interaction between the measuring device and the system, so that the environment is modified. The λ parameter must be assumed to be large, so that the quantity

$$\frac{h}{2\pi i} \cdot \sum_s H_{rs} c_s$$

can be neglected in comparison.

(b) When $t > \tau$, i.e., after the interaction is switched off, one has $\lambda \sim 0$, and (44) tends to become (46), i.e., the theory of Bohm and Bub tends to Schrödinger's theory.

4. Other problems present themselves. *They are problems which have to do with the "violations" of QM, connected with the nearly classical Bell inequality and with other inequalities.*

Let us look in detail at the former, recalling Bell's theorem which states: "A local theory of hidden variables cannot reproduce all the statistical predictions of QM." However, if we apply the analysis to the "spin functions" which describe the state of the system of two particles through the singlet state wave function (rotational invariance), whe have:

According to quantum mechanics, if we call $\bar{\sigma}(1)$ and $\bar{\sigma}(2)$ the spins of the two particles, the probable value of the product of the related spin observables is

$$P_{qm}(\bar{a},\bar{b}) = \langle \bar{\sigma}(1) \cdot \bar{a}\bar{\sigma}(2) \cdot \bar{b} \rangle = -\bar{a} \cdot \bar{b} \ , \tag{47}$$

with $\bar{a}$ and $\bar{b}$ arbitrarily chosen unit vectors. By contrast, in the case of singlet states with direct correlation, i.e., $P(\bar{b}',b^-)=1$ for $\bar{b}'=-\bar{b}$, Bell states, considering three vectors $\bar{a},\bar{b},c^-$, the inequality

$$\left| P_{hv}(\bar{a},\bar{b}) - P_{hv}(\bar{a},\bar{c}) \right| \leq 1 + P_{hv}(\bar{b},\bar{c}) \ , \tag{48}$$

where the subscript hv stands for "hidden variables" and qm for quantum mechanics.

From Bell's inequality one derives the following for small values of $|\bar{b}-\bar{c}|$; one has

$$\left| P_{hv}(\bar{a},\bar{b}) - P_{hv}(\bar{a},\bar{c}) \right| \sim \left| \bar{b}-\bar{c} \right|,$$

whence, under the said conditions,

$$P_{hv}(\bar{b},\bar{c}) \geq \left| \bar{b}-\bar{c} \right| - 1,$$

which contradicts the equality:

$$P_{qm}(\bar{b},\bar{c}) = - \bar{b} \cdot \bar{c} = - 1;$$

(-1 = stationary value, minimum), which holds in quantum mechanics. In the case of "imperfect correlation," i.e., when

$$P(\bar{b}',\bar{b}) = 1-\delta \; , \; (0 \leq \delta \leq 1, \; \Psi\bar{b}', \; \Psi\bar{b}) \tag{49}$$

(notice that when the correlation is good, δ is almost zero), the following inequality has been formulated by Clauser, Horn, Shimony, and Holt:

$$\left| P(\bar{a},\bar{b}) - P(\bar{a},\bar{c}) \right| + P(\bar{b}',\bar{b}) + P(\bar{b}',\bar{c}) \leq 2 \tag{50}$$

Other inequalities have been formulated by Selleri, for instance, the following:

$$\left| P(\bar{a},\bar{b}) - P(\bar{a},\bar{c}) \right| + \left| P(\bar{d},\bar{b}) - P(\bar{d},\bar{c}) \right| \leq 2 \; . \tag{51}$$

Selleri, Capasso, and Fortunato have also examined the problem of establishing for what state functions, among which the singlet states considered by Bell are particular instances, inequalities (48) and (51) can be violated through a convenient choice of state vectors.

5. Now, among *non orthodox theories providing an interpretation* of quantum mechanics, also the *stochastic interpretation, originated in Fürth's 1933 statement, according to which Schrödinger's equation of motion is analogous* to Smolukowski's equation for Brownian motion. Later, in 1952, Fényes showed how Schrödinger's equation and Heisenberg's uncertainty principle can be considered analogous to the ones ruling Markoffian stochastic processes and stated, in opposition to the orthodox interpretation of quantum mechanics, that Heisenberg's uncertainty relations had to be considered *independent* of the perturbation exerted from observations and measurements on the particle. However, Fényes himself did not believe that Schrödinger's equation could be related to the existence of hidden variables[1].

Another theory that deserves to be mentioned is the one due to Nelson and L. de la Pena-Auerbach.

Starting from a "subquantal" viewpoint, one hypothezises that quantal phenomena are the result of stochastic interaction which occurrs between the system investigated and the whole surrounding medium. Under such circumstances it can be demonstrated that Schrödinger's time dependent equation essentially "generates" a system of two differential equations. The first of them - which was also deduced by Bohm - is a Hamilton-Jacobi type of equation. The second one is essentially an equation of Brownian motion and concerns the probability density $\rho = \Psi\Psi^*$. It must be noticed that Nelson and de la Pena-Auerbach have utilized the stochastic formalism to study the problem of a particle moving inside an electromagnetic field, and they have shown that the overall force acting on the particle is not only consisting of a Lorentz' term, but contains terms of a stochastic nature as well.

De la Pena-Auerbach and coworkers have elaborated an interesting theory of motion for the satellite electron within the hydrogen atom and have

---

(1) Another contribution within the stochastic framework is due to Weizel's theory of "zerous," which appears difficult to accept.

shown that it is just one of the terms of stochastic nature, if the intrinsic force field of the moving electron is considered, which is responsible for the largest contribution to the shift of quantum S levels of the hydrogen atom. It must has been extended by these authors to the case of a system of more interacting particles and to the case of spin-endowed particles, as well as to the relativistic treatment of the motion of spin 1/2 particles in which case the Dirac equation was obtained, and finally to the similar problem of particle of arbitrary spin, where the equation of Feynman and Gell-Mann resulted.

The formalism of the stochastic theory of L. de la Peña-Auerbach is based on the following layout. Let x(t) be a set of casual variables that define the stochastic description of the motion of one point in configuration space. The probability density $\rho(x,t)$ in x(t), an essentially positive quantity, can be represented in the form:

$$\rho = e^{2R} , \quad R = R(x,t) \text{ a real function.} \tag{52}$$

The conservation of total probability is expressed by a continuity equation

$$\frac{\partial \rho}{\partial t} + \text{div } \bar{j} = 0 , \tag{53}$$

which is transformed, through straight forward substitutions, into

$$\frac{\partial R}{\partial t} = - \frac{1}{2} (\nabla \cdot v) - (v \cdot \nabla) R . \tag{54}$$

Letting $v = \alpha \nabla S$, with α constant and $S = S(x,t)$ a real function, and letting

$$\psi = e^{(R+iS)}, \tag{55}$$

we obtain the following differential equation for the wave function ψ (such that Born's probability amplitude is $\rho = |\psi \psi^*|$):

$$\frac{\partial \psi}{\partial t} - i \frac{\partial S}{\partial t} = \frac{i\alpha}{2} \Delta_2 \psi - \frac{i\alpha}{2} \left[\Delta_2 R + (\nabla R)^2 - (\nabla S)^2 \right] \psi . \tag{56}$$

Now, introducing the function V defined by

$$V = - \frac{\partial S}{\partial t} + \frac{1}{2} \alpha \left[\Delta_2 R + (\nabla R)^2 - (\nabla s)^2 \right], \tag{57}$$

The Schrödinger's equation of motion results:

$$i \frac{\partial \Psi}{\partial t} = - \frac{\alpha}{2} \Delta_2 \Psi + V\Psi, \quad \alpha = \frac{1}{2\pi m} \tag{58}$$

In other words, Schrödinger's equation derives from two hypotheses which are generally assumed in order to obtain the differential equations of Markoffian processes.

6. Let us now consider the quasi-classical theory of radiation (Jaynes and coworkers). The solution of Schrödinger's equation, in its spinless nonrelativistic version, is obtained starting from an Hamiltonian function

$$H = H_o + W , \tag{59}$$

where H_o is the term describing the atomic field in the absence of any electromagnetic field and W is the interaction term due to the presence of fields. In the case of the radiating electron it assumed that the electromagnetic field acting on the electron is made up of two parts: the first consists of the field of the impinging radiation, the second of the intrinsic field generated by the electron investigated. This intrinsic field is the most important characteristic of Jaynes' theory and its calculation containes a view of the electron as a space distribution of electric charge of density $\rho \propto \Psi\Psi*$, i.e., a return, in a sense, to the original interpretation given by Schrödinger in the wave function Ψ. The difficulties related to such interpretation reappear also. What is worse, in Jaynes' theory the commutation rules for the electron do not appear to be conserved with time. Actually, the spontaneous emission theory originated from Jaynes' line of thinking shows violations of Heisenberg's uncertainty relations although the Schrödinger wave equation is assumed to hold.

Other difficulties are still present, and many authors draw attention to them; nevertheless, presently quantum mechanics appears to be dominating the scene as theory for the description of the microobjects of the physical world.

7. A peculiar viewpoint, which can lead to natural philosophy types of interpretation of quantum mechanics and is at variance with that of the orthodox Copenhagen school, is one which can be identified as the "statistical interpretation." It takes its lead essentially from Einstein's thinking, and has evolved to a large extent within the Russian scientific milieu (a reflection also of their adverse stand towards idealistic philosophical theories) and eventually merges in with the line of thinking which encompasses personalities such as Popper and Landé. We will briefly consider a few essential aspects of this doctrine. In the URSS school, a fundamental work is undoubtedly Blokhinzev's *Foundations of Statistical Mechanics*. The author strictly rejects the "orthodox" interpretation given by the Copenhagen school and gives a statistical interpretation of quantum mechanics.

According to Blokhinzev, modern quantum mechanics is not a theory of "micro-processes" but rather something which studies their properties on the basis of statistical ensembles, which are in turn described in terms of concepts, as it were, "borrowed" from macroscopic classical physics, such as energy, momentum, coordinate. The measuring devices are interpreted as "spectral analyzers" of quantal systems: analyzers which select, out of the ensemble given, certain sub-ensembles, in agreement with the nature of the instrument, or subdivide an ensemble (pure state) into a "mixture of sub-ensembles" (state mixture). To assign a new wave function to one of the subensembles quoted amounts to what is usually called reduction of the wavepacket. And from a physical point of view, it must be remarked that such "reduction" belongs, after the measurement is taken, to a new, pure ensemble. Blokhinzev further clarifies his statistical interpretation. He defines ensembles as infinite sequences of identical micro-systems, each of which belongs to the same ensemble of macroscopic objects. The "microenvironment" somehow determines the state of motion or, simply, the state of the micro-system. The system is considered quantal whenever the uncertainty relation between the Lagrangian coordinates q and the conjugate momenta p, taken as an average over the ensemble, is satisfied. The Bohr complementarity principle is considered by Blokhinzev simply as an "exclusiveness" principle in the sense that the dynamical variables are subdivided in mutually exclusive group that cannot coexist in quantal systems. There is, nevertheless, a quantity which completely characterizes the quantal system, in the sense that its knowledge allows us to calculate all the different probability *measurements*. This is the wave function Ψ_M, which describes the quantal ensemble both in terms of coordinates and of conjugate momenta. A point of view analogous to Blokhinzev's is illustrated by L. I. Mandelstam. By contrast, an opposite stand is taken by Blockhinzev since, according to him, the wave function represents a something which is potentially possible but not actually realized (the language is clearly Aristotelian). In V.A. Fock's view, the concept of statistical ensemble in quantum mechanics is acceptable up to the point where it refers to the results of measurements, wince every ensemble is associated with a specific experimental device rather than with the micro-objects themselves.

8. The epistemological positions of Karl Popper and Alfred Landé must certainly be taken into account. The former states first that certain relationships of quantum mechanics hold over certain intervals of "distribution" or "variancy" or "dispersion," and calls them "statistical distribution relations." Popper also states that measurements having an higher degree of accuracy than allowed by the Heisenberg uncertainty principle are *not* incompatible with the system of formulae of quantum mechanics itself or with the statistical interpretation of this body of doctrine. If it were possible to take measurements with a higher precision than allowed for by the uncertainty principle, quantum mechanics would not be invalidated; this would rather imply that the precision limits stated by Heisenberg constitute a separate or added assumption, rather than a logical consequence of the quantal mathematical formalism; actually, according to Popper, the added assumption of Heinseberg *in fact contradicts* the formulae of quantum mechanics whenever they are interpreted in statistical terms.

As far as causality goes, Popper writes verbatim that he "dissents from indeterministic methaphysics". He adds, too, "what distinguishes it from determinist metaphysics is not so much its greater lucidity, but rather its greater sterility."

Let us turn now to Landé. In his imposing scientific production, he constructs a "unitary interpretation of the particle," which is at variance both with Schrödinger's wavelike unitary interpretation and with Bohr's dualistic wave-particle interpretation. Landé rejects the theory of hidden variables and denies that the microphysical probability can be reconducted to classical determinism and seeks a foundation for its theory in what could be termed the "principle of cause-effect continuity." Landé states verbatim that the "uncertainty due to discontinuous transitions among the states of an object, the heart of quantum theory, is truly the necessary counterpart of a principle of wide generality, the principle of continuity for the deterministic cause-effect relations."

After the comparative consideration of the most important positions concerning the great debate on the interpretation of the wave-function , we will try to construct an objective view of the theory of micro-objects, based on a clear ontological distinction between the categories of causality and physical interpretation of the differential equations of quantum mechanics, and will conclude our work with the proposal of a *metaquantal theory*.

PART II. AN OBJECTIVISTIC-ONTOLOGICAL INTERPRETATION OF QUANTUM MECHANICS

(1) From the foundation and the development of quantum mechanics in the orthodox sense of the Copenhagen school, the following characteristics of the doctrine stand out clearly:
(a) the theory has a postulate-type of character, i.e., it is a hypothetic-deductive body of doctrine, the way mathematical theories are;
(b) quantum mechanics itself has an operative character;
(c) the fundamental quantities of quantum mechanics, above all the wave functions, have a probabilistic, rather than deterministic, character.
It must be added that the logical tool of quantum mechanics itself is three-value logic. Let us keep in mind that, in classical two-value logic, one usually defines as "logical formula" a combination of propositions which, for every value of truth of elementary propositions, has a "truth value". Such a formula is called a "tautology." It is necessarily true and, on the other hand, void since it does not supply us with any information on the values of the truth of any elementary proposition. The value of tautologies consists exactly in their being void, beside being necessary.

The tautological formulas can always be added to physical statements; indeed they don't add any empirical content, and it is necessary to add them if one wants physical statements to have consequences. Essentially, the physicist finds in the construction of

tautologies a very powerful tool for the derivation of new consequences
and new statements. Mathematics must be considered within the domain of
tautologies. The principle of identity (a = a), the rule of double
negation ($\bar{\bar{a}}$ = a), the rule "tertium non datur" (a ∪ $\bar{a}$), and the
principle of contradiction (a ∩ $\bar{a}$) must be viewed as tautologies.

The formulae of two-value logic can easily be verified by analyzing
the single cases. Every formula which contains both values "true" and
"false" is called a "synthetic formula." A formula which, for a given
proposition, only contains values "false," that is, a formula which is
invariably false, is a "contradiction." A synthetic formula states an
empirical truth, and it must be remembered that every physical law and
every physical statement are synthetic.

A few strongholds of quantum physics can be interpreted on the basis of
a "three-value" logic (true, false, undetermined). The construction of a
three-value logic is based on the idea according to which there exists a
"meta-language" of the language considered, one which can be viewed as
belonging to a two-value logic. The operations of three-value logic are
a generalization of those of two-value logic, and they have been
introduced by E.L. Post (with the exception of the "complete negation,"
of the "alternative implication," and of the "alternative equivalence,"
which are originated for the very requirements of quantum mechanics.

One must remember this proposition: What is required in the first place
of an implication that it allows the process of "inference," which is
represented by the following rule: if A is true and A implies B is true,
then B is true. Now, in three-value logic, one has three implications:
the standard implication, the alternative implication, and the
quasi-implication, and all three satisfy the rule stated above.

The fundamental point of quantum mechanics where three-value logic is
necessarily introduced, is the Heisenberg uncertainty principle. This is
a fundamental physical law which is valid for all possible physical
states. It implies - as we have already recalled above - an essential
perturbation of the object, due to the measurement. What happens is the
following: Statements on simultaneous value of complementary quantities,
such as position coordinates and corresponding kinetic momenta of a
particle, can be introduced only by means of definitions. In the
physical world there appear therefore, first of all, the physical
phenomena, which can be "inferred" from observations; there further
appear *inter-phenomena*, which can be introduced only by means of an
interpolation based on definitions. And one has a fundamental law, named
the "principle of anomaly." On its basis, the integration of the world
of phenomena by means of the world of inter-phenomena cannot be given
without incurring anomalies.

Rather than talking about "structures," one is led to talk about
the "languages" through which such world can be described. One has then
the "language of observation" and the "quantal language." The former
does not present anomalies, except for a few non-verifiable
implications. The latter is presented in the corpuscle version, the wave
version, and the neutral version, respectively, and again involves the
so-called inter-phenomena.

Now, the corpuscle- and wave-like versions of the quantum language
contain statements of "causal eliminable anomalies" since they do not

appear at mutually corresponding places. The "neutral" language, which is neither corpuscle-like nor wave-like, contains "causal anomalies" in connection with the fact that the statements concerning inter-phenomena have as truth value the third logical value (of three-value logic), that is, the undetermined value. This "linguistic" aspect of indeterminacy in the structure of the physical world must be considered in the light of the rules which "hold together" - so to speak - the proposition of three-value logic. But the observation we want to point out is the fact that quantum mechanics is essentially *a system founded on postulates*.

2. We want to also stress a few other points. It is well known that de Broglie and Vigier were involved, a few years later, in intersting attempts to construct the theory of the "double solution", which has since been abandoned. That theory was supposed to allow a reconciliation of probabilistic views with an objective and "causal" view of the wave-corpuscle dualism. This aim was pursued with a new conception, in which the structure of the system considered was seen as described by a wave having a singularity, as opposed to the axiomatic quantal feature in which the wave function is characterized essentially by the fact of *not having* singularities. Thus, before the double solution theory, quantum axiomatic yields a vision which is "globally" equivalent to a deeper-rooted theory (so to speak).

The second observation is the following. The statistical interpretation (such as, for instance, the one of Blokhinzev quoted above) presents itself as having a concept of the wave function that is representative of a "collection" of similar systems, rather than identical systems, which amounts to assuring an "underlying" theory of the system which we will call "individual." In other words, the wave functions represent a "collective object, an ensemble of analogous systems, the differences among which are not detectable with our observational means (measuring devices?), but which are sufficient to justify the different behavior of such systems in further experiments.

In this connection, we must carefully rethink the deduction of the wave function effected by Feynman. Here the wave function expresses the probability of finding the system in a given time and in a given space configuration such as to appear as the results of all the possibilities to reach them via different processes.

The statistical interpretation of quantum axiomatics agrees with the fact that experimental possibilities are limited (this point has been utterly overlooked in the arguments in favor of the purely probabilistic vision of the mechanics of quanta, for which the uncertainty has been viewed for a long time as essential and deep-rooted). In other words, apparatuses represent an observation system with "limitations" in its capability of "distinguishing" and of "separating." They leave in a apparently identical status systems which are only similar and which can be represented by the same wave function in the model at hand. It must be kept in mind that the "limitation" is present at all "levels" of investigation of the physical world. When one chooses the level of quantum mechanics, one is confronted with what is termed the "ultimate interaction" (which is the more "violent" the "closer" it is) between the system observed and the observation device, in line with Heisenberg's uncertainty principle and Bohr's complementarity principle. This so called "ultimate interaction" is the reason for viewing probabilism as the definitive stage of knowledge of micro-objects. We stress: *definite stage of knowledge*. This is the place to point out that the non-determination characteristic of quantum mechanics concerns the *gnoseological* aspect, while it does not contradict causality, which is typical of the *ontological world*, i.e., of *being*.

Indeed the wave function, with its statistical character, represents the status of *our knowledge* on the system, rather than the status of the system itself. We must, furthermore, remember that theory of micro-objects which causes, so to speak, determinism to itself to a space time description. One considers "causality" as absorbed within the relativistic space time so that the usual quantities of the three-dimensional world are like optical illusions. This "absorption" (if we may say so) is based on the belief that what is responsible for the uncertainty is the "Weltanschaung", which is typical of non-relativistic physics. An attempt in this direction (see M. Pierucci) is in line with the mechanistic, i.e. deterministic and causal character, of Einstein's thinking.

At this point it might be worthwhile to re-examine the stand taken by de Broglie, Bohm, and Vigier. The micro-objects are invariably, at one and the same time, waves and corpuscles. One defines the classes of motion related to the corpuscular part of micro-objects, and one considers their corpuscular aspects as a "singularity region" within the continuous solution of the usual equations. This point of view, which involves both the nonrelativistic and relativistic domains, is in the latter outlined as follow: The starting point are Einstein's field equations

$$R_{\alpha\beta} - \frac{1}{2} R g_{\alpha\beta} = -8\pi \frac{\gamma}{c^2} T_{\alpha\beta} \tag{1}$$

where all symbols have their usual meanings, and it can be demonstrated that a particular solution of (1) can be found, such that the "particle singularity" is associated with an extended field of its own. This is responsible for the fact that the particle must follow the trajectories predicted by the causal theory. The statistical behavior of the ensembles of micro-objects in causal theory is, in short, the following: If we assume that the quantal ensembles which are to be found in a given state are defined by the identity of their own real extended fields ϕ, then the fundamental Bohm-Vigier stochastic proposition holds: "The conservation of the proper extended field in the middle of the fluctiations due to external conditions causes the singularities to distribute themselves with the density predicted by the probabilistic interpretation, i.e.

$$|\psi|^2 = K |\phi|^2, \quad K \text{ constant.} \tag{2}$$

This is equivalent to the statement that the "density" of the

corpuscular aspects of a real statistical ensemble of micro-objects
tends to the probabilistic distribution. As concerns the systems
containing N interacting particles, which involve N corpuscular aspects
and N waves propagating in real space, they can be represented by means
of fictious wave — which satisfies Schrödinger equation — in
configuration space. The square of the module of this fictious wave
function yields the statistical distribution imposed by the
probabilistic interpretation.

As far as the problem of "measurement" is concerned, David Bohm
shows that the well known occurrence of discrete eigenvalues is due to
the effective interaction between the instruments utilized and the
"causal" model.

3. At this point we want to emphasize that a causal vision of the
world of micro-objects, that is a causal interpretation of the mechanics
of quanta, an interpretation which concerns essentially the ontological
world rather than the gnoseological one, is not necessarily connected
with a concept. Let us try to justify our statement. First of all, it
must always be remembered, with a sort of detachment, if you will, that
quantum mechanics is a *system of postulates*. Such system, as we recalled
in the first section, declares to begin with the existence of the wave
function, endowed with a few essential properties; the existence of a
linear Hermitean operator A <——> a, corresponding to any observable a,
the connection between the two being provided by the wave function
itself; it also states that ψ at t=0 is completely determined by a
"maximal observation," and, finally, that the vicissitudes in time of
the state of the system are described by the fundamental time evolution
equation. Furthermore, one must not forget the proposition of necessity
and sufficiency for the compatibility of two physical observations,
namely the commutation of the respective operators.
Now, within the system of axioms of quantum mechanics one encounters the
problem of *non-contradictoriness* and *completeness*, in exactly the
Hilbert connotation. When constructing any mathematical theory, the
requirements one places on every system of axioms are the following:
(a) the independence of every postulate,
(b) their completeness, and
(c) their non-contradictoriness.
The requirement of independence implies that it must not be
possible to demonstrate one of the axioms starting from the others.
Hence it is a fundamental principle not to insert in axiomatic
foundations anything that is not strictly necessary. Now it is well
known that it is possible to demonstrate *the independence* of an axiom
from the remaining propositions of a system (by means of the
construction of appropriate models, such as Klein's model in geometry).
Let us now come to *non-contradictoriness*. The latter, in geometry, is
reduced both to the one of arithmetic. Analytical geometry associates
with the points of plane geometry the real number pairs (x, y)
(coordinates); it associate straight lines with linear equations, and so
on. Now only two alternatives are possible: Either geometry is void of
contradictions and then - according to Hilbert - so is arithmetic. The
non-contradictoriness of geometry is thus proved in a "relative" way.

As far as completeness is concerned, we will say this much: The axioms of a theory are said to possess completeness if all the truths of that doctrine can be deduced from the axiomas. As far as arithmetic is concerned, Kurt Gödel did the following:

(a) He demonstrated that it is possible to construct an arithmetic formula G which represents the meta-mathematical proposition: "the G formula is not demonstrable." Such formula thus states its own undermonstrability.

(b) He proved that G is demonstrable if, and only if, its formal negation ~ G is demonstrable; hence, if one formula and its formal negation are both formally demonstrable, arithmetical calculus is contradictory. Due to this situation, if calculus is to be self-consistent, neither G nor ~ G are deducible from the axioms of Arithmetics.

(c) He himself also proved that, although G is not formally demonstrable, nevertheless it is a "true" arithmetic formula.

(d) Hence, since G is true and at the same time formally undecidable the axioms of arithmetics are not complete and furthermore arithmetics is essentially incomplete.

(e) Gödel described the way one must construct an arithmetic formula A which represents the meta-mathematical proposition: "Arithmetic is self-compatible," and he showed that A is *non-demonstrable*. Thus the self compatibility of arithmetic cannot be assessed with arguments representable in formal arithmetical calculus.

In summary, if S is an arithmetic system of postulates and is non-contradictory, then the non-contradictoriness of such system S cannot be demonstrated with the means of the system itself.

More generally, the following statement gains credit: Just as the whole of mathematics consists of systems of postulates, and the compatibility of the systems of postulates concerned with it cannot be demonstrated from the inside of the systems themselves, thus quantum mechanics is *a body of doctrine essentially based on postulates. Therefore it is necessary, also for quantum mechanics, to gain access to a meta-quantum theory; i.e., to an external (and more extended) system, in order to check the non-contradictoriness, the compatibility, the very completeness of the postulates of the doctrine, i.e., of the postulate concerning the wave functions* Ψ.

This "meta quantal space", where the axiomatic system of quantum mechanics must be observed in depth, can lend itself to the "accusation" of being a metaphysical concept. This should not frighten us. Perhaps (this is a fundamental question) even the forced coincidence of the categories of determination and causality, which occurs in the "ultimate-orthodoxy" of von Neumann, depends on the fact that one does not take into account the wider meta-quantal system, which contains the quantal system, which amounts essentially to taking a stand in opposition to accepting the greater amplitude of natural-philosopy thinking, and especially philosophical thinking, than physical thinking.

Here the profound need for an interrelation suitable for the construction of a consistent theory of the hidden structures, of which phenomena are the outer manifestation, appears evident.

We will not ignore *the question, posed by Jean Dieudonné, of the*

*discussion of whether metamathematics itself is non-contradictory or
not*, which appears to be critical example pursued to its extremes.
That's all right. But the following problem now presents itself: Either
the critique is brought to the point where one accepts an infinite
succession of searches for "environments" of demonstration of
non-contradictoriness of "preceding systems" contained in "successive
systems" (and here the limitation of quantum mechanics appears very
clear), or one accepts to go back to a more serene consideration of a
few intuitive evidences which give back some balance to the exasperated
thinking itself. Beyond these evidences and equilibrium positions, one
encounters a region of continuous doubt on rationality or even of
negation of the same, which is actually caused by a supreme, desperate
attempt toward, so to say, "total" rationality. The meta-quantal
position, independent of any possible formulation, is essentially a
general philosophical vision of the problem. The following
considerations present themselves. Any neo-deterministic vision of the
mechanics of elementary particles requires much more than a causalistic
vision. *Indeed causalism, that is the ontological statement of the
necessity of existence of a sufficient reason for anything that happen*
(here it does not make sense to distinguish between the macroscopic
world and the microscopic world), *is invariant even before a solution of
the problem of the interpretation of quantum mechanics in a
probabilistic sense.*
 The probabilistic meaning of Ψ and of the probability density $|\Psi|^2$
is a question that concerns essentially the world of knowledge, which is
different from that of being, best it becomes impossible to construct a
science valid for all scientists. And in the gnoseological world, a
distinction between the macroscopic world and the microscopic world has
meaning and full substance.
 In this sense, what clearly emerges is an *epistemological
proposition* such as follows: *causalism concerns positions and ways of
thinking which are not related to physical determinism by mutual
dependence. The two categories causalism and determinism* (hence
probabilism excludes determinism) *are related, the first to the
ontological world and the second to the gnoseological world.* (It must be
noticed that quantal indeterminism arises from the inadequacy of a
model, namely the corpuscular model). But we must pursue our analysis
further. Let us reconsider the fundamental equation of quantum
mechanics, namely the equation of motion in the Hamiltonian form:

$$H(\bar{r},p)\ \Psi(\bar{r},t) = \frac{ih}{2\pi}\ \frac{\partial \Psi(\bar{r},t)}{\partial t}. \tag{3}$$

Let us first remark that no restriction of validity is placed on the
following reasoning by considering the nonrelativistic version of the
equation, since the category "causalism" is not altered when going from
three-dimensional space to relativistic space-time. Actually,
relativistic theories are typically causal, so it is important to stress
the value of causality in ordinary space, in classical Euclidean
geometry, in our opinion, *rather than* in relativistic space-time. This

being taken into account, it appears fundamental that a differential equation such as (3) establishes a relation, the essence of which is that it naturally expresses a causality principle.

If the meaning of the solutions of the differential equation considered is probabilistic at a gnoseological level, it is causal at an ontological level, since causality implies a connection between probabilistic distributions. *And it appears more than a conjecture to think that, in correspondence with a cause identically equal to zero, every hypothetical state probability distribution at time t cannot be different from the probabilistic distribution of the initial time* (unless one accepts the creation of a phenomenon as a spontaneous effect). Consequently we state:

I. In the correspondence represented by a differential equation - or by an equivalent relation - to a *void ensemble of cause*, there must correspond a *void ensemble of effects*.

II. Between initial and final probabilistic distributions there exists, because of the wave equation, a *causal connection*.

III. The indeterminism expressed by Heisenberg's principle is *not* a condition suitable for the negation of causalism; however causalism (on the ontological level) and indeterminism (on the gnoseological level) are *compatible*, and the uncertainty is expressed by the meaning of $|\psi|^2$ according which $|\psi|^2 dS = dP$, where dP is the probability of finding the micro-object considered in the space element dS, and one has $\int_S |\psi|^2 dS = 1$, i.e., the normalization condition for ψ expresses the certainty of finding the micro-object in the whole space S. (If someone should object that the state observed as final state could actually be other than that, one can retort that, if that happens, then the final state can be subject to ordering inversion. *What we mean is that one may envisage a transformation of causalism into finalism*).

IV. Questions such as the EPR paradox or relations like Bell's inequalities do not appear by themselves apt to modify the probabilistic interpretation of ψ in a gnoseological sense, that is, they *cannot* shake the foundations of quantum mechanics. We stress again that the question of causalism is fundamentally an ontological question.

V. The choice of a philosophical (environment also cuts short the question of the (potentially infinite) succession of searches for doctrine schemes suitable for demonstration of the non-contradictoriness of *preceding* systems and of *contained* systems.

VI. The choice of this philosophical "milieu" also causes the interruption of what could bécome a chain of searches for "metalanguages." Actually, from the point of view of the philosophy of language, one finds oneself within those systems in which causality is considered an indestructible, indispensable category.

VII. *Inter-phenomena* belong then in the ordinary class of *phenomena*.

VIII. Three-valued logic becomes two-valued Aristotelian logic when· I $\longleftrightarrow$ F or T $\longleftrightarrow$ F, that is, when the value of uncertainty becomes the value of truth or value of falsehood. Two-valued logic can indeed be considered as a particular instance of three-value logic or, more generally, of many-value logics. It is formally so. But we must remark the following: Starting from the fact that probabilism, as mentioned

above, must be considered as an essentially gnoseological category, it
is the very *two-valued logic* (where the value I disappears), the one
corresponding to the *ontological world*, which containes the category of
causality.

   IX. The identification of ontological world and gnoseological world
is *not admissible*, due to the general, unequivocal validity of the laws
of nature.

BIBLIOGRAPHY

J. von Neumann, *Mathematische Grundlagen der Quantenmechanik* Springer,
Berlin, 1932; "Allgemeine Eigenwerttheorie Hermitescher
Funktionaloperatoren," *Math. Ann.*, $\underline{102}$, 49 (1929).
E. C. Kemble, *The Fundamental Principles of Quantum Mechanics* (Mc Graw
Hill, New York, 1937).
G. Ludwig, *Die Grundlagen der Quantenmechanik* (Springer, Berlin, 1954).
E. Wienholtz, "Bemerkungen über elliptische Differentialoperatoren",
Arch. Math. $\underline{10}$ *(1959)*.
E.C. Titchmarsh, *Eigenfunctions Expansions Associated with Second-order
Differential Equations*. II. (Oxford, 1958).
G. Hellwig, *Differentialoperatoren der mathematischen Physik* (Berlin,
1964).
S. Bell, "On the Einstein-Podolsky-Rosen paradox", *Physics* $\underline{1}$, 195
(1964).
F. Clauser, M.A. Horne, A. Shimony, and R.A. Holt, "Proposed experiment
to test local hidden-variables theories", *Phys. Rev. Lett* $\underline{23}$, 880
(1969).
F. Selleri, "A stronger form of Bell's inequality", *Lett. Nuovo Cimento*,
3, 581 (1972).
V. Capasso, D. Fortunato, and F. Selleri, "Sensitive observables of
quantum mechanics", *Int. J. Theor. Phys.*, $\underline{7}$, 319 (1973).
M. Jammer, *The Philosophy of Quantum Mechanics* (Wiley, New York, 1974).
E. Nelson, "Derivation of the Schrödinger Equation from Newtonian
Mechanics", *Phys. Rev.* $\underline{150}$, 1079 (1966); *Dynamical Theories of Brownian
Motion* (Princeton University Press, Princeton, 1967).
L. de la Peña-Auerbach, "A simple derivation óf the Schrödinger equation
from the theory of Markov processes", *Phys. Lett.* $\underline{24A}$, 603 (1967); "New
formulation of stochastic theory and quantum mechanics", *ibid.* $\underline{27A}$, 594
(1968).
L. de la Peña-Auerbach, E. Braun, L.S. Garcia-Colesi, "Possible
interpretation of quantum mechanics", *J. Math. Phys.* $\underline{9}$, 916 (1968):
"On the generalized Schrödinger equation", *Rev. mex. Fis.* $\underline{16}$, 221
(1967): "A new formulation of stochastic theory and quantum mechanics",
ibid., $\underline{17}$, 327 (1968).
L. de la Peña-Auerbach, "New formulation of stochastic theory and
quantum mechanics", *J. Math. Phys* $\underline{10}$, 1620 (1969): "Stochastic quantum
mechanics for particles with spin", *Phys. Lett.* $\underline{31A}$, 403 (1970).
L. de la Peña-Auerbach and M.A. Cetto, "Self interaction in a
nonrelativistic stochastic theory of quantum mechanics", *Phys. Rev.* $\underline{D3}$,
795 (1971).
P. Caldirola, *Dalla Microfisica alla Macrofisica*, 2nd ed., EST, Milano

(1977).

D.I. Blokhintsev, *Osnovy Kvantovoi Mekhaniki* GITTL, Moscow (1949); "Printsipalnye Voprossy Kvontovoi Mekhaniki (Nauka, Moscow, 1966); *The Philosophy of Quantum Mechanics* (Reidel, Dordrecht, 1968).

L.I. Mandelstam, *Polnoje Sovrannje Trudov*, V, 356 (1950) (Academy of Science USSR, Moscow.

V.A. Fock, "0 tak jaryvajemykh asambljakhv kvantovoi mekhaniki", *Voprossy Filosofii*, 170 (1952).

K. Popper, *Logica della Scoperta Scientifica*, tr. by M. Trinchero (Einaudi, Torino, 1970).

A. Landé, *Quantum Mechanics* (Pitman, London, 1952); "Continuity, a key to quantum mechanics", *Phil. Sci.* 20, 101 (1953); "Quantum indeterminacy, a consequence of cause-effect continuity", *Dialectica* 8, 199 (1954); "Quantum physics and philosophy", *Current Science* 27, 81 (1958); "New Foundations of Quantum Mechanics" (Cambridge University Press, Cambridge 1965); "Quantum Theory without dualism", *Scientia* 101, 208 (1966); "Quantum Mechanics in a New Key" (Exposition Press, New York, 1973); "Albert Einstein and the quantum riddle", *Am. J. Phys.* 42, 459 (1974).

E.L. Post, "Introduction to a general theory of elementary propositions", *Am. J. Math.* 43, 163 (1921).
cfr. anche: J. Lucasiewicz, *C.R. Soc. Sci. Varsovie* 23 (cl. III), 51 (1930).

J. Lucasiewicz, and A. Tarski, *ibid.*, p. l.

H. Reichenbach, *Philosophic fondations of Quantum Mechanics* (University of California Press, Berkeley, 1954).

P. Février, "Les relations d'incertitude de Heisenberg et la Logique", *C.R. Acad. Sci.* 204, 481 (1937).

L. de Broglie, *La Physique quantique restera-t-elle indeterministe?* (Gauthier-Villars, Paris, 1953).

R.P. Feynman, "Space-time approach to non relativistic quantum mechanics", *Rev. mod. Phys.* 20, 367 (1948).

H. Freistadt, "The causal formulation of quantum mechanics of particles (the theory of de Broglie, Bohm and Takabayas)", *Suppl. Nuovo Cimento*.

M. Pierucci, "Alcune considerazioni sulle particelle e sui campi", *Suppl. Nuovo Cimento* (1956), and references therein.

K. Gödel, "Über formal unentscheidbare Sätze der *Principia Mathematica* und verwandter Systeme. I", *Monatschr. Math. Phys.* 38 (1931).

H. Meschkowski, *Wandlungen der mathematischen Denkens* (Vieweg, Braunschweig, 1960).

E. Nagel, and J.R. Newman, *Gödel's Proof* (New York University Press, New York, 1958).

A. Pignedoli, "Sui fondamenti della Meccanica quantica". *Atti Accad. Naz. Sci. Lett. Arti*, Modena, 1972); "Meccanica classica e meccanica quantica di fronte alla filosofia del linguaggio", *ibid.*, XIX, 1977; *Alcune teorie meccaniche superiori* (CEDAM, Padova, 1969, and references therein).

B. d'Espagnat, "Conceptual Foundations of Quantum Mechanics" (Benjamin, New York, 1976).

G. Tarozzi and A. van der Merwe eds., *Open questions in Quantum Physics* (Reidel, Dordrecht, 1985) and references therein.

PERSPECTIVES OF PHYSICAL DETERMINISM

Nicola Cufaro-Petroni
Istituto Nazionale di Fisica Nucleare
Dipartimento di Fisica dell'Università
Via Amendola 173
70126, Bari,Italy

ABSTRACT

   A short review of the problems posed to the physical determinism by quantum mechanics and relativity is presented, along with the suggested ways to circumvent them.

1. INTRODUCTION

   Generally speaking the so called "impossibility theorems" cannot be considered as absolute impossibility statements. In fact, it is trivial to remark that, if we look at the hypotheses on the ground of which these theorems are proved, they fix rather the price we must pay in order to get some result, than the fact that this result cannot be reached at all.
   In that sense we think that the von Neumann theorem and the no-inte_ raction theorem can be considered as impossibility theorems about the problem of the construction of a causal interpretation of the quantum mechanics. In fact, they were for a long time considered as a crucial obstacle on the way to a definition of a deterministic theory of quantum phenomena.
   However, recent and less recent developments have shown that it was not completely impossible to get such a deterministic description. Of course, the theorems are correct, but it is possible, by looking at their hypotheses, to find a way to circumvent them. The aim of this paper is to sketch a possible path between the obstacles represented by the von Neumann and the no-interaction theorems toward the final result of a causal description of nature.

G. Tarozzi and A. van der Merwe (eds.), The Nature of Quantum Paradoxes, 197–205.
© 1988 by Kluwer Academic Publishers.

2. VON NEUMANN'S THEOREM AND BEYOND IT

It is well known that the first difficulty for a causal interpretation of the quantum theory was posed by the existence of the uncertainty
relations. They could be viewed as a consequence of the formal property
of the Fourier transform characterizing the wave function of a microscopic system, and thus, in some sense, as an inherent behavior of the
microscopic world, in which the observables are not always well defined
(so that sometime they can lack their physical meaning). However, in
another sense, the uncertainty principle could be considered as the
effect only of the disturbance induced by a measurement of the observables of the physical system. Of course, in this case the observables
always mantain their meaning, but the description of the system given by
the wave function must be regarded, in some sense, as an incomplete one.
It is in this context that we can pose the problem of the existence
of the "dispersion free" states, namely the states (or wave functions)
for which all the observables have a well defined value without dispersion.

Regarding this question, it is well known that the von Neumann
theorem[1] states the impossibility of the existence of dispersion free
states in the framework of a theory reproducing the results of the quantum mechanics. We do not enter here into a discussion of this theorem,
whose analysis can be found in an already vast and well-known literature.
We, however, allow ourselves the remark that, despite the proof of von
Neumann, dispersion free states were explicitely constructed in a causal
model proposed by Bohm[2] which reproduces all the features and the
results of non-relativistic quantum mechanics.

This causal theory, based on earlier works of Madelung and de
Broglie,[3] considers the wave function as a physical field and, by means
of a separation of the real and imaginary parts of the Schrödinger equation, furnishes a continuity equation and a gereralized Hamilton-Jacobi
equation, describing the dynamics of a classical particle subjected to
a field of force exerted by the wave function. For such a classical
particle all the physical observables are well defined.

The field of force of ψ was represented by the socalled quantum
potential defined as

$$U(\vec{r}) = - \frac{\hbar^2}{2m} \frac{\nabla^2 R}{R} ; \qquad R^2 = |\psi|^2 . \tag{2.1}$$

But how was it possible to get around the von Neumann theorem?
We must remark that, among the hypotheses of this theorem, there
is the statement, directly generalized from ordinary probability
theory, that the expectation values of observables are linear, in the
sense that

$$E(aA + bB) = aE(A) + bE(B) \tag{2.2}$$

for A, B observables and a, b real numbers. This statement is revealed
as an excessively restrictive hypothesis which could be relaxed if one
takes into account the existence in quantum mechanics of incompatible
(noncommuting) observabes. In this case, following a remark of Bell,[4]

we can immediately see that in the Bohm model, for example, the relation (2.2) is not always verified when A and B are incompatible observables.

Thus the Bohm model and its subsequent analysis indicated that a causal interpretation of quantum mechanics could not be considered absolutely impossible on the basis of the von Neumann theorem. This important remark opened the way to a number of investigations on this argument: modifications, generalizations of the original model, stochastic interpretations, etc.

3. QUANTUM POTENTIAL AND CAUSAL ANOMALIES

However, if the road to building a causal interpretation of the quantum formalism appeared open from the standpoint of the von Neumann theorem, another important difficulty lurked in the idea of quantum potential itself.

If we consider two particles 1 and 2, described by a nonfactorizable ψ (1,2), and if we decompose the wave equation into its real and imaginary parts, we find that the two particles are subjected to quantum potentials respectively of the form

$$U_1(1,2) = - \frac{\hbar^2}{2\mu} \frac{\nabla_1^2 R}{R} , \quad U_2(1,2) = - \frac{\hbar^2}{2\mu} \frac{\nabla_2^2 R}{R} ; \quad R^2 = |\psi|^2 . \qquad (3.1)$$

It is evident that these are two nonlocal potentials, as Bohm himself pointed out immediately[2], in the sense that the force acting on 1 (or on 2) instantaneously depends on the position of 2 (or of 1). Of course there was no problem in a nonrelativistic theory, but it is possible to carry out the same analysis on the relativistic Klein-Gordon equation; one obtains as relativistic nonlocal quantum potentials,

$$U_1(1,2) = - \frac{\hbar^2}{2\mu} \frac{\Box_1 R}{R}, \quad U_2(1,2) = - \frac{\hbar^2}{2\mu} \frac{\Box_2 R}{R} . \qquad (3.2)$$

On the other hand, it is well known that the nonlocal potentials between directly interacting particles, such as (3.2), are generally viewed as carriers of instantaneous signals between the two particles and that this possibility introduces the socalled "causal anomalies." Classical examples of these (signals travelling backward in time, effects preceding causes, etc.) can be found in the standard literature on relativity theory.[5]

It thus seems that the idea of a quantum potential, introduced in order to get a coherent causal interpretation of quantum mechanics, directly implies the breakdown of causality in another sense. Indeed, the existence of signals (or particles) travelng faster than light is not in direct contradiction with the formalism of the theory of the relativity: a number of investigations in this field (and about tachyons) are based on this possibility. But the difficulty aries when, besides the relativity principle, we admit a suitable idea of causality. In this sense, the existence of nonlocal quantum potentials could be incompatible with a causal theory of the microscopic phenomena.

In fact, a suitable Lorentz transformation applied to a superluminal signal, could allow it to propagate backward in time, so that the

succession of causes and effects will depend on the choice of the reference frame and, as a direct consequence, we could influence our past
history.

First of all we must say that there exists a general opinion today
that it is impossible to send macroscopic superluminal signals by means
of quantum potentials or, for example with an EPR apparatus.[6] That
will depend on the "uncontrollable" (macroscopic) character of this
subquantum nonlocality, which manifests itself in the fact that we never
can get any signal from an EPR apparatus if we look only at one polarizer. We must always consider coincidences of measurements in order to
appreciate changes or signals in the apparatus. Accordingly, we must
in any case wait for the knowledge of the results of both polarizers
and never can receive a signal by looking at only one of them.

However, going beyond this preliminary remarks, we need here a
deeper discussion about the idea of causality.[7] In some sense we must
distinguish between two concepts of causality, depending on whether we
are dealing with closed or non closed systems:

(a) The first concept is a development of the ideas of the classical mechanics: For a closed systems we speak of causal or predictive
description when the knowledge of the initial conditions, at t_o , determines the evolution in time by means of the equations of motion (Cauchy
problem).

(b) For nonclosed systems we speak of the propagation of a signal
from one event (cause) to another (effect) by means of an interaction of
the given system with an external perturbation: In this case "causality"
deals with statements about an invariant succession of causes and effects such that anomalies can be avoided.

These two concepts are not equivalent, and the causality idea discussed in connection with the principle of relativity and the quantum
potentials is of type (b). Hence, from the contradiction between (b)
and the existence of superluminal signals, we cannot deduce anything
about the relation between (a) and the nonlocal quantum interactions.

In this sense, the possibility of a theory of a direct nonlocal
interaction, in the scheme of a relativistic predictive mechanics, is
not influenced by the preceding analysis of the causal anomalies, if
we clearly state that we always are dealing with closed dynamical systems.

As an example,[8] let us consider the case of two particles 1 and
2 interacting via a nonlocal potential. We can solve our dynamical
equations and determine their world lines L_1 and L_2. Until now, no
causal anomalies can arise. However, we can perturb L_1 so that the
disturbance will be propagated instantaneously via the nonlocal interaction. Of course, in order to do this we must use something external to
the system 1 + 2: for example a particle 3 acting on 1. As long as 3
remains an "external" agent with the ability "to be or not to be ,"
there is place for causal anomalies. But, if the world is a unique set
of events in which the particle 3 is (and cannot "choose" anything else),
we need only consider the enlarged closed system 1 + 2 + 3 in order to
get a new perfectly causal system in a predictive sense, completely
free of any causal anomaly.

4. THE NO-INTERACTION THEOREM ...

Following the analysis of the preceding section, we now consider
the possibility of existence of a predictive theory of closed systems
in the presence of a direct nonlocal interaction.

To this end we must start with a more clear discussion about the
idea of covariance.[9] We can distinguish two meanings of this expression: The first is the socalled "manifest covariance" which deals with
the use of a mathematical apparatus (tensor calculus) that exploits the
spacetime symmetry. Another, nonequivalent meaning concerns the group
of inhomogeneous Lorentz transformations. In this sense relativistic
covariance requires that the linear space of the states of our physical
system be a representation space for the Lorentz group. This approach
was first developed for the relativistic quantum theories, where the
ambiguity of the concept of the "position" of a particle in spacetime
makes it difficult to directly use the scheme of manifest covariance.
In fact, this idea of relativistic covariance is concerned only with
the Lorentz group as expressing the relationship between physical phenomena as viewed by different observers. Accordingly, the fundamental
demand is that the vector space for the states of our system (defined,
for example, by means of a homogeneous linear differential equation) be
the ground on which we can define a linear representation of the Lorentz
group.

In this sense the problem of defining the dynamics of our system is
that of identifying the ten fundamental infinitesimal generators of the
group that satisfy the Lie bracket relationship characteristic of their
algebra:

$$[P_i, P_j] = 0, \quad [P_i, H] = 0, \quad [J_i, H] = 0,$$

$$[J_i, J_j] = \varepsilon_{ijk} J_k, \quad [J_i, P_j] = \varepsilon_{ijk} P_k,$$

$$[J_i, K_j] = \varepsilon_{ijk} K_k, \quad [K_j, H] = P_j,$$

$$[K_i, K_j] = -\varepsilon_{ijk} J_k, \quad [K_i, P_j] = P_j; \tag{4.1}$$

$$i, j, k = 1, 2, 3 .$$

This program, developed in relativistic quantum theories, has an
extension to classical mechanics that was pioneered by Dirac.[10] In
fact, the Lorentz symmetry in a classical Hamiltonian theory of a fixed
number of particles can be introduced by means of functions of their
canonical variables. This can be done by requiring that the generators
of the canonical transformations connected to the Lorentz group satisfy
the usual relations characteristic of the group structure, i.e., as
(4.1) in the quantum case. Of course, they being the generator functions of the
canonical variables, the Lie brackets will be the classical anolog of
the quantum commutators, viz. the Poisson brackets.

Until now nothing has been said about the particular way in which
some specific physical quantities transform in the theory. If, for
example, we require [9,10] that the spacetime coordinate Q_j of a parti-

cle transform, in the familiar manner, according the Lorentz transformation formula, we find that they will satisfy a set of Poisson bracket equations with the ten generators of the group:

$$[Q_j, P_j] = \delta_{ij}, \quad [Q_i, J_j] = \epsilon_{ijk} Q_k,$$
$$[Q_i, K_j] = Q_j[Q_i, H] ; \quad (i,j,k = 1,2,3). \tag{4.2}$$

These relations, which must be postulated in addition to the algebraic relations (4.1) among the generators, are connected with the canonical character of the Q_j and with the question of the invariance property of the world lines.

It is in this context that the no-interaction theorem was developed.[11] This theorem shows that, by combining the demand of relativistic symmetry (as expressed by the relations between the generators of the group) with the requirement of the manifest covariance for the space time positions of the particles of our system, we get a very restrictive set of conditions on the form of the generators. More precisley, it can be proved that the above conditions are compatible only if we are dealing with a set of noninteracting particles. If indeed we calculate the acceleration of each particle, we find that it always is zero, and hence all the velocities must be constant.

In this sense, in a Lorentz symmetric, classical, Hamiltonian theory, the requirement of the ordinary manifest covariance of the space time positions rules out any interaction. This important theorem was considered for a long time as a proof of the impossibility of constructing a classical theory of directly interacting particles and, consequently, as the main reason for introducing a quantized field theory in order to describe interacting systems.

5. . . . AND BEYOND IT

In a precise sense, the fact that, in a relativistic Hamiltonian theory, one cannot use covariant particle coordinates as canonical variables, indicates the formal "cost" of a relativistic theory with action at a distance, rather that the impossibility of it. Indeed, a few forms of a relativistic dynamics of directly interacting particles have emerged during last years.[7,12] Of course, all these approach must take into account the above mentioned results of the no-interaction theorem, so that in general we will deal with noncanonical positions or noninvariant world lines, and so on. However, these attempts show that, beyond some conpatibility condition imposed on the form of the interaction, a theory of relativistic particles interaction via action at a distance is possible and works. Thus the problem, outlined at the end of Sec. 3, of the construction of a covariant dynamics for interacting closed systems admits solutions, even if not the more simple conceivable ones.

Here we will briefly sketch one of these approaches, based on the socalled constraint formalism for classical mechanics,[13] that present, in a new suggestive form, the essential results of the preceding predictive theories.

In a phase space with simplectic form (Poisson's brackets) there is
a way to define trajectories without reference to any time parameter.
Thus these trajectories will be manifestly covariant. The theory will
also be Lorentz symmetric if it constitutes a linear representation of
the Lorentz group. To accomplish this we consider k functions $K_i(q,p)$
$(i = 1,\ldots,k)$ satisfying the relations

$$\{K_i, K_j\} = \sum_m \lambda_{ij}^m K_m, \qquad \lambda_{ij}^m \text{ const.} \tag{5.1}$$

The set of equations $K_i = 0$ will constitute a constraint which defines
a hypersurface in phase space. If now we consider canonical transformations generated by K_i, it is evident that the constraint hypersurface
will be left invariant. Thus, starting with an arbitrary point on the
hypersurface, the iteration of the canonical transformation generated
by K_i defines a trajectory lying on the hypersurface.

Now it is possible to define generalized Hamiltonian equations of
motion by establishing the functional dependence between the q_i^0 and the
remaining canonical variables. However, this passage is not necessary in
the constraint formalism, which is totally algebraic: No explicit time
dependence of variables needs to enter into its statements, so that it
will remain covariant and also manifestly covariant. In this sense the
Hamiltonian formalism and the constraint formalism are closely related
but not identical.

In this formalism we can also introduce direct interactions, for
example, in the case of two particles, by means of a scalar potential
$V(r)$, where r, a scalar, is the spatial separation of the two particles
in the rest frame of the center of mass. Thus we get a well defined
covariant system of directly interacting particles. Of course, not all
forms of $V(r)$ are permitted.

However, the price to be payed for the no-interaction theorem is
that we need in this theory a deeper analysis of the idea of observable.
Formally we can introduce in phase space the socalled "syntactic"
observables, as functions of p_i and q_i, which commute with the K_i on the
constraint hypersurface: They are constant along each phase space trajectory. But these observables are not coincident with the physical
"semantic" observables, namely: times, distances, velocities, etc. as
measured by rulers and clocks in some frame of reference. The main difficulty of the constraint formalism lies in the connection between syntactic and semantic observables. In fact, in order to obtain the space-
time orbits, we must associate semantic observables with the phase space
coordinates. But this association is ambiguous, in the sense that it is
unique only in a preferred frame of reference: the center of mass rest
frame. In this frame, the association of the q^0 with the semantic time
gives rise to auxiliary conditions and to a unique identification of
Lorentz covariant orbits in spacetime.

6. CONCLUSIONS

The preceding brief discussion shows that the road to a causal interpretation of quantum theory cannot be considered closed on the ground

of the von Neumann theorem and of the no-interaction theorem. In fact, these theorems state what we can do and what we cannot do in order to build a deterministic theory of the microscopic phenomena. On this question preliminary works exist[14] showing that, at least for the scalar-particle case, the quantum potential between two correlated particles satisfies the compatibility conditions for a predictive mechanics, so that it can be used in the construction of a "classical" mechanics for these systems. Of course, this is only a first step along the way leading to the definition of the description of a deterministic world. However, nothing has been said till now about the problem, outlined in Sec. 5, of the canonicity of the coordinates or of the identification of the physical observables, which items will constitute nothing less than an exciting program for the future.

A final remark: In order to avoid any possible causal anomaly, the world must be treated in a nonlocal theory as a unique, closed, predictive system, in the sense sketched in the Sec. 3. This implies that we must accept, in some sense, a completely deterministic point of view about the evolution of the world. Of course, such a choise could not be made without a discussion concerning fields and arguments (epistemology, philosophy, ethics, etc.) not directly connected with physics. We cannot eneter here in to a debate on these questions; we recall only that in the past centuries the idea of a deterministic universe was widely considered a real possibility and that a long chain of thinkers and physicists, from Spinoza through Diderot and Laplace up to Einstein himself, was in various forms concerned with this "Weltanschauung." Perhaps the time has come to reopen the debate on these problems.

REFERENCES

1. J. von Neumann, *Mathematical Foundations of Quantum Mechanics* (University Press, Princeton, 1955).
2. D. Bohm, *Phys. Rev. 85*, 166, 180 (1952).
3. E. Madelung, *Z. Phys. 40*, 332 (1926).
    L. de Broglie, *C. R. Acad. Sci. 183*, 477 (1926); *184*, 273 (1927); *185*, 580 (1927).
4. J. S. Bell, *Rev. Mod. Phys. 38*, 447 (1966).
5. C. Møller, *The Theory of Relativity* (University Press, Oxford, 1960).
6. J. Jarrett, "Bell's Theorem, Quantum Mechanics and Local Realism", Ph. D. theisis, Chicago, 1983.
    A. Shimony, Communication to the International Symposium on the Foundations of Quantum Mechanics, Tokyo, 1983.
    N. Cufaro Petroni and J. P. Vigier, *Lett. Nuovo Cimento 25*, 151 (1979)
7. P. Havas, "Causality and Physical Theories", Wane State University, 1973.
    R. P. Gaida, *Sov. J. Part. Nucl. 13*, 179 (1982).
8. N. Cufaro Petroni, in "Open Questions in Quantum Physics", G. Tarozzi and A. van der Merwe, eds.(D. Reidel Dordrecht, 1985).
9. L. L. Foldy, *Phys. Rev. 122*, 275 (1961).
10. P. A. M. Dirac, *Rev. Mod. Phys. 21*, 392 (1949).
11. D. G. Currie, *J. Math. Phys. 4*, 1470 (1963).

    D. G. Currie, T. F. Jordan, E. C. G. Surdarshan, *Rev. Mod. Phys. 35*, 350, (1963).

12. L. H. Thomas, *Phys. Rev. 85*,
    B. Bakamjian and L. H. Thomas, *Phys. Rev. 92*,
    Ph. Droz Vincent, *Phys. Scrip. 2*, 129 (1970); *Ann. Inst. H. Poinceré 27*, 407 (1977).
    L. Bel, *Ann. Inst. H. Poincaré 3*, 307 (1970); *Phys. Rev. 18D*, 4770 (1978).

13. I. T. Todorov, preprint E2-10125, JINR, Dubna, 1976.
    V. V. Molotov and I. T. Todorov, *Comm. Math. Phys. 79*, 111 (1981).
    A. Komar, *Phys. Rev. 18D*, 1881, 1887, 3617 (1978).

14. N. Cufaro Petroni, Ph. Droz Vincent and J. P. Vigier, *Lett. Nuovo Cimento 31*, 415 (1981).

DETERMINISM, SEPARABILITY, AND QUANTUM MECHANICS

S. Notarrigo

Dipartimento di Fisica, Università di Catania
Corso Italia, 57 I 95129 Catania, Italy

ABSTRACT. It is shown that the mathematical formalization of
the concept of determinism and separability in the derivation
of Bell's inequality is too restrictive to express the standard
meaning of these concepts as used in classical physics. In fact,
classical physical systems which are non-problematic and non-
paradoxical are exhibited which violate Bell's inequality. For
this reason it is concluded that Bell's inequality, as it has
been used so far in the physical experiments related to the
socalled Einstein-Podolsky-Rosen paradox cannot exclude a
realistic interpretation of quantum mechanics. New kinds of
experiments are proposed to test the validity of these concepts
in the physical world.

1. INTRODUCTION

The Einstein-Podolsky-Rosen (EPR) paradox,[1] as
reformulated by Bohm-Aharonov,[2] raises serious doubts about
the possibility of maintaining both determinism and separability
in the physical world if the standard interpretation of quantum
mechanics is accepted.

Such a conclusion is usually derived from the comparison of
Bell's inequality,[3] with the results of recent experiments,[4]
and it, of course, presupposes a preliminary definition of terms
like determinism and separability, which in philosophical debates
do not usually have very stable meanings. Moreover, having
defined these terms, it is necessary to ascertain whether the
mathematical representations of the related concepts actually
possess their intended meaning and, in turn, if the physical
experiments actually test the stated hypotheses.

G. Tarozzi and A. van der Merwe (eds.), The Nature of Quantum Paradoxes, 207–214.

I intend to show in the present paper that, if we accept the most usual formulation of the foregoing concepts, then their formalization, as presupposed in the derivation of Bell's inequality, is not adequate for expressing them. In other words, the mathematical hypotheses, needed for the derivation of the final theorem, are too restrictive for a deterministic and separable physical system; in fact, many classical separable mechanical systems, non-problematic and non-paradoxical, can be identified which violate Bell's inequality.

2. DETERMINISM AND SEPARABILITY

The most usual definition of the concept of determinism implies that "everything happening in nature must be predetermined by previous happenings in the physical word". [5]

EPR added to this the socalled principle of reality, [6] formulated by asserting the existence of "elements of reality" corresponding to quantities of which we can predict with certainty (i.e., with probability equal to unity) their values without in anyway disturbing the physical system to which they pertain. [1]

The two previous principles leads to the assertion of the principle of "statistical determinism," according to which:

> "If two statistical ensembles of physical systems are
> submitted to identical treatments, and if subsequent
> observation reveals significant statistical
> difference between them, the implication is that the
> two ensembles were not identical at the start," [6]

i.e., the systems belonging to different ensembles possess different elements of reality. Bell's mathematical formalization of this principle becomes the assertion of the existence of a set Λ of hidden variables (elements of reality) λ with a given probability distribution over the Borel subsets of Λ. Moreover (in the usual interpretation connected with the experimental situation), each value of λ determines unequivocally the registration or not of a count (in a fixed interval of time) in either of two detectors in any direction in which we choose to set the two polarizers (which fix the two local variables being measured by the two counters); it also determines unequivocally the registration of a coincidence between the two detectors. [7] The paradigm of such a set Λ is the phase space of a classical dynamical system over which is given the probability measure defining the macroscopic state in classical equilibrium statistical mechanics.

Now we briefly examine the notion of separability. Suppose we have two physical systems that are no longer interacting, then the separability criterion may be expressed by asserting that

> "the factual situation of one system is independent
> of what is done with the other system spatially
> separated from the former. [8]"

Here one intends to exclude any physical interaction between
the two systems during the time of observation, but not to
exclude correlations brought about by any interaction that the
two systems may have undergone in the distante past.

Bell's mathematical formalization of the separability
concept consists in the forbidding of any dependence of the local
variables measured in one system on the parameters controlled
in the other system. [3]

This formalization, as will be shown, goes ouside the limits
set by the previous formulation when applied to certain
experimental situations.

3. OBSERVABLES AND DYNAMICAL VARIABLES

For our subsequent discussion, let us recall the
formalization of the concept of a *Kolmogorovian probability
measure*.

Given a triple S,F,p with S a space of "outcomes," F a
family of subsets of S called "events," p a set function that
assigns a positive real number to each event, then the structure
of F is such that

$$\forall a, b \in F \Longrightarrow a \lor b \in F \text{ and } a \land \bar{b} \in F;$$

$$S \in F;$$

$$\forall a_i \in F, \ i \in N \Longrightarrow \bigcup_{i=1}^{\infty} a_i \in F;$$

$$\forall a_i \in F, \ i \text{ and } j \in N, \ a_i \land a_j = \phi \text{ for } i \neq j \Longrightarrow p(\bigcup_{i=1}^{\infty} a_i) = \sum_{i=1}^{\infty} p(a_i);$$

$$p(S) = 1.$$

In order to apply this concept to physical situations, we
usually build up particular statistical structures.

In classical equilibrium statistical mechanics we identify
the space S of the outcomes with a particular subset of the
phase space of the dynamical system (e.g., in the construction
of the microcanonical esemble, we take the hypersurface of
constant total energy), the events are generally constituted by
all its Borel subsets, and the probability measure α over the
events defines the macroscopic state of the system.[9] Further,

for each dynamical variable A, defined as a Borel function on
the phase space of the given dynamical system, we set up a
probability distribution $p(A,\alpha,E)$ over the Borel subsets E
(events) of the range of values of the variable A, which in
general coincides with the real line R.[9]

Because of the peculiar properties of the variables A we can
always set[9]

$$p(A,\alpha,E) = \alpha(A^{-1}(E)) = \alpha(\omega), \tag{1}$$

where α is the probability measure which defines the macroscopic
state of the system and $\omega = A^{-1}(E) \subset S$ is the inverse image, over
the space S, of the set $E \subset R$. Because of this relation, all the
probability distributions $p(A,\alpha,E)$ for all the dynamical
variables are reconduced to the unique probability distribution
α which defines the macroscopic state of the system. In
particular, if E is a Borel subset of the range of the possible
values of A and E' is a Borel subset of the range of A', then
the probability of having simultaneously $(A \in E)$ and $(A' \in E')$ is
given by

$$P = \int_{\omega \wedge \omega'} Q_E^A(n) Q_{E'}^{A'}(n) \ (dn), \tag{2}$$

where Q_E^A and $Q_{E'}^{A'}$ are the indicators of the sets $\omega \equiv A^{-1}(E)$ and
$\omega' \equiv A'^{-1}(E')$, respectively.

All these properties are presupposed for the set Λ of the
hidden variables in the derivation of Bell's inequality.[3]
But now we show that in classical systems there exist observables
which cannot be considered as dynamical variables in the
restricted sense defined above.

In fact, consider a single trajectory in phase space of a
single classical dynamical system, and let A be a dynamical
variable as defined above; during the evolution of the system,
A takes the values $A(x_t)$ with $x_t = T_t x_0$, wherein T_t is the evolution
operator of the system and x_0 the initial phase point. We then
introduce the function of time

$$f(t) \equiv A(x_t)$$

and note its zero crossings, that is, the points t_n such that

$$f(t_n) = 0.$$

Assume one has a situation in which everything is fixed except
the initial time phase, which we may treat as a random variable,
so that a stochastic process $f(t-\tau)$ with τ random variable,
obtains. If the trajectory in phase space is sufficiently
entangled and the stochastic process $f(t-\tau)$ is stationary,[10]
one is free to fix t and consider different physical systems

with different τ, or one may fix τ, and hence the system, and consider different times of observation t.

Let $Q_E^A(\tau)$, with $E \equiv [t, t+\Delta t]$, be a binary random variable which takes the value 1 if at least one zero crossing of A lies in E, and the value 0 otherwise. $Q_E^{A'}(\tau)$ a similar variable for A'. Assume $O_1 \equiv Q_{E,E'}^{A,A'}(\tau)$ is a binary random variable which takes the value 1 if a zero crossing t_A of A lies in E and, simultaneously, a zero crossing of A' in $E' \equiv [t', t'+\Delta t']$, with the condition $t'=t_A$. This is the observable measured by means of a time-to-height converter,[11] with the zero crossing of A as "start" and the zero crossing of A' as "stop." In a standard coincidence circuit the measured observable[11] is

$$O_2 \equiv Q_{E,E'}^{A,A'}(\tau) + Q_{E',E}^{A',A}(\tau)$$

Both O_1 and O_2 are quite different from the observable $O_3 \equiv Q_E^A(\tau) Q_{E'}^{A'}(\tau)$ appearing in formula (2) and assumed by Bell.

Our observables O_1 or O_2 cannot obey both hypotheses of Bell. Either we take O_1 or O_2 and apparently violate Bell's formalization of the principle of locality, or we take O_3 and will in general be unable to find a density $\rho(\lambda)$ of hidden variables that could simulate, by an averaging process, the experimental result which is expressed by $<O_1>$ or $<O_2>$, depending on whether we use a time-to-height converter or a standard coincidence circuit. In other words, there exists neither a set Λ of hidden variables nor observables with all the previously stated properties, in spite of the fact that our system obeys the principle of "statistical determinism" and the criterion of separability as formulated above.

4. A CONCRETE PHYSICAL EXAMPLE

We now give a simple concrete example. Consider a physical system of two mass points whose respective motions relative to a rectangular Cartesian coordinate system are identical uniform circular motions in the xy plane to which are added uniform displacements of the same speed in opposing z directions. The two mass points do not physically interact with one another, but their motions of course are strongly correlated by virtue of their identity when projected on the xy plane. These projected motions are described by the equations

$$x = A \cos(\omega t + \phi),$$
$$y = A \sin(\omega t + \phi),$$

which means that the projection of the radius vector along a direction θ in the xy plane is given by

$$r_\theta = x\cos\theta + y\sin\theta = A \cos(\omega t + \phi - \theta).$$

Now imagine that at random instants of time one prepares
identical pairs of such correlated mass points and allow them to
start moving at z=0 with random initial phases ϕ, but with the
same amplitude A and angular speed ω.

Let us suppose that, at symmetrical distance $\pm$ L from the
origin, two observers count the number of zero crossings of the
projection r_θ and the coincidences between the two counters,
which are at distance 2L a part, during a fixed interval of time
T. We further assume that the coincidences are measured by a
time-to-height converter, with the "start" furnished by the
nearby counter, after a fixed delay of $2L/c-R/2$, where c denotes
the speed of the signal carrying information from the distant
counter to the "stop" and R is the time range of the time-to-
height converter.

Let $n(\theta)$ denote the number of counts registered by one
counter when we observe r_θ; it is proportional to the probability
of the event

$$Q_E^A(\tau) = 1; \quad A \equiv r_\theta, E \equiv [t, t+\Delta t] \ .$$

Let, furthermore, $n(\theta_1, \theta_2)$ be the number of coincidences when
we observe r_{θ_1} in one counter and r_{θ_2} in the other counter; this
number is proportional to the probability of the event

$$Q_{EE'}^{AA'}(\tau)=1; \quad A \equiv r_{\theta_1}, \quad A' \equiv r_{\theta_2}, \quad E \equiv [t, t+\Delta t] \ , \quad E' \equiv [t_A-R/2, \ t_A+R/2].$$

Of course, we have supposed that the intensity of the incoming
pairs is so low that the dead-time distortion can be neglected[11]
Finally choose three directions $\theta_1=\theta$, $\theta_2=3\theta$ and $\theta_3=2\theta$ such that

$$\frac{\theta}{\omega} < \frac{R}{2} < \frac{2\theta}{\omega} \ .$$

Then one way of expressing Bell's inequality in term of our
observables is: [7]

$$n(\theta_1,\theta_3) + n(\theta_2,\theta_3) < n(\theta_3) + n(\theta_1,\theta_2). \tag{3}$$

In our simple case, one gets

$$n(\theta_1,\theta_3) = n(\theta_2,\theta_3) = n(\theta_3) = \text{integral part of } T\omega/\pi.$$

and $n(\theta_1,\theta_2) = 0$

because of the particular selection of the range R. Thus, the
Bell inequality is grossly violated, by much more than the quantum
mechanical margin; in fact, the ratio between first and second
term of (3) is 2 in our case, as against 6/5 in quantum mechanics
(see ref.7).

This allows us to devise a physical system that will

faithfully reproduce the correlation curve of quantum mechanics
[7]: It is only necessary to complicate the motion of the systems
by replacing the two mass points by two separated arrays of
coupled linear oscillators which reproduce formally the
polarization space of the two photons in the Bohm-Aharonov
experiment [2].

5. CONCLUSIONS

The foregoing discussion demonstrates that there are
observables of classical systems that do not behave like the
dynamical variables of classical equilibrium statistical
mechanics and behave more like the observables of quantum
mechanics [7]. So it may not be impossible that an appropriate
generalization of our considerations could favor Einsten's idea
that quantum mechanics can be completed. [8] After all, the
square modulus of the cross-correlation [12] $R_{AB}(0)$ for two
dynamical variables defined over two separated but correlated
physical systems A and B, for instance

$$A(t) = f_A(t)\cos\theta_1 + g_A(t)\sin\theta_1,$$

$$B(t) = f_B(t)\cos\theta_2 + g_B(t)\sin\theta_2,$$

would give the correlation curve of quantum mechanics, i.e.,
the $\cos^2(\theta_1 - \theta_2)$ law, if

$$R(0)_{f_A g_B} = R(0)_{f_B g_A} = 0 \text{ and } R(0)_{f_A f_B} = R(0)_{g_A g_B}.$$

A sufficient condition for the latter is that one has two
identical rotating vectors in a plane whose motion is a
superposition of an arbitrary number of uniform circular
motions of arbitrary parameters, which implies identical angular
momentum and parity for the two vectors.

Of course one obtains the $\sin^2(\theta_1 - \theta_2)$ law in the case of
opposite angular momenta. Such an observable is connected with
the number of coincidences between the zero crossings of the two
local observables $A(t)$ and $B(t)$ and, in simple particular cases,
this connection can be explicitly exhibited; [12] an example may
be the one described in ref.7.

In passing we note that the algebra of random variables is
just the algebra of a linear vector space with an inner product,
[10,12] like the algebra of the states of quantum mechanics; so
it cannot be the formal structure of quantum mechanics that
forbid its realistic interpretation.

It may also be that the paradoxes, as quite often in the
history of science, derive from overly strict conceptual
barriers which are more or less arbitrarily imposed on nature.

But, independently of future possible developments in the interpretation of quantum mechanics, our considerations lead us to the conclusion that all the experiments thus far conducted in order to test quantum mechanics in connection with the EPR paradox,[1] including those experiments that use particular devices for testing nonlocality,[13] are not suited for testing determinism and separability in the more general sense formulated above. In order to test nonlocality, it would be necessary either to check beforehand the invariance of the statistical distribution of the coincidence time intervals (life-time measurements) against the variation of the relative angle of the two polarizers or, better, to devise an experiment in which the coincidences are taken by fixing a time window at instants of time predetermined before a count takes place in any of the two detectors.

REFERENCES

1. A. Einstein, N. Rosen, and B. Podolsky, *Phys.Rev. 47*, 777 (1935).
2. D. Bohm and Y. Aharonov, *Phys.Rev. 108*, 1070 (1957).
3. J.S. Bell, *Physics 1*, 195 (1964).
4. J. F. Clauser and A.Shimony, *Rep. Progr. Phys. 41*, 1881 (1978).
5. F.J. Belinfante, *A Survey of Hidden-Variables Theories* Pergamon Press,(Oxford, 1973).
6. B. D'Espagnat *Conceptual Foundations of Quantum Mechanics,* 2nd edn. Addison-Wesley, Benjamin, (Reading, Massachusetts, 1976).
7. S. Notarrigo, *Nuovo Cimento 83B*, 173 (1984).
8. A. Einstein, in *Albert Einstein: Philopher, Scientist,* P.A. Schilpp, ed. (Open Court, La Salle, Illinois, 1970).
9. G.W. Mackey, *Mathematical Foundations of Quantum Mechanics* (Benjamin, New York, 1963).
10. J.S.Bendat, *Principles and Applications of Random Noise Theory* (Krieger, New York, 1977), corrected and expanded edition of original book (Wiley, New York, 1958)
11. G. Faraci, S. Notarrigo, and A.R. Pennisi, *Nucl. Instrum. Meth. 165,*325 (1979).
12. A. Papoulis, *Probability, Random Variables and Stochastic Processes* (McGraw-Hill, New York , 1981).
13. A. Aspect, J. Dalibard, and G.Roger, *Phys. Rev. Lett.* 49 1804 (1982).

PART 4

QUANTUM THEORY AND GRAVITATION

CAN SINGULARITIES AND MULTIDIMENSIONAL SPACES INFLUENCE THE
EVOLUTION OF QUANTUM MECHANICAL SYSTEMS?

S. Bergia

Dipartimento di Fisica, Università di Bologna
I.N.F.N., Sezione di Bologna, Italy

ABSTRACT. Some results of quantum gravity suggest possible
violations of quantum mechanics in the form of transitions from
pure to mixed states. A "quantum violating evolution equation"
put forward by Ellis, Hagelin, Nanopoulos and Srednicki is used
in the case of rotationally invariant density matrices describing
a system of two spin-$\frac{1}{2}$ particles. It is shown that the equation
may account for the transition from a pure singlet state to a
mixture of maximal entropy discussed by Baracca, Bohm, Hiley and
Stuart in 1975 in connection with Bohm's version of the EPR
experiment. It is argued that similar effects on the evolution of
microscopic systems could also arise as a consequence of extra
spatial dimensions, such as those introduced in multidimensional
unified theories.

1.   The general theme of this meeting does not seem to indicate,
at first sight, any role for astrophysics, quantum gravity, and
multidimensional theories. My intention is to point out that,
quite to the contrary, fairly recent developments in these fields
play, or may play, an important part with respect to the issues
that are here being debated: In particular, the developments in
question seem to indicate the possibility of violations of
quantum mechanics, although in regions still far outside the
realm of current experiments.
    My talk will be divided into three parts. The first part is
the most important, but, I am afraid, the less original. In it I
will try to expose, in a condensed form, some results of black
hole theory and quantum gravity which bear on our subject. My
justification for reviewing other people's achievements is
twofold. On the one hand, I believe there has been a lack of

G. Tarozzi and A. van der Merwe (eds.), The Nature of Quantum Paradoxes, 217–229.
© 1988 by Kluwer Academic Publishers.

communication between people working on the foundations of quantum mechanics and those occupied with the general aspects of the theory of gravitation. I therefore take the opportunity offered by this meeting to try to convince the former party of the relevance of some of the latter's results for the problems they are working on. The propagandist may hopefully be forgiven where the reviewer would probably be faulted.

On the other hand, this meeting is unusual inasmuch as it adresses a wider audience than the one formed by the participants. This fact, too, can perhaps justify the giving of a not strictly technical account on general results by someone who has not directly taken part in their derivation. I shall make evident the role to which I confine myself in presenting the introductory review by very often quoting almost verbatim the authors I am referring to.

The second part of my presentation refers to personal work and will be concerned with a direct application of the general ideas discussed in the first part to a specific problem much debated at this meeting.

My third part is largely speculative. I would like to devote it to the exposition of some ideas, not yet developed in full, concerning the possibility that effects characterizing the evolution of microscopic systems, similar to those analysed in the first part, would arise not only in connection with the occurrence of singularities, but also as a consequence of extra spatial dimensions, such as those introduced in multidimensional unified theories.

2.   ON BLACK HOLES, QUANTUM GRAVITY, AND A QUANTUM-VIOLATING EQUATION

The conceptual problems relating to the conclusion that nothing would stop the gravitational collapse of a star that has a sufficiently high mass were first faced in the sixties, starting from a paper by Penrose. The analysis of this author showed that the indefinite collapse of such objects can lead to singularities in which physically measurable quantities will become infinite. This seems to be a nonsensical conclusion. Because what is meant, for instance, by matter concentrated in a point with infinite density? Various ways out were proposed: modifications of general relativity (Brans-Dicke and Einstein-Cartan), "cosmic censorship," or "nature abhors naked singularities," meaning that any singularity developing from a gravitational collapse shall be hidden from the sight of an observer at great distance by an event horizon. This hypothesis, true in a simple case, but unproven in general, would allow us to disregard the breakdown of physics in connection with space-time singularities, since they would produce no detectable effects for observers "careful enough not to fall into a black hole," to use

Hawking's expression (1). Quantum effects, expected to be important in the very strong field near a singularity, might also prevent the singularity from occurring (*ibid.*).

However, in 1974 Hawking discovered the socalled black hole evaporation, according to which black holes create and emit particles at the expense of their mass. The central question is then: What is the final stage of the black holes evolution? From the point of view of a semiclassical theory, based on a non-quantal theory of gravitation, the conclusion is that a black hole would indefinitely continue to lose mass and essentially disappear, leaving behind it a naked space-time singularity.

For reasons to which I shall briefly return in a while, this approach is expected to break down once the black hole's mass got down to a value of the order of Planck's mass, a limit at which one does not have any reliable quantitative theory. As discussed by Hawking (2), however, the most probable hypothesis is that the black hole would continue to "evaporate," carrying with it all the information about the black hole states. It is here that quantum mechanics gets involved, although apparently in a very formal and abstract way. Let us see why. Suppose we are given a quantum state associated with a given complete set of commuting observables. What the occurrence of an event horizon can do is to "swallow" a subset of the observables, say those belonging to a subsystem. In Hawking's words,

> "the system would still be in a pure quantum state, but an observer at infinity could measure only part of the state; he could not even in principle measure what fell into the hole. Such an observer would have to describe his observation by a mixed state which was obtained by summing with equal probability over all the possible black hole states. One could still claim that the system was in a pure quantum state though this would be rather metaphysical because it could be measured only by an angel and not by a human observer." (2).

In his 1976 paper, Hawking insisted that the above

> "represents a fundamental limitation to our ability to predict the future, a limitation that is analogous but additional to the limitation imposed by the normal quantum-mechanical uncertainty principle", and introduced the term "principle of ignorance" (2).

For our purposes, I would like to stress the following points: Firstly, the effect here described, the transition from a pure to a mixed state, to the extent that it is "spontaneous," violates the quantum mechanical evolution law for the density operator; secondly, the formal reason for the transition is the averaging one has to carry out over the states of a subsystem, just as in the quantum mechanical theory of measurement.

Not withstanding the formal resemblance, one must admit that, up to this point, Hawking's analysis does not seem to have much to do with the familiar situation of two-body decays which can be controlled in a laboratory.

There is, however, a development of the idea which leads us much further in the desired direction. In order to introduce this development, I will first rapidly recall the basic qualitative features of quantum gravity.

The first point is that gravitation, which dominates at the scale of celestial and cosmic phenomena, should become prominent as a quantum feature also at the ultramicro scale of Planck's length, $(hG/c^3)^{1/2} \sim 10^{-33}$ cm, the length which characterizes a theory based on the constants h, G, and c. As everyone knows, quantum gravity is a largely speculative theory: Twenty orders of magnitude separate its domain of validity from that of the elementary particles. There is however a diffuse agreement on the opinion that the metric should manifest violent and rapid fluctuations at the scale of the Planck's length and Planck's time $(hG/c^5)^{1/2}$.

It is these fluctuations that will interest us here. Already in his article of 1976, Hawking stressed that

"in any quantum gravitational situation there is the possibility of "virtual" black holes which arise from vacuum fluctuations and which appear and then disappear again" (1).

In his 1982 paper he made the more explicit statement
"... a decay of a pure quantum state into a mixed state can occur with a macroscopic configuration such as a black hole, it also ought to occur on a microscopic elementary particle level because of quantum fluctuations of the metric which could be interpreted as virtual black holes which appear and disappear again" (2).

Ellis, Hagelin, Nanopoulos and Srednicki have taken this indication very seriously in a recent article (3). They maintain explicitly:

"It seems inescapable that space-time should have a foamy structure on the Planck scale, with Planck radius black holes appearing and diseappearing on a Planck time scale. Hence space-time should be topologically complicated with many evanescent microhorizons... therefore we believe that our microphysical laws should also accomodate violations of quantum mechanics such as the evolution of pure states into mixed states." (3).

An interesting aspect of the paper by Ellis *et al.* is that they propose an evolution equation for the statistical operator ρ modified with respect to the quantum mechanical equation. The quantum mechanical evolution equation for ρ is

$$\dot{\rho} = L_o \, \rho \, , \tag{1}$$

with L_o, the Liouville operator, given by

$$L_o = i \, \left[\quad , H \right] \, , \tag{2}$$

where H is the Hamiltonian. The equation proposed by Ellis *et al.* is

$$\dot{\rho} = L\rho \, , \tag{3}$$

with

$$L = L + L' \, , \tag{4}$$

in which L' does not have in general the form (2). If the state space is finite-dimensional and density operators are expressed in a given basis, Eq.(3) admits a solution, satisfying the initial condition

$$\rho(o) = \rho_o \, , \tag{5}$$

of the form

$$\rho(t) = \$ \, \rho_o \, , \tag{6}$$

where $\$$, the so-called *superscattering operator*, is a tetradic form; it factorizes into the product of two ordinary evolution matrices for pure states if $L' = 0$. It is easily checked, even if the authors have not stressed it, that $\$$ factorizes *only* if $L' = 0$.

A general feature of the approach of Ellis *et al.* stressed by these authors is that invariance principles are no longer equivalent to conservation laws, as it is in quantum mechanics. To see how this comes about, just consider the case of a system formed of two spin-1/2 subsystems. In quantum mechanics the system must necessarily be in a state of given angular momentum, a pure state. For example, in the state of zero total angular momentum, the singlet state, one has:

$$\rho = 1/4 \, (I - \underset{\sim}{\sigma}_1 \cdot \underset{\sim}{\sigma}_2) \, , \tag{7}$$

where $\underset{\sim}{\sigma}_1$ and $\underset{\sim}{\sigma}_2$ are the Pauli operators for the subsystems.

From the point of view of possible quantum-mechanics violating effects here discussed, a system like this can evolve to a mixed state. On the other hand, the dynamics producing the violation can still be rotationally invariant, so that it will lead to a rotationally invariant mixed state. The most general rotationally invariant form of the density matrix for a system formed of two spin-1/2 subsystems is

$$\rho' = 1/4 \; (I - \beta \underset{\sim}{\sigma}_1 \cdot \underset{\sim}{\sigma}_2) \; . \tag{8}$$

A density matrix of the form (8) describes a mixture of singlet and triplet states ($\beta = 1$ gives the pure singlet state). A process in which $\rho \longrightarrow \rho'$ therefore does not conserve the quantum mechanical angular momentum.

Another way of looking at the problem consists in computing the average value of the observable $(\underset{\sim}{\sigma} \cdot \hat{a})(\underset{\sim}{\sigma} \cdot \hat{b})$, with $\hat{a}$ and $\hat{b}$ unit vectors along arbitrary directions, in a system described by the density matrix (8). Quantum mechanics gives the value -1, corresponding to perfect anticorrelation. Now, since quantum mechanics is in any case violated in the above transition, one may wish to connect this violation with the one foreseen by local theories, and wonder whether a sensible local theory would permit the transition, while still conserving perfect anticorrelation. If this turned out true, one could argue that, although quantum mechanical angular momentum conservation is violated along with quantum mechanics itself, a naive idea of conservation survives in an alternative (partial) scheme in the form of constant "deterministic" alignment along opposite directions. However, this is not the case. In a recent paper (4) it was shown that, at least if one restricts oneself to local theories guaranteeing the existence of quantum mechanical single particle states with the correct properties under rotation, one has

$$(\underset{\sim}{\sigma}.\hat{a}) \; (\underset{\sim}{\sigma}.\hat{b}) = - 1/3 \; ,$$

hence a large violation.

We take this result as and indication that, in the transition $\rho \longrightarrow \rho'$, conservation of angular momentum is dropped in the widest possible sense.

The next question becomes what attitude one should take with respect to such violations (similar problems arise also with respect to energy and momentum (5)).

As stressed by Ellis *et al.*, their evolution equation has the structure of a Markovian master equation, of the type describing, in particular conditions, (ref. (6)), a system plunged in a thermal bath. If the analogy is taken seriously, one could argue that the fluctuating sea of microhorizons provides some kind of thermal bath for quantum systems. Energy-momentum and angular momentum exchanges would then be allowed. The authors, however, state quite clearly that their philosophy is different: They stress that, in Hawking's approach, the conservation laws are imposed axiomatically on the super-scattering operator and choose to retain them in their own. For

systems like the one described above, angular momentum conservation should then, according to their views, be put in by hand as
a statistical constraint. At any rate, they chose to discard as a
possible field of application any system described by a statisti-
cal operator for which the transition $\rho \longrightarrow \rho'$ would imply mixing
states of different angular momenta. This excludes classic probes
of quantum mechanics, like EPR-type experiments, from the set of
problems in which the authors intend to check their approach.

In the next section I would like to explore the perspectives
arising from applying the form of quantum violating evolution
equation proposed by Ellis *et al.* just to such cases. (For the
cases considered suitable by these authors, the reader is
referred to the original paper.) This is personal work (7), for
which -- it must be stressed -- the authors quoted till now do
not share any responsibility, since the attitude taken here is
alien to their views.

3. A SOLUBLE MODEL FOR IRREVERSIBLE QUANTUM PHENOMENA

Our attitude (7) is to assume the overall presence of a
thermal bath related to the fluctuations of the metric according
to the above. The first step would then be a derivation of Eq.
(3); this problem is however only at a preliminary stage (8).
Here we will take Eq. (3) as it stands and apply it to density
matrices of the general form (8).

The interest of this approach for the set of problems
discussed at the present meeting is evident. To quote but one
aspect, I will recall that Baracca, Bohm, Hiley, and Stuart (9)
expressed the opinion a few years ago that Bell's theorem could
be considered as a test of the hypothesis that the wave function
of a many-body system tends to factorize into a product of
localized states of the constituent particles. The "spontaneous
transition," envisaged by the authors, from the state of the
second kind, describing a composite system, to a mixture would
get rid of one of the most disconcerting features of quantum
mechanics, *viz* that of the unlimited space-time extension of the
quantum states of composite systems, or, as it may be called, of
the *inseparability* of such states. Baracca *et al.* set themselves
also the problem of a mechanism which would be able to produce a
transition of this kind. Without going into any detail, it can be
mentioned that the authors thought that the transition could
arise from the system being actually non-isolated. The question
then becomes what it is that provides the surroundings, and this
is where the mechanisms discussed previously comes in. From the
above, it is clear that the transition here envisaged is bound to
violate the conservation of angular momentum. The situation in
this respect is neither better nor worse but just the same as in
our use of the evolution equation, but not the philosophy, of
Ellis *et al.*

The application of the evolution equation (3) to density
matrices of the form (8), with β dependent on time, gives rise to
a trivial dynamical system. The desired behavior arises if a
further condition is imposed on the system. Quantum mechanics
forbids the decay of pure states ($\text{Tr } \rho^2 = 1$) into mixtures ($\text{Tr } \rho^2$
< 1) and therefore implies $d(\text{Tr } \rho^2)/dt = 0$. This condition must
be released and replaced by

$$\text{Tr } \rho\dot{\rho} \leq 0 , \tag{9}$$

which allows the decay of a pure state into mixtures, while still
forbidding the reverse process.

Then it can be shown that all "states," either mixed or
pure, evolve exponentially to the mixed state of maximal entropy
(ref. (7)).

Incidentally, this result appears satisfactory as a
corollary to the analysis by Ellis *et al.*, inasmuch as one could
have wondered whether the proposed evolution equation was general
enough to be able to produce this effect.

We have thus indicated, in principle, a mechanism and a
formalism which are able to produce the kind of transition
foreseen by Baracca *et al.*

The question of the proper time scale for this type of
process is of course of primary importance. We lack a theory
capable of producing quantitative predictions. Some observations
of a very general character will however permit us to conclude
that the scale can only be cosmological and that, as a consequence, according to the approach here discussed, the evolution of
pure into mixed states would in principle take place in the
universe, but would not be testable in the laboratory. Indeed, it
must be stressed that the notion of entropy, to which is here
referred, arises from considering a set of equivalent systems in
the sense the term has in quantum mechanics. As a consequence,
the possibility of a transition from a pure to a mixed state
described here would concern any system described by quantum
mechanics. Against this possibility speak, on the one hand, the
stability on a cosmological scale of the atomic levels, on the
other hand, the invariance under T and CPT (10), apart from the
neutral K-meson system. (The time scale of a possible violation
of these principles would also be cosmological.)

On this basis, the negative outcome of the experiments, as
analysed in papers by Wilson, Lowe, and Butt and by Bohm and
Hiley (11) was to be expected.

Before leaving this section, I will also mention possible
applications of the formalism analyzed so far in the quantum
theory of measurement. Transitions from pure to mixed states
arise in this field and could be accounted for by an equation of
the form (3). In this case the physical meaning of the second
term in Eq.(4) would be completely different. One could also try
to describe the preparation of a pure state. A linear evolution

equation would however not do, because it would not produce the needed saturation mechanisms. Mathematical models which could produce the desired effects in simple cases are being studied (12).

As a final comment to this specific item, it should be made clear that the idea put forward here is not claimed to be more than a *pedagogical model* for the measurement process; for a discussion of *theories*, see the report by Ghirardi in this volume.

4. ON THE POSSIBILITY THAT EXTRA SPATIAL DIMENSIONS COULD INFLUENCE THE QUANTUM MECHANICAL EVOLUTION LAW

This third, and more speculative, part of my contribution deals with the possibility that extra spatial dimensions may play a role, similar to that discussed in the preceeding sections, in influencing the time development of quantum states.

I am here referring to the set of ideas and procedures going under the name of Kaluza-Klein theories. As is well known, the initial idea was put forward by Theodor Kaluza (13) in 1919, who tackled the problem of a unified geometrical formulation of gravitation and electromagnetism by coming, in his own words, "to the strange decision to take recourse to a fifth dimension of the world (14)." A further impulse to the development of this idea was given by O. Klein in 1926 (15), coincident with the advent of quantum mechanics (16). (I shall briefly return to some of his ideas further on.) Fallen into near oblivion for a few decades, the scope of Kaluza-Klein-type theories has recently grown enormously in connection with attempts to build a unified theory embracing all known interactions: For this purpose, more spatial dimensions must be added to Kaluza's fifth dimension, so that one speaks more properly nowadays of *multidimensional unified theories*, or MUT's for short.

This is not the place for entering into any detail. For immediate reference, I shall only quote a few results and speculations. In the original five-dimensional case, the fifth dimension is rolled up to form a tiny circle, a "fiber"; the electric charge is the momentum conjugate to the fifth coordinate; quantization of charge follows then from the periodicity of space-time in the fifth dimension; the value of the fine structure constant fixes the length of the fiber at roughly 10 times Planck's length; for the scalar case, the theory can accomodate charged particles with a mass of the order of one tenth of Planck's mass (10^{19} GeV) and a "stack" of particles with multiple charges and with masses which, in the simplest hypothesis, are also multiples of Planck's mass. All such particles would be inaccessible at accelerator energies.

Similar but much more complex conclusions hold for the multidimensional case. No general agreement exists on a definite

MUT, although there are arguments which favor a formulation with
seven compact extra spatial dimensions (17) (forming, in one of
the versions, a 7-sphere). Several problems plague the attempts
to formulate a realistic theory, among them that of the masses
and group representations of the Fermion states. In the spinor
case, charged states with vanishing mass (zero-modes) can exist.
The general tendency is to associate both quarks and leptons with
the zero modes (17). For sure, particles with a mass of the order
of a microgram cannot be identified with any existing particle.

An alternative possibility which has been considered (18) is
to' think of the leptons as zero modes, but to associate the
quarks to the states of the first level of the stack. In this
way, one would turn one of the major drawbacks of the theory into
an interesting perspective. The proposal can succeed if one can
envisage a mechanism through which massive colored modes can
combine to form colorless hadrons. Such a mechanism is at least
compatible with the general frame and, if operating, could
produce the following result: If a hadron is approximately
described in terms of a point-like product of quark fields, then
the component quarks behave as if they had exactly vanishing
masses. To a finite dimension $\sim m^{-1}$ of the hadron would then
correspond an effective mass of the quark $\sim m$. A rigorous confinement would thus be replaced by an imperfect confinement (tight
binding, (ref. (18)), which leaves open the possibility of free
quarks with a mass of the order of Planck's mass.

Let me go back to the question formulated in the first
paragraph of this section: Are there reasons for concluding that
the existence of further spatial dimensions would produce a
modification of quantum mechanics of the type produced by the
metric fluctuations arising in quantum gravity?

My feeling is that one cannot speak of compelling reasons,
but that there are hints suggesting such a conclusion.

The first hint is related to the "unobservability" of the
extra dimensions. The notion can be given various semi-philoso-
phical meanings, but by it I mean here simply that there is no
pratical way of localizing objects within distances of the order
of ten Planck's lengths, where extra dimensions, if they really
existed, would manifest themselves. The current view is that
ordinary quantities arise from averaging over the extra space
coordinates.

From the point of view of the structure of the Hilbert space
of quantum states, this seems to imply the following. One introduces, to begin with, further space variables. A basis in Hilbert
space can then be built in general as a direct product of an
ordinary coordinate basis and a basis of eigenvectors of the new
space observables, which leads to a direct product structure of
the Hilbert space. The above unobservability and the related
averaging procedure imply that the actual density operator for a
quantum system would be obtained by forming the trace of the
density matrix referred to a basis for the direct product space

over the eigenstates of the new variable. This process reduces pure to mixed states.

The above argument points to the possibility that extra spatial dimensions could imply a transition of the desired type. But it does not exactly establish this possibility; and it does not properly work. Indeed, according to it, any quantum state would necessarily be a mixed state. And transitions do not just come about. So it was perhaps a hint, but we must look farther.

The second hint is introduced by the following chain of arguments. Let us first observe that an ordinary Hamiltonian would arise by averaging over the extra space coordinates. It therefore would not have nonvanishing matrix elements between "zero-mass" and higher-mass states. For that we need a dependence on the extra coordinates, or, in other terms, the off-diagonal elements of the Hamiltonian should correspond to "charged" spurions. This means that there is no ordinary pertubation theory involving the higher states, and that these are always to be considered as a manifestation of the extra dimensions. Note, on the other hand, that a dependence of the Hamiltonian on the extra coordinates emerges naturally in a Kaluza-Klein scheme if one takes the extra dimensions seriously enough by considering fluctuations of the metric about its background value (this has a counterpart in any quantity, such as the curvature scalar, computed from the metric). The off-diagonal spurion part of the Hamiltonian would then act as a constant perturbation, causing transitions by transferring energy to and from vacuum fluctuations of the metric in multidimensional space-time. The possibility of these processes would then lead to a modified evolution equation of the ordinary states, in connection with the "leakage" to and from the states of the particle stack. The basic mechanism is akin to those which give rise to non-Markovian master equations.

A question related to the above concerns what would be the fate of a higher-mass state once created at the expense of the energy localized in the vacuum. An intriguing possibility is that, as an object with a mass of the order of Planck's mass and possibly with the same size-to-mass relation as the known particles, it would give rise to a mini black-hole. (19) (This, by the way, could also concern non-tight-bound quarks in the view expressed in Ref. (18).) An outcome of this eventuality would be an apparent (extremely rare) disappearance of the known particles with no visible final products.

Fortunately, for those who would be inclined to think that all this comes too close to science fiction, there is no compelling theoretical framework; as established as for the cases discussed in the first section, in which the above arguments and partial conclusions necessarily hold. Moreover, it might well be that they would not withstand criticisms of same loopholes which were overlooked in the above. Here is an example of criticism which falls very short of invalidating the whole argument at its root. Let us, for simplicity, refer once more to the five-dimensional situation. For this case, as we mentioned, the charge is the fifth component of the momentum; but then it seems one could treat charge as *the* further observable compatible with the ordinary space, spin, etc. variables. If this conclusions were indeed unavoidable, the entire discussion of unobservability would become empty and, with it, all the subsequent analysis.

The argument can however (perhaps) still be saved by the observation that, in the second track suggested above, one does not average over the extra coordinates, implying a limited accessibility to them, corresponding to an incomplete determination of charge.

As far as a formal approach it concerned, it need not differ in principle from the one set up for the general case of a quantum system plunged in a thermal bath. Before any definite step is taken in this direction, one should however define a theoretical scheme, founded on a self-consistent set of physical ideas, in which the above hints are elevated to compelling reasons.

Finally, I would like to recall that Klein started speculating about the fifth dimension by wondering if an ordinary wave equation in five dimensions would produce a quantum mechanical wave equation when projected onto ordinary space-time (20). It was actually in this way that he obtained what we now know as the Klein-Gordon equation. This does not imply that he derived quantum mechanics from a holistic description in terms of pure waves: He just started with waves and ended up with waves.

It seems however legitimate to ask whether one could move farther along similar directions: The aim would be then to produce not just, or not only, violations of quantum mechanics, but quantum mechanics itself. It is known that Einstein contemplated for a while this possibility (21). The idea might be completely hopeless, but it seems to pose a very worthy challenge for people working on the foundations of quantum mechanics.

ACKNOWLEDGMENTS

During preparation of this report I benefited from conversations with R. Balbinot and F. Cannata, and what in this paper is both nontrivial and correct is probably due to them. Needless to say, they do not share any responsibility for any of my speculations or inferences.

REFERENCES

1. S. W. Hawking, "Breakdown of predictability in gravitational collapse", *Phys. Rev. D14*, 14 (1976).
2. S.W. Hawking, "The Unpredictability of quantum gravity",

Commun. Math. Phys. 87, 395 (1982).

3. J. Ellis, J.J. Hagelin, D.V. Nanopoulos, and M. Srednicki, *Nucl. Phys. B244* 125 (1984).

4. S. Bergia, F. Cannata, and V. Monzoni, *Found. Phys. 15*, 145 (1985).

5. T. Banks, L. Susskind, and M.E. Peskin, *Nucl. Phys. B244*, 125 (1984).

6. V. Gorini, A. Frigerio, M. Verri, and A. Kossakowski, E.C.G. Sudarshan, *Rep. Math. Phys. 13*, 149 (1978).

7. S. Bergia, F. Cannata, B. Giorgini, and V. Zamboni,*Lett. Nuovo Cimento, 43*, 113 (1985).

8. S. Bergia, and B. Giorgini, in progress; based on an adaptation of the analysis carried out in G.G. Emch and G.L. Sewell, *Journ. Math. Phys. 9*, 946 (1968).

9. A. Baracca, D.J. Bohm, B.J. Hiley, and A.E.G. Stuart, *Nuovo Cimento 28B*, 453 (1975).

10. For a general discussion of this aspect, as well as of general features of the matter discussed in Sec. 1, apart from Ref.(2), see: R.M.Wald, *Phys. Rev. D21*, 2742 (1980); R.M. Wald, "Black holes, thermodynamics, and time reversibility, in *Quantum Gravity 2*, C.J. Isham, R. Penrose, and D.W. Sciama, eds. (Oxford University Press, Oxford, 1981); D.N. Page, *Gen. Rel. Grav. 14*, 299 (1982); F.J. Tipler, *Gen. Rel. Grav. 15*, 1139 (1983); and references therein.

11. A.R. Wilson, J. Lowe, and D.K. Butt,*J. Phys. 42* , 613 (1976); D.J. Bohm, and B.J. Hiley, *Nuovo Cimento 35B*, 137 (1976).

12. S. Bergia, and F. Cannata. "Non linear models for quantum mechanical measurements", presented at the Conference "Microphysical Reality and Quantum Formalism", Urbino, Sep. 25-Oct. 3, 1985.

13. T. Kaluza, *Sitzungsber. preuss. Akad. Wiss.*, p. 966 (1921).

14. C. Hoenselaers, English translation of Ref.13, in *Unified Field Theories of More than Four Dimensions*, (Proceedings of the International School of Cosmology and Gravitation, Erice, 1982), V. De Sabbata and E. Schmutzer, eds. (World Scientific, Singapore, 1983).

15. O. Klein, *Z. Phys. 37*, 895 (1926).

16. C. Hoenselaers. English translation of Ref.15, the Erice Proceedings of Ref.14.

17. See, in particular, E. Witten, *Nucl. Phys. B186*, 412 (1981).

18. S. Bergia, C.A. Orzalesi, and G. Venturi, *Phys. Lett. 123B*, 205 (1983).

19. F. Cannata, private communication.

20. M. Kargh, *Am. J. Phys. 52, 1024 (1984)*.

21. *See, e.g., A. Pais, "Subtle is the Lord: The science and life of Albert Einstein"*, (Oxford University Press, Oxford, 1982).

SUBQUANTUM MECHANICS AND GRAVITON BLAS

G.F. Cerofolini
SGS Microelettronica
20041 Agrate MI
Italy

ABSTRACT. Subquantum mechanics (a theory which contains quantum mechanics
as limiting case, has no measurement paradox, and is able to separate two
spacelike separated events) is tested for consistency and completeness.
The physical mechanism responsible for subquantum behavior, the graviton-
particle interaction, can be invoked to explain gravitation as well as gravitational instability of white holes and during the Big Bang.

1. INTRODUCTION

In a now classic paper,[1] Einstein, Podolsky and Rosen (EPR) opened in 1935
the debate on completeness of quantum mechanics. The rebuttal by Bohr [2]
of Einstein's criticism was judged so convincing that the Copenhagen interpretation is still accepted by most physicists as the (orthodox) interpretation. In last years, however, much work has been done on the foundations of quantum mechanics, and the results obtained have led to the foundation of new theories containing quantum mechanics as limiting case.
One of them, subquantum mechanics, has no measurement paradox, succeeds in
separating two spacelike-separated events and can be substantiated by a microscopic model of physical vacuum.

2. AXIOMATIC FOUNDATIONS OF QUANTUM MECHANICS

Quantum mechanics is based on an analogy with classical mechanics and necessarily presumes the latter as the starting point. Assuming that the classical mechanics of an elementary particle is known, its quantum mechanics
is based on the following set of axioms.

2.1. Axioms of Representation

The set of these axioms maps the physical concepts of particle, state, and

231

G. Tarozzi and A. van der Merwe (eds.), The Nature of Quantum Paradoxes, 231–254.
© 1988 by Kluwer Academic Publishers.

Table 1: Correspondence between classical and quantum mechanical representations of physical concepts

Concept	Mathematical Quantity	
	Classical	Quantum
particle	phase space S	Hilbert space $\mathcal{H}$
state	point of S	direction in $\mathcal{H}$
observable	function on S	linear operator in $\mathcal{H}$

observable in terms of mathematical quantities.

(QM1) A Hilbert space $\mathcal{H}$, whose explicit representation can be obtained with the aid of the other axioms, is associated with any elementary particle.

(QM2) A direction in the Hilbert space of the particle is associated with any possible state of the system.
This axiom means that if a is an arbitrary complex number, $|\Psi\rangle$ and $a|\Psi\rangle$ describe the same state; the vector zero is excluded because it is common to all directions. Hence attention can be confined to normalized vectors, $\langle\Psi|\Psi\rangle = 1$. This condition is not yet sufficient for a one-to-one correspondence between states and suitable vectors of $\mathcal{H}$ because $|\Psi\rangle$ and $\exp(i\,\rho)\,|\Psi\rangle$ (ρ, an arbitrary real number) represent the same state.

(QM3) A linear Hermitean operator acting on the Hilbert space of the elementary particle is associated with any observable A.
This axiom ensures that the eigenvalues of A are real and their spectrum is complete.

The task of transforming quantities of classical physics into quantum-mechanical quantities has then been exhausted (see Table 1).

2.2. Axioms of Dynamics

This family of axioms permits: the possible states of the particle to be constructed (by means of the superposition principle), the time evolution of the actual state to be determined (by the Schrödinger equation) and the explicit expression of operators associated with classical observables to be found.

The superposition principle states:

(QM4) If $|1\rangle$ and $|2\rangle$ are in correspondence with possible states of the particle, then the linear combination

$$| \Psi \rangle = \Psi_1 | 1 \rangle + \Psi_2 | 2 \rangle$$

(with Ψ_1, Ψ_2 arbitrary complex numbers) also defines a possible state of the particle.

In naive language, the superposition principle states that the dynamic state of the particle is the result of a superposition of all possible states. This axiom is the one in strongest contrast with the classical point-of-view.

(QM5) The time evolution of the state vector $| \Psi \rangle$ representing the actual state of the particle is given by the Schrödinger equation

$$i\hbar \frac{d | \Psi \rangle}{dt} = H | \Psi \rangle \tag{1}$$

where $\hbar$ is a universal constant (the reduced Planck constant) and H is the operator associated with the classical Hamiltonian of the particle. If $| \Psi \rangle_0$ is the state-vector at time $t = 0$, Eq.(1) specifies the state vector $| \Psi \rangle_t$ at time t,

$$| \Psi \rangle_t = \exp(-iHt/\hbar) | \Psi \rangle_0$$

the operator acting on $| \Psi \rangle_0$ being a continuous, unitary operator.

In order to find the expression for the operators associated with classical observables, I strictly follow the Dirac procedure [3]:
The Poisson bracket $\{A, B\}$ of two operators A and B is not uniquely defined unless their commutator, $[A, B] = AB - BA$, is given by

$$[A, B] = i\hbar \{A, B\}. \tag{2}$$

We advance the following axiom:

(QM6) The quantum Poisson bracket has the same value as the corresponding classical Poisson one (at least in particular reference frames).

If this axiom is accepted, then the commutation relations between various observables can be obtained from classical mechanics. The commutation relations (2) are sufficient for establishing the analytical expressions for the operators of quantum mechanics.

2.3. Axioms of Measure

The previous scheme allows us to find the actual state of the system from the initial state by solving the Schrödinger equation. Nothing, however, has been said about the initial state and the result of a measure process. The following axioms make up for this omission.

(QM7) If a measurement of the observable A is performed, the possible values that can be obtained are the eigenvalues of A. If the value observed is a_n, the actual state immediately after the measurement is $| a_n \rangle$, irrespective of the state vector $| \Psi \rangle$ immediately before the measurement.

The state vector before the measurement influences only the probability of obtaining a particular eigenvalue:

(QM8) The probability of obtaining the eigenvalue a_n as a result of a measure process is given by

$$P(a_n) = |\langle a_n \mid \Psi \rangle|^2 . \tag{3}$$

It is obvious that the law of evolution (QM7) of the state vector contradicts (QM5): The Schrödinger equation gives a continuous, deterministic evolution, while (QM7) states that $\mid \Psi \rangle$ changes, as a consequence of a measurement, in a sudden and discontinuous way; besides, (QM8) gives to such a discontinuous evolution a character of randomness which cannot be avoided in quantum mechanics.

3. CONSISTENCY

Just as any other formal theory, quantum mechanics must be tested for consistency and completeness. Consistency occurs when every theorem of quantum mechanics is in agreement with experimental facts (within a given interpretation). Completeness occurs when all experimental facts (expressed in terms of observables) are consequences of quantum mechanics.

If $\mid \Psi \rangle$ is interpreted as an actual physical quantity, the theory outlined in the previous section contains an inconsistency: In fact, axiom (QM5) states that $\mid \Psi \rangle$ evolves in a deterministic, continuous way during an unobserved motion, while axiom (QM7) gives an erratic, discontinuous evolution during a measurement. To avoid an animistic view of nature (for which the particle knows whether or not it is observed) and to make up for the inconsistency, Bohr proposed a restrictive interpretation of $\mid \Psi \rangle$ - the orthodoxy was established.[4]

Though Bohr did never formulate explicitly the orthodox interpretation, I believe that the following formulation is shared by most physicists: Macroscopic objects, which can be described by common-sense words, obey classical physics (including relativity). Macroscopic events are changes of state of macroscopic objects. Some rare, carefully prepared events, referred to as measurements (e.g., the ''tick'' of a Geiger counter, a track in a Wilson chamber, and so on) are conveniently ascribed to the effects of microscopic particles.

These particles are necessarily described by common-sense words (motion, proximity, etc.) and by quantities of classical physics (mass, charge, velocity, etc.).

This description is, however, experimentally found to be inadequate - classical physics does not describe the behavior of microscopic particles.

Quantum mechanics is an algorithm, a machinery, which is superposed to classical mechanics for the considered microscopic particle and is capable of retaining the part of an earlier measurement useful for predicting the results of a later measurement.

   This is the core of the Copenhagen interpretation. It succeeds in removing the inconsistency of the double behavior of nature at the microscopic
level, but the cost of such an operation is to reduce $|\Psi\rangle$ to a mere math-
ematical tool without any physical reality.

   All theories which consider the state vector as actually associated with
the particle will be called *etherodox*. All etherodox theories find a difficult obstacle just at their beginning - the inconsistency between (QM5)
and (QM7). Two approaches are possible to remove this inconsistency: The
first assumes one kind of evolution as fundamental and derives the other
as a consequence, while the second assumes both evolutions as simultaneously present, their relative weight being specified by the degree of interaction with the measuring apparatus.

4. HIDDEN-VARIABLE THEORY

This theory assumes that axioms (QM5) and (QM7) hold true simultaneously
and independently. Obviously, since (QM5) and (QM7) are contradictory, neither one nor the other can hold true exactly, but both must be limiting cases
of a more complex evolution.

   Bohm and Bub [5] developed a general theory, the ``hidden-variable theory'', along this line. Their theory describes the elementary particle by
two normalized state vectors, $|\Psi\rangle$ and $|\lambda\rangle$:

$$| \Psi \rangle = \sum_n \Psi_n | a_n \rangle$$

and

$$| \lambda \rangle = \sum_n \lambda_n | a_n \rangle$$

relative to a suitable complete orthonormal system $\{|a_n\rangle\}$. The time evo-
lution of the state vector $|\Psi\rangle$ of quantum mechanics is postulated as given
by

$$\frac{d\Psi_n}{dt} = -\frac{i}{\hbar} \sum_n H_{nm}\Psi_m + \gamma\Psi_n \sum_m \left[\frac{|\dot{\Psi}_n|^2}{|\lambda_n|^2} - \frac{|\Psi_m|^2}{|\lambda_m|^2} \right] |\Psi_m|^2 \qquad (4)$$

where $H_{nm} = \langle a_n | H | a_m \rangle$ and γ is a coupling constant between particle
and measuring apparatus.

   In absence of coupling, $\gamma=0$ and Eq.(4) is reduced to the Schrödinger equation (1). For strong coupling, the second term dominates and the evolution is described by

$$\frac{d\Psi_n}{dt} = \gamma\Psi_n \sum_m \left[\frac{|\Psi_n|^2}{|\lambda_n|^2} - \frac{|\Psi_m|^2}{|\lambda_m|^2} \right] |\Psi_m|^2 .$$

This equation can easily be solved if $\forall n :$ λ_n is supposed slowly variable
with time. In this case, all Ψ_n go to zero in a time of the order of $1/\gamma$,

except one component, say $\Psi_{\bar{n}}$, which goes to 1. This component is the one for which $|\Psi_{\bar{n}}|^2 / |\lambda_{\bar{n}}|^2$ is maximum at the time of measurement.
This evolution fits the collapse evolution postulated by (QM7), and the probability distribution (3) is produced if all λ_n are uniformly distributed over their possible values.
In conclusion, the Bohm-Bub theory gives a deterministic equation of motion of the state vector $|\Psi\rangle$ if the hidden vector $|\lambda\rangle$ is known. This evolution is the same as that of the Schrödinger equation in the absence of coupling with the measuring apparatus, while, during a measurement, the state vector collapses onto an eigenvector with the same probability as predicted by quantum mechanics, provided that the hidden vector has a kind of statistical distribution.

Bohm and Bub's theory allows the prediction of new phenomena in disagreement with quantum mechanics if two subsequent measurements are performed within a lapse of time shorter then the time required by the hidden vector to randomize. Bohm and Bub's estimate of this time, of the order of $\hbar/k_B T$, where k_B is the Boltzmann constant and T the absolute temperature of the measuring apparatus, leads to observable phenomena.
A measurement to verify a Bohm-Bub prediction gave however results in agreement with quantum mechanics [6]; though the disagreement between hidden-variable theory and experience can be removed by choosing a shorter randomization time,[7] the initial failure of the hidden-variable theory has stifled the interest in this beautiful theory.

5. QUANTUM THEORY OF MEASUREMENT

In the other context two possibilities can be considered: Either the collapse evolution can be deduced from the Schrödinger equation, or vice versa.
The idea that the collapse evolution can be explained in terms of the Schrödinger evolution goes as far back as the Jordan [8] hypothesis. This idea was reconsidered by Ludwig [9] and developed by Daneri, Loinger, and Prosperi [10, 11] to assume the dress of a ''quantum theory of measurement and ergodicity conditions.''
This theory is first concerned with the foundations of a quantum theory of macrosystems, which theory would, in principle, replace classical mechanics as fundamental theory at the macroscopic level. In this context, the measuring apparatus is regarded as a macrosystem for which macrostates can be defined in a suitable way within the conceptual framework of quantum mechanics.
The development proceeds by observing that the detection of the state of a microsystem by a macroscopic apparatus always involves an amplification process which can be seen as the irreversible transition of the amplifying apparatus to a condition of stable equilibrium. This irreversible transition permits the linear superposition of Hilbert-space vectors, rep-

resenting the quantum state of the ''microsystem + macrosystem'' after the measurement, to be replaced by a statistical distribution of quantum microstates.

The aim of the Daneri-Loinger-Prosperi theory is to show that this statistical distribution coincides with the distribution of eigenstates predicted by the algorithm of quantum mechanics.

This approach avoids the characterization of a macrosystem by a state vector as made by von Neumann [12] in his regressive analysis of the measure process, and confers a sound physical ground on the Copenhagen interpretation. In fact, Bohr's discrimination between classical and quantum objects, described by classical and quantum physics, respectively, is replaced in the quantum theory of measurement by the distinction between microscopic and macroscopic objects, both ruled by quantum mechanics, the macroscopicity of the measure apparatus being accounted for by a description in terms of quantum statistical mechanics.

In spite of these merits, orthodox people see the quantum theory of measurement merely as a restatement of quantum mechanics with no possible test to verify it (and, in absence of a direct test, why not accept Bohr's simpler view ?), while etherodox people judge as too loose the link between microsystem and state vector.

''Thus, although the analyses of Daneri, Loinger and Prosperi and other such treatments do help to clarify the measurement process to a certain extent by emphasizing the existence of an amplifying stage in the process, the basic problem is still unresolved. ... The question of the behavior of an individual system ... is not answered.''[5]

6. SUBQUANTUM MECHANICS

The last possibility is that the collapse evolution is proper on the microscopic level, the Schrödinger evolution being there reduced to it.

This possibility, whose logical analysis is necessary to complete the study of all the possibilities for etherodox theories, was proposed by me under the name of ''subquantum mechanics''.[13]

The basic idea to overcome the inconsistency between (QM5) and (QM7) is the following: *While quantum mechanics is concerned with two kinds of evolution of the same vector, subquantum mechanics is characterized by two vectors, each with its proper law of evolution.* The first one, the ''true state vector'', is seen during the measurement process, while the second one, the ''statistical state vector'', is a statistical description of the first and evolves according to the Schrödinger equation.

Like quantum mechanics, subquantum mechanics (SQM) associates a Hilbert space with any particle:

(SQM1) = (QM1),

appropriate vectors of the Hilbert space with the states of the particle:

(SQM2) = (QM2),

and linear Hermitean operators with classical observables:

(SQM3) = (QM3).

Among the other quantum mechanical axioms, from (QM4) through (QM8), (QM7) has the most direct and immediate experimental evidence.

Subquantum mechanics therefore accepts this axiom:

(SQM4) If a measurement of a observable A has taken place, the possible values which can be obtained are the eigenvalues of A. If the observed value is a_n, the actual state vector immediately after the measurement is $|a_n\rangle$.

A linear superposition of eigenstates can never be observed: It is rather strange, but true, that the main entity of quantum mechanics (a theory built upon the criticism to the concept of observability) is a quantity that cannot be observed. It is therefore tempting to use, instead of the unobservable state vector $|\Psi\rangle$ subject to the superposition principle, another vector $|\xi\rangle$ obeying the *principle of internal exclusion*:

(SQM5) For any observable A a true state vector $|\xi\rangle^A$ can be defined. The true state vector $|\xi\rangle^A$ coincides at any time with one, and only one, of the eigenstates $|a_n\rangle$ and switches among them according to proper dynamical laws, at present unknown, whose statistical features are given by (SQM7).

If the eigenvalue a_n is degenerate, the associated eigenstate is in a one-to-one correspondence with the linear manifold having a_n as eigenvalue.

It is somewhat natural to postulate that during a measurement of A at time t one observes the true state vector $|\xi\rangle_t^A$. This can be stated in a way very similar to (QM8):

(SQM6) The probability of obtaining the eigenvalue $|a_n\rangle$ as a result of a measurement process is given by

$$P(a_n) = |\langle a_n | \xi \rangle^A|^2$$

where $P(a_n)$ can however assume only the value 0 or 1.

The true state vector $|\xi\rangle^A$ can again be written as a linear combination of eigenstates

$$|\xi\rangle^A = \sum_n \xi_n^A |a_n\rangle$$

where the at present unknown time evolution of the coefficients $\xi_n^A = \xi_n^A(t)$, is such that they satisfy at any time the conditions

$$\xi_n^A(t)\xi_{n'}^A(t) = \delta_{nn'}\xi_n^A(t) \tag{5}$$

and

$$\sum_n \xi_n^A(t) = 1 \tag{6}$$

required by the principle of internal exclusion.

Condition (5) ensures that, at any time, at most one of ξ_n^A is equal to

1, while condition (6) states that surely one among all ξ_n^A is
equal to 1.

Formulas (5) and (6) describe a completely new law of motion: At a given
time one and only one of ξ_n^A is equal to 1 and remains constant for a length
of time; then ξ_n^A falls discontinuously to 0 and another coefficient, $\xi_{n'}^A$,
simultaneously rises to 1. This phenomenon can be described as the switch-
ing of $|\xi)^A$ from the state $|a_n\rangle$ to the state $|a_{n'}\rangle$.

The dynamic law of the coefficients ξ_n^A is the cardinal point of the theory. Though such a law is missing, its statistical features are easily formulated. Let me define the temporal mean values of the coefficients ξ_n^A:

$$\overline{\xi}_n^A = \frac{1}{\tau_A} \int_t^{t+\tau_A} \xi_n^A(t')dt' = \frac{1}{\tau_A}\text{m supp } \xi_n^A(t') \quad t \le t' < t + \tau_A \tag{7}$$

where ''m supp'' denotes the measure of support of $\xi_n^A(t)$, and the lapse of
time τ_A is so long that the process of average is meaningful, but so short
that $\overline{\xi}_n^A(t)$ does not vary appreciably in the interval $(t, t+\tau_A)$; the time τ_A
can in principle depend upon the observable A. It is evident that events
occurring within a duration τ_A cannot be described statistically, while a
statistical description is possible for events separated by a lapse of time
longer than τ_A. A tentative estimate τ_A will be proposed later.

The mean values of the coefficients satisfy the normalization condition,

$$\sum_n \overline{\xi}_n^A(t) = 1 \tag{8}$$

which directly follows from (6) and (7).

In the framework here outlined, if a measurement of A is carried out in
the time interval $(t, t + \tau_A)$, the value a_n is obtained with a probability
$\overline{\xi}_n^A(t)$:

$$P(a_n \mid t) = \overline{\xi}_n^A(t) \tag{9}$$

This probability is associated with a single particle, not with the ensemble of replicas of the particle, each prepared in the same way (see the discussion by Bohm and Bub [5] on this logical necessity).

The average value of the observable A, resulting from a set of measurements performed in the lapse of time $(t, t+\tau_A)$, will be

$$\overline{A}(t) = \sum_n a_n P(a_n \mid t) = \sum_n a_n \overline{\xi}_n^A(t)$$

I assume the following axiom:

(SQM7) If a system is prepared at time 0 in an eigenstate $|b_m\rangle$ of the
observable B, then at time t one has

$$\overline{\xi}_n^A(t) = |\langle a_n \mid \exp(-iHt/\hbar) \mid b_m\rangle|^2 \tag{10}$$

This axiom gives the statistical evolution of the true state vector.

Finally, I accept that the operators of subquantum mechanics are the same as in quantum mechanics:

(SQM8) = (QM6).

In quantum mechanics there is one true state vector per observable, though in general only its statistical description

$$\overline{|\xi\rangle}^A = \sum_n \overline{\xi}_n^A(t) \mid a_n\rangle$$

is known. The law of evolution of $\overline{|\xi\rangle}^A$ from a known initial condition, *i.e.* axiom (SQM7), is however very complex. It is therefore more convenient to define a ''statistical state vector''

$$\mid \psi\rangle_t^A = \sum_n \langle a_n \mid \exp(-iHt/\hbar) \mid b_m\rangle \mid a_n\rangle$$

which is strictly linked with $\overline{|\xi\rangle}_t^A$ because

$$\overline{\xi}_n^A(t) = \mid \langle a_n \mid \psi\rangle_t^A \mid^2 . \tag{11}$$

The statistical state vector $\mid \psi\rangle_n^A$ contains the same statistical information as $\overline{|\xi\rangle}_t^A$, but has some very interesting formal features:

1. $\forall A, B : \mid \psi\rangle^A = \mid \psi\rangle^B = \mid \psi\rangle$ [this follow from the completeness of $\{\mid a_n\rangle\}$];

2. $\mid \psi\rangle$ evolves according to Eq.(1) [from (10)];

3. $\mid \psi\rangle$ is normalized [from (8)];

4. the probability of observing the eigenvalue a_n is given by $P(a_n) = \mid \langle a_n \mid \psi\rangle \mid^2$ [from (9) and (11)].

These features show that the statistical state vector of subquantum mechanics has the same formal behavior as the state vector of quantum mechanics: *quantum mechanics can therefore be seen as the statistical description of subquantum mechanics.* This conclusion, however, requires that further analysis be attempted. In fact, since $\mid \psi\rangle$ has a physical existence, macroscopic properties must be explained in terms of $\mid \psi\rangle$ and hence *the quantum theory of measurement must hold true in subquantum mechanics.*

This is a serious proof of consistency that subquantum mechanics overcomes because of the validity of the Daneri-Loinger-Prosperi theory. Paradoxically, subquantum mechanics is a theory requiring the validity of a competing theory.

Obviously the previously considered theories do not exhaust all the attempts proposed to remove the measurement problem. For instance, I recall Wigner's theory (essentially a modification of the Copenhagen interpretation, with the difference that a measurement is considered to have taken place when an animate observer could obtain the result, at least in principle [14]) about which Zweifel [15] writes:

''...we are not suggesting that this description of nature has any particular merit, although it is certainly no more bizzarre than other theories which have been advocated, presumably seriously (Everett [16])''.

But Ross-Bonney [17] writes:

''Wigner has the most mystical explanation: he suggests that the re-
gression stops at human consciousness. This should be an exciting thought
to serious students of ESP.''

7. THE UNCERTAINTY PRINCIPLE

Subquantum mechanics states that, at any time, all observables have well
defined values even if they are taken in n-tuples. This property holds true,
in particular, for two non-commuting observables A and B.
The following condition seems to me a *necessary condition* for physical reality: A physical quantity has an element of physical reality only if it can
be measured with absolute precision.

Obviously, the attribute ''absolute'' is used here in the same sense as
in quantum mechanics - the degree of precision is that due to intrinsic limits of the theory and not to actual, but contingent difficulties.

If the above necessary condition is applied to the true state vectors
$|\xi\rangle_t^A$ and $|\xi\rangle_t^B$, we find that these vectors have an element of physical reality only if A and B can be measured simultaneously with absolute precision. If $[A, B] \neq 0$, the uncertainty principle seems to state that the
necessary condition is not satisfied. This violation, however, is only seem-
ing, not actual. The demonstration is carried out in two steps.

1. Heisenberg principle is formulated in pre-quantum mechanics as a corollary of the wave-particle dualism and of the complementarity principle.
Quantum mechanics contains a theorem (due to Robertson [18]) anti-correlating
the expectation values of the standard deviations of two observables, ΔA
and ΔB (where $\Delta A = \left\langle \psi \,|\, (A - \langle A \rangle)^2 \,|\, \psi \right\rangle^{\frac{1}{2}}$ and a similar definition holds for
ΔB), at the same time, in the following way

$$\Delta A \Delta B \geq \frac{1}{2} \,|\, \langle [A, B] \rangle \,| \tag{12}$$

In this form, the so called uncertainty principle does not state anything
about the measurement of two non-commuting observables, but rather states
that, if the observable A has been measured within ΔA, the value of B at
the same instant cannot be predicted with precision better than that allowed by inequality (12).
This discussion shows that the formalism of quantum mechanics is not in-
compatible with the necessary existence condition for $|\xi\rangle_t^A$ and $|\xi\rangle_t^B$.

2. The second step is the consideration of a particular example showing that A and B can be measured with absolute precision even if $[A, B] \neq$
0.
Consider a bank of counters used to detect the position of a particle and

the usual single slit diffraction experiment. I shall show that for a class
of events the error associated with momentum measure, δp_x, and the error
for position measure, δq_x, do not satisfy the Heisenberg relationship

$$\delta p_x \delta q_x \geq \frac{\hbar}{2} \tag{13}$$

To this end, we examine the diffraction experiment from a single slit,
where particles of momentum $p = p_z$ pass through a slit of width a or are
stopped by a screen. The uncertainty δq_x of the position measurement is roughl
the size w of the window of the counter with which the particle
interacted.
The knowledge of the position q_x of the counter allows the x component of
momentum, $p_x = p \sin \vartheta = p q_x / d$, to be determined with an uncertainty

$$\delta p_x = (a + w) p \cos^3 \vartheta / d$$

where d is the distance between the screen and the bank of counters and ϑ
is the angle subtended by the counter which gave the tick seen from the slit.
The product of uncertainties is thus given by

$$\delta p_x \delta q_x = w(a + w) p \cos^3 \vartheta / d \tag{14}$$

so that a whole class of events, those for which $w(a + w) p \cos^3 \vartheta / d < \hbar/2$,
does not satisfy Heisenberg's inequality. A single process of measurement
is described by (14), not by (13), and in some cases it allows the simultaneous measurement of p_x and q_x with absolute precision.

By contrast, Heisenberg's principle applies to what will happen immediately after the diffraction from the slit: Localizing the particle with
uncertainty $\delta q_x \simeq a$ does not permit the momentum p_x to be predicted with
an accuracy greater than $\delta p_x \simeq \hbar/2a$.
This uncertainty does not allow us to foresee *a priori* which counter will reveal the particle: Quantum mechanics can only associate a probability with
any counter. The simple example cited above shows that Heisenberg's principle limits the predictability of immediately future measurements and not
the possibility of simultaneous measurements of position and momentum.

8. THE EPR PARADOX

In the previous sections, I have described three etherodox consistent theories containing quantum mechanics as a limiting case and avoiding the restrictive orthodox interpretation. The debate about which, if any, of the
considered theories is the ''true'' theory, shifts therefore from consistency to completeness.
The quantum theory of measurement is unable to predict anything different from that of quantum mechanics. The degree of completeness of the former theory is therefore the same as that of quantum mechanics.

The hidden-variable theory, on the contrary, predicts phenomena different from those foreseen by ordinary quantum mechanics. The original predictions of Bohm and Bub's theory have however been disproved by experiment. This disagreement can be ascribed to an overestimation of the randomization time as originally proposed by Bohm and Bub. Though the value
of this time does not influence the validity of the hidden-variable theory [7], when the randomization time becomes comparable with the characteristic time of the Schrödinger equation ($\approx \hbar/\langle H \rangle$), the already-complex
law of motion (4) becomes totally unmanageable, thus forbidding in prac-
tice any prediction, because the components of the hidden vector $|\lambda\rangle$ can
no longer be considered constant with time. Though this difficulty is not
fundamental, but only of a computational nature, it nonetheless makes it
difficult to test the hidden-variable theory. This computational difficulty is associated with the second term of the right hand side of Eq.(4),
having this term been introduced to allow $|\Psi\rangle$ to collapse onto $|a_n\rangle$ dur-
ing the measurement.

Since subquantum mechanics does not suffer from this difficulty (because
the true state vector does not collapse, but always is in a one of the eigenstates and switches among them), there is a hope that this theory is capable of a more complete description of nature than quantum mechanics.

The usual check for completeness is the analysis of the Einstein-Podolsky-
Rosen ideal experiment.

8.1. The EPR Experiment in Quantum Mechanics

Consider two systems R and S which interact for a certain finite lapse of
time and then separate; an example might be a scattering experiment involv-
ing short-range forces. Denoting with $|\Psi\rangle_{RS}$ the state vector of the com-
plete system after the interaction and with $\{|a_n\rangle_R\}$ a complete orthonor-
mal system of the system R alone, it is possible to expand $|\Psi\rangle_{RS}$ in the
following way

$$|\Psi\rangle_{RS} = \sum_n |\Lambda_n\rangle_S \, |a_n\rangle_R$$

If, in particular, $\{|a_n\rangle\}$ is the set of eigenvectors of the observable A,
then after the measurement of A on the system R alone one or the other of
the eigenvalues a_n will be obtained with a certain probability.
At the same time, the state vector representing the system R alone will col-
lapse onto $|a_n\rangle$, $|\Psi\rangle_R \rightsquigarrow |a_n\rangle_R$ and $|\Psi\rangle_{RS} \rightsquigarrow |\Lambda_n\rangle_S \, |a_n\rangle_R$.
This means that, by performing a measurement on the system R alone, auto-
matically and simultaneously a measurement on the dinstant system S is performed:

$$|\Psi\rangle_R \rightsquigarrow |a_n\rangle_R \Rightarrow |\Psi\rangle_S \rightsquigarrow |\Lambda_n\rangle_S$$

This process implies the existence of an action at a distance, contradicting the principles of relativity, and the only way to avoid antinomies becomes the placing of strong limits on the concept of physical reality. In
particular, a condition as weak as the EPR one (''if, without in any way

disturbing a system, we can predict with certainty - *i.e.*, with probability
equal to unity - the value of a physical quantity, then there exists an element of physical reality corresponding to the physical quantity'') is not
a *sufficient condition*, as assumed by Einstein, Podolsky and Rosen, for physical reality in orthodox quantum mechanics.

8.2. The EPR Experiment in Subquantum Mechanics

Since the EPR paradox in quantum mechanics is strictly linked to the paradoxical description of the measurement itself, the subquantum mechanical
description of the EPR experiment does not suffer from paradoxes.
In fact, if the measurement of A on the system R alone yields the eigenvalue a_n, this means that at the time of the measurement the true state
vector $| \xi \rangle_R^A$ coincided with $| a_n \rangle_R$.
The statistical law of motion is such that the specification $| \xi \rangle_R^A = | a_n \rangle_R$
restricts any observable X of S to be described by a true state vector
$| \xi \rangle_S^X$, the statistical description of which is $| \Lambda_n \rangle_S$. With this kind of
evolution, the predictions of subquantum mechanics are the same as that of
quantum mechanics and Bell's inequality is violated, in agreement with an
increasing body of experimental evidences. There is no paradox because,
while quantum mechanics interprets the experiment stating that measuring
A in R *projects* $| \Psi \rangle_S$ onto $| \Lambda_n \rangle_S$, subquantum mechanics asserts that having
measured a_n in R, the law of motion is such that $| \xi \rangle_S^X$ *is* statistically de-
scribed by $| \Lambda_n \rangle_S$. In general, however, the statistical law of motion is
insufficient for a complete specification of $| \xi \rangle_S^X$, unless the system S is
forced into an eigenstate by some conservation principle (energy, momentum, or angular moment).

In conclusion, not only does subquantum mechanics admit a physical interpretation of the state vector and have no measurement paradox, but also
succeeds in separating two spacelike separated events.

9. TIME DOMAINS OF SUBQUANTUM MECHANICS

Subquantum mechanics presumes the existence of three time domains:
1. the duration $\theta_n^{(k)}$ of each of the r_n ($k = 1, 2, ..., r_n$) intervals forming the support of $\xi_n^A(t')$ for $t \leq t' < t + \tau_A$;
2. the duration τ_A over which the average is taken;
3. durations longer than τ_A.

If two events are separated by a lapse of time longer than τ_A, their statistical features are correctly predicted by ordinary quantum
mechanics.
Denoting by Δt the time interval separating two events involving A, the time
domain of ordinary quantum mechanics is given by $\{\Delta t : \Delta t > \tau_A\}$. By contrast, if $\Delta t \leq \tau_A$, then predictions obtained by quantum mechanics cannot
be applied. Consistency considerations require that the statistical dis-

tribution of $|\xi\rangle^A$ over the various states $|a_n\rangle$ does not vary appreciably with time within the duration τ_A. Hence the law of evolution of $|\psi\rangle$ (that gives the statistical distribution of $|\xi\rangle^A$) must be characterized by the existence of a time τ_0 within which $|\psi\rangle$ remains practically unchanged. The duration τ_0 is an upper bound to τ_A.

The estimation of τ_A follows from the analysis on the uncertainty principle for time and energy. Indeed, time in quantum mechanics is a parameter, not an observable, and Robertson theorem [18] does not apply. The typical time within which the observable A (not depending explicitly upon time) has a large variation is given by

$$\tau_A = \Delta A / |\langle \dot{A} \rangle|$$

From Robertson theorem [Eq. (12)], one has

$$\Delta A \Delta H \geq \frac{1}{2} |\langle [A, H] \rangle|$$

and, remembering that the time derivative of the observable A is given by $i\hbar \dot{A} = [A, H]$, it follows that

$$\Delta A \Delta H \geq \frac{\hbar}{2} |\langle \dot{A} \rangle|$$

which, in view of the definition of τ_A, gives

$$\tau_A \Delta H \geq \hbar/2$$

i.e.,

$$\tau_A \geq \hbar/2\Delta H \tag{15}$$

independent of A.

If the hypothesis $\tau_A = \tau_0$ is accepted, then the probabilities of two events, the first taking place at time t' and the second at time t'', are correctly foreseen by quantum mechanics provided that $t'' - t' \geq \hbar/2\Delta H$; by contrast, if $t'' - t' \leq \hbar/2\Delta H$, the average considered in (7) has no meaning and the probabilities of the events cannot be predicted by quantum mechanics. The time domain of this theory is therefore given by the condition

$$\Delta H \Delta t \geq \hbar/2 \tag{16}$$

Inequality (16) is just the uncertainty principle for time-energy parameters.

Durations obeying the inequalities $\theta \leq \Delta t \leq \tau_0$, θ_A being the upper bound to $\theta_n^{(k)}$, define the time domain within which quantum mechanics cannot be applied and the two measurements are said to be simultaneous in the Heisenberg sense.

Finally, I consider durations shorter than θ_A. The general expression of the support of $\xi_n^A(t')$ is

$$\text{supp } \xi_n^A(t') = \bigcup_{k=1}^{r_n} \left(t_n^{(k)}, t_n^{(k)} + \theta_n^{(k)} \right) \quad t \le t' < t + \tau_A$$

If two measurements are carried out in the time interval $\left(t_n^{(k)}, t_n^{(k)} + \theta_n^{(k)} \right)$ (small but finite, at least for discrete spectra), then the results will be the same eigenvalue with certainty in both observations. This result contradicts the law evolution (QM5) and in principle should allow subquantum mechanics to be falsified.

10. CONSISTENCY OF SUBQUANTUM MECHANICS

In Bohm and Bub's formulation of hidden-variable theory, the measurement was supposed impulsive and so strong that the Hamiltonian of the unperturbed particle can be neglected during the interaction with the measuring apparatus. This interaction is responsible for the asymptotic collapse, with a time constant τ_c of the order of $1/\gamma$, of the state vector $|\Psi\rangle$ onto the eigenstate $|a_n\rangle$. Later, Tutsch [7] has shown that such an evolution is characteristic of Eq. (4) too, provided that γ is independent of position.

Bohm and Bub's theory is associated with three characteristic times:

1. the time τ_A considered in the previous section;

2. the collapse time τ_c, i.e., the time constant required by $|\Psi\rangle$ to collapse asymptotically onto $|a_n\rangle$;

3. the randomization time τ_r, i.e., the time necessary in order that $|\lambda\rangle$ be uniformly distributed over all its possible values.

Two of them, τ_c and τ_r, are parameters of the theory subjected to the consistency constraint

$$\tau_c \ll \tau_r \tag{17}$$

Here I wish to show that with the choice

$$\left. \begin{array}{l} \tau_c = \hbar/mc^2 \quad (\gamma = mc^2/\hbar) \text{ for any } A \\ \tau_r = \tau_0 \end{array} \right\} \tag{18}$$

the hidden-variable theory is reduced to subquantum mechanics.

First of all, I observe that, in atomic or subatomic phenomena, choice (18) is consistent with Papaliolios's experimental result, showing that Bohm and Bub's early proposal of τ_r is at least 75 times larger than the actual value.

Second, I show that choice (18) is consistent with (17). In fact, Eq. (4) obviously applies to the non-relativistic range, i.e., to the range $\langle H \rangle \ll mc^2$. Since $\Delta H \le \langle H \rangle$, the previous inequality ensures that condition (17) is satisfied.

Third, because of Tutsch result, Eq. (4) with choice (18) is characterized by a collapse evolution; from a given initial value $|\Psi\rangle_0$, the state vector collapses asymptotically (with a time constant τ_c of the order of $\hbar/mc^2$) onto the eigenvalue $|a_{\bar{n}}\rangle$ and remains very close to it until $|\Psi_{\bar{n}}|^2 / |\lambda_{\bar{n}}|^2$ remains maximum; the evolution of $|\lambda\rangle$ and the Schrödinger evolution of $|\Psi\rangle$ eventually force $|\Psi\rangle$ to switch (again in a time of the order of τ_c) from $|a_{\bar{n}}\rangle$ to the eigenstate $|a_{\bar{n_1}}\rangle$ for which $|\Psi_{\bar{n_1}}|^2 / |\lambda_{\bar{n_1}}|^2$ is maximum, and so forth. A regression shows that $|\Psi\rangle_0$, too, is an eigenstate.

Fourth, the staying time close to each eigenstate is given by $|\Psi_n|^2$, provided that the fundamental ergodic hypothesis of Bohm and Bub's theory $(|\lambda\rangle$ is uniformly distibuited over all its possible values in a randomization time) is satisfied.

If the staying time close to the eigenstate of Bohm and Bub's theory is identified with the sojourn time in the eigenstate of subquantum mechanics, the above behavior of $|\Psi\rangle$ is the same as that described by subquantum mechanics. Hidden-variable theory, endowed with choice (18), can therefore be thought of as a model for (i.e., has the same consistency as) subquantum mechanics.[21]

The quasi-equivalence (the prefix "quasi" is related to the previous identification of the "staying time" with the "sojourn time") between the two theories is somewhat formal because, in their original formulation, Bohm and Bub felt that only occasionally $\gamma \neq 0$, while here I have demonstrated the quasi-equivalence in the hypothesis that γ is always non-null and very large compared with $\langle \Delta H \rangle / \hbar$ - a clue for the subquantum medium.[22]

11.   TOWARDS CLARIFYING THE SWITCHING MECHANISM BETWEEN EIGENSTATES

Quantum mechanics, i.e. the statistical description of subquantum mechanics, is possible only if $\theta \ll \tau$. The time τ is overestimated by τ_0, as given by (15). If θ does not depend on H, then as soon as the energy of the particles increases, τ_0 decreases eventually approaching θ (the arbitrary additive energy constant is eliminated as in relativity, i.e., is put equal to the rest energy). Hence it follows that *quantum mechanics cannot remain true at high energy.*

Now, it is well known that, at high energies, relativistic effects arise, quantum mechanics does not apply and must be replaced by its relativistic extensions.
It is tempting to ascribe relativistic effects and subquantum effects (both responsible for the breakdown of quantum mechanics) to the same cause. Accordingly, as soon as relativistic effects appear, subquantum phenomena emerge at an observable level. This idea allows both the estimation of the duration θ and the elucidation of the nature of the switching mechanism.

Let me consider first the relativistic extension of the Schrödinger equation for spin-$\frac{1}{2}$ particles, i.e., the Dirac equation

$$\left(i\hbar\partial/\partial t - c\underset{\sim}{\alpha} \cdot \mathbf{p} - \beta mc^2\right) \Psi^{\dagger} = 0$$

where $\underset{\sim}{\alpha} = (\alpha_1, \alpha_2, \alpha_3)$ and β denote four Hermitean operators that are independent of p and q, acting on spin variables alone, m is the particle mass, c the velocity of light, and $\Psi^\dagger$ the four-row state vector depending on q and t.

The components of $\underset{\sim}{\alpha}$ do not commute and $c\underset{\sim}{\alpha}$ represents the instantaneous velocity v of the particle.[3] Since each component of $\underset{\sim}{\alpha}$ has, in all, eigenvalues +1 and -1, the motion is seen in quantum mechanics as the superposition of a direct motion, with velocity + c, and of a reverse motion with velocity - c (*Zitterbewegung*).

Zitterbewegung appears so strange in quantum mechanics that Messiah [23] states that it is ''an additional indication that the classical picture of phenomena should not be taken too seriously.'' In contrast to this view, I shall assume that *Zitterbewegung* is a phenomenon where the subquantum nature of matter emerges at a quantum level. Since the instantaneous speed of the Dirac particle is always $\pm c$, *subquantum mechanics implies that the true state vector for the velocity component* $| \xi \rangle^{v_j}$ *randomly switches between* $|+c\rangle$ *and* $|-c\rangle$.

In this picture there are two velocities: the instantaneous velocity (always equal to $\pm c$) and the drift velocity u (related to particle momentum by the usual relation $u = p/m$), which concepts are in line with the Dirac description.

A reasonable estimation of the mean sojourn time in each of two eigenstates is the period of oscillation between them, as follows from the Dirac equation. In the static limit (null drift velicity), this period is given by $\theta = h/2mc^2$ ($h = 2\pi\hbar$).

Adopting a point of view close to the Brownian-motion theory of quantum mechanics, [24, 25] we can ascribe the random switching between eigenstates to the interaction of the particle with the subquantum medium or alternatively, with subquantum particles. Such particles have been postulated to exist (Weizel's zerons [26]), but their nature has never been clarified. In the following I shall show that, under a reasonable assumption, it is possible to specify the nature of the subquantum particles.

The fundamental hypothesis is the following [27]: *Zitterbewegung represents the physical reality of microsystems and is due to the collisions of the particle against the subquantum particles. The frequency of these collisions is equal to the frequency of the Zitterbewegung.*

If Φ is the flux (supposed to be uniform and isotropic) of subquantum particles and Σ their cross section for the interaction with the particle, the above hypothesis gives

$$\Phi\Sigma = 2mc^2/h \qquad (19)$$

in the static limit.

This result permits the physical nature of subquantum particles to be clarified. Consider, in fact, two different spin-$\frac{1}{2}$ particles, with mass m_1 and m_2, respectively. Applying Eq.(19) to them we find $\Sigma \propto m$, irrespective of the nature of the particle (hadron or lepton, charged or neutral). But the only interaction with this feature is the gravitational interaction, so that *subquantum particles must be viewed as gravitons.*[27]

Again from (19) it follows that $h \propto \Phi^{-1}$; in this view h is no longer a universal constant, but rather is a local property of the universe: The larger the graviton flux, the smaller h. This conclusion permits the physical meaning of the limit $h \to 0$ to be clarified. In fact, $h \to 0$ means $\Phi \to \infty$, and in this limit the statistical fluctuations of the graviton flux become negligible; since quantum behavior is due to these fluctuations, it vanishes as $\Phi \to \infty$, i.e., as $h \to 0$. The same result is obtained considering the limit for $m \to \infty$.

But the particle-graviton interaction is a property of all massive particles, irrespective of their spin. Therefore, a phenomenon similar to *Zitterbewegung* must characterize not only the Dirac particle, but also all other particles (such as Klein-Gordon and Proca-Yukawa particles) in the relativistic limit.

The *Zitterbewegung* of the Dirac particle can be removed by the Foldy-Wouthuysen representation [28]; in this representation, however, the eigenfunctions for the position operator are no longer Dirac δ distributions, but bell-shaped functions with approximate width $h/2mc$. The same fact characterizes the Klein-Gordon and the Proca-Yukawa particles. In the present view, the impossibility of localizing an elementary particle in a space region with size smaller than $h/2mc$, i.e. to within its Compton wavelength, is due to the collisions against gravitons.

12.   CONSTRUCTIVE FOUNDATION OF SUBQUANTUM MECHANICS

Having identified the cause responsible for the erratic motion in quantum mechanics, our next objective is a description of particle dynamics in relation to the switching mechanism; i.e., a constructive foundation of subquantum mechanics. This goal requires the first principles of subquantum mechanics, which principles must concern:

   1. the particle dynamics,
   2. the graviton-particle interaction, and
   3. the statistical features of this interaction.

The particle dynamics is concerned with the laws describing the motion of the particle under an external force $\mathbf{F}$ in the absence of collisions against gravitons.

The graviton-particle interaction is concerned with the graviton scattering on the particle; in practice, this process can be adequately, though statistically, described in terms of a differential cross section for scattering at an angle Θ.

Because of graviton scattering, the path given by the dynamic laws of motion will be broadened by the collisions against gravitons. Since the description is in terms of cross section, it is therefore necessary to have a statistical law to describe how the dynamical trajectory is broadened by these collisions.

The following assumptions appear to be resonable answers to points 1 and

3:

(CSQM1) The laws of classical mechanics hold true in subquantum mechanics, provided that (classical) velocity means drift velocity. The instantaneous velocity is not defined in the realm of classical mechanics, but only in that of subquantum mechanics.

(CSQM2) The statistical evolution is given by a Boltzmann equation for a probability density which is a function of position, instantaneous velocity, and time.

Since the particle is in *Zitterbewegung*, its instantaneous velocity in any direction is always $\pm c$. This fact requires some modifications of the Boltzmann transport equation. In particular, limiting the attention to a particle in one dimension (1D), rather than the function $f(x, v_x, t)$ two functions of x and t only must be considered: $f_+(x, t) = f(x, v_x = +c, t)$ and $f_-(x, t) = f(x, v_x = -c, t)$.

It is possible to show [29, 30] that a particle in 1D, in the absence of external forces, is described by an equation, the 1D Aron equation [31, 32] for null external forces, which is equivalent to the Dirac equation in the hydrodynamic representation.

This result shows that the objective of a constructive foundation of subquantum mechanics can be thought of as accomplished. This success is only partial because the derivation is intrinsically based on the restriction to 1D and cannot be extended to 3D.

Further progress in the constructive foundation of subquantum mechanics requires the accurate consideration of point 2., *i.e.* a description of the graviton-particle interaction.

13. GRAVITON BLAS

I shall postulate the following axioms to describe the graviton-particle interaction.

(CSQM3-1) Gravitons are massless particles moving at the velocity of light. The Graviton dispersion law is given by $E = cp$.

(CSQM3-2) The particle-graviton interaction takes place through the temporaneous capture of the graviton by the particle.

(CSQM3-3) The particle's instantaneous velocity is the same as that of the captured graviton.

(CSQM3-4) A captured graviton is released when and only when another graviton is captured.

(CSQM3-5) The global capture-release process is a Compton scattering of the graviton on the particle.

(CSQM3-6) In this process, most events take place without momentum exchange.

(CSQM3-7) Momentum is exchanged only when the graviton is in resonance with the particle, *i.e.* when the graviton energy E_Γ equals the particle rest mass mc^2 within the energy width Δ_{mc^2}.

(CSQM3-8) On the average, momentum is released from the graviton to the particle (or *vice versa*) when the graviton energy is greater (lower) than the particle energy.

Because of axioms (CSQM3-1) and (CSQM3-2), the instantaneous velocity of the particle is always $\pm c$ and therefore the particle is in *Zitterbewegung*.

Because of axioms (CSQM3-1), the graviton rest mass is null, which implies that the cross section for the graviton-graviton interaction is null, too. Gravitons do not interact mutually and form a kind of gas.

Since the concept of gas is intrinsically associated with the occurence of two-particle interactions, while gravitons never collide with one another, I shall use a different word to denote a set of non-interacting particles - blas [1]. Two other examples of blas are: the blackbody radiation at 3 K, and a rarified gas in a system where the molecular mean free path is much larger than the system size.

In the above hypotheses it is possible to demonstrate that two particles embedded in a graviton blas manifest a gravitational effect.[27, 33, 34] If the temperature of the graviton blas is higher than that of matter, *i.e.*, if the particle kinetic energy is lower than the graviton energy, the gravitational interaction is attractive. Otherwise, the reverse Compton scattering can occur and the gravitational interaction becomes repulsive - matter at very high temperature is gravitationally unstable.

A rough estimate of the temperature T at which a neutron star becomes unstable is the following. When the thermal kinetic energy $k_B T$ is of the order of $m_n c^2$ (m_n, neutron rest mass), $k_B T \approx m_n c^2$, it is resonable to admit that the reverse Compton effect prevails.This estimate gives $T \approx 10^{13}$ K. As the temperature exeeds this value, matter becomes gravitationally unstable; this fact suggests a mechanism for the Big Bang and for white holes.

In passing, I remark that, in my model, the Big Bang took place at a temperature of the order of 10^{13} K. This limit makes the universe younger by a remarkable amount, about 2×10^{-2} s. Though this time may appear extremely short, especially in comparison with the age of the universe (of the order of 3×10^{17} s), it however removes a serious difficulty of the standard Big Bang theory - the absence of magnetic monopoles in the universe. Indeed, quite irrespective of details, all grand unification theories predict that magnetic monopoles are stable particles with a huge rest mass of the order of 10^{16} proton mass. These particles should have been formed in the early universe, when its temperature exceeded 10^{29} K, and, being stable, they should now be distribuited throughout the space. However, experiments to detect the magnetic monopole have given negative results. In my theory this contradiction does not exist because the Big Bang took place at a temperature of 10^{13} K.

Finally, I wish to emphasize that in my view there are only three kinds of interaction: weak, electromagnetic, and strong, possibly unified. The so-called gravitational interaction appears as a second order effect of quan-

Table 2: Current physical theories as presumed limiting cases of subquantum mechanics

Classical mechanics takes into account neither kinematic effects
nor graviton fluctuations.

Quantum mechanics takes into account only graviton fluctuations.

Relativistic mechanics takes into account only kinematics effects.

Relativistic quantum mechanics takes into account both kinematic effects
and graviton fluctuations in a statistical way.

General relativity takes into account kinematic effects and only large
fluctuations of graviton density.

---

tum behavior. This fact is the cause of the numerous unsuccessful attempts
at quantization of the gravitational field.[35] Indeed, just as the quantization of the phonon field is not carried out starting from the sound wave
equation (because sound represents a macroscopic statistical behavior of
phonons), so it is impossible to quantize the gravitational field starting from its statistical macroscopic properties as described by the Einstein equations.

Table 2 shows how current theories should be obtained as limiting cases
of subquantum mechanics.

14.   CONCLUSIONS

Subquantum mechanics is very close to de Broglie view of nature according
to which a particle is thought of as

''subjected to a kind of Brownian motion that forces the particle to jump
from one trajectory to another ... . This obliges us to think that the
subquantum medium is a kind of heat reservoir ... . The present situation [ordinary quantum mechanics] recalls the one existing when classical
thermodynamics constituted a rigorous formal system capable of exact predictions, but based on principles widely considered arbitrary and *à priori*.
It was only because of the statistical interpretation of thermodynamics,
mainly due to L. Boltzmann and J.W. Gibbs, that the very nature of entropy
was understood and both prediction and the explanation of new phenomena were
possible.''[36]

In subquantum mechanics, $|\psi\rangle$ plays a role similar to that of the partition function in statistical mechanics: Though the partition function does not describe in detail the actual state of a macrosystem, it however contains the information required for the calculation of most observable properties. Similarly, though $|\psi\rangle$ does not describe completely the actual state $|\xi\rangle^X$ of a microsystem, it however is capable of correct predictions. The description in terms of $|\psi\rangle$ fails during single events (such as a measurement process) or in the lapse of time during which a statistical description has no meaning. The graviton blas is de Broglie's ''kind of heat reservoir'' responsible for the jump from one eigenstate to another.

NOTE

[1] The words ''gas'' and ''blas'' were coined by van Helmont, an iatrochemist of the XVII Century, to denote any ferment governing life functions and the spirit responsible for all life processes, respectively.

REFERENCES

1. A. Einstein, B. Podolsky, and N. Rosen, *Phys. Rev.* **47**, 777 (1935).

2. N. Bohr, *Phys. Rev.* **48**, 696 (1935).

3. P.A.M. Dirac, *The Principles of Quantum Mechanics* (Oxford University Press, Oxford, 1958).

4. N. Bohr, *Atomic Theory and the Description of Nature* (Cambridge University Press, Cambridge, 1934).

5. D. Bohm and J. Bub, *Rev. Mod. Phys.* **38**, 453 (1966).

6. C. Papaliolios, *Phys. Rev. Lett.* **18**, 622 (1967).

7. J.H. Tutsch, *Rev. Mod. Phys.* **40**, 232 (1968).

8. P. Jordan, *Phil. Sci.* **16**, 269 (1949).

9. G. Ludwig, *Die Grundlagen der Quantenmechanik* (Springer-Verlag, Berlin, 1954).

10. A. Daneri, A. Loinger, and G.M. Prosperi, *Nucl. Phys.* **33**, 297 (1962).

11. A. Daneri, A. Loinger, and G.M. Prosperi, *Nuovo Cimento B* 44, 119 (1966).

12. J. von Neumann, *Mathematical Foundations of Quantum Mechanics* (Princeton University Press, Princeton, New Jersey, 1955).

13. G.F. Cerofolini, *Nuovo Cimento B* **58**, 286 (1980).

14. E.P. Wigner, *Am. J. Phys.* **31**, 6 (1963).

15. P.F. Zweifel, *Int. J. Theor. Phys.* **1**, 67 (1974).

16. H. Everett, *Rev. Mod. Phys.* **29**, 454 (1957).

17. A.A. Ross-Bonney, *Nuovo Cimento B* **30**, 55 (1975).

18. H.P. Robertson, *Phys. Rev.* **34**, 163 (1929).

19. J.S. Bell, *Physics* **1**, 195 (1964).

20. L. Mandelstamm and I. Tamm, *J. Phys. (USSR)* **9**, 249 (1945).

21. G.F. Cerofolini, *Lett. Nuovo Cimento* **35**, 457 (1982).

22. D. Bohm and J.P. Vigier, *Phys. Rev.* **96**, 208 (1954).

23. A. Messiah, *Quantum Mechanics*, Vol.II (North Holland, Amsterdam, 1970).

24. D. Kershaw, *Phys. Rev. B* **136**, 1850 (1964).

25. E. Nelson, *Phys. Rev.* **150**, B 1079 (1966).

26. W. Weizel, *Z. Phys.* **134**, 264 (1953).

27. G.F. Cerofolini, *Lett. Nuovo Cimento* **23**, 509 (1978).

28. A.O. Barut and S. Malin, *Rev. Mod. Phys.* **40**, 632 (1968).

29. G.F. Cerofolini, *Lett. Nuovo Cimento* **34**, 424 (1982).

30. G.F. Cerofolini, *Nuovo Cimento B* **79**, 59 (1984).

31. J.C. Aron, *Found. Phys.* **9**, 163 (1979).

32. J.C. Aron, *Found. Phys.* **11**, 77 (1981).

33. G.F. Cerofolini, *Lett. Nuovo Cimento* **26**, 125 (1979).

34. G.F. Cerofolini, *Lett. Nuovo Cimento* **29**, 305 (1980).

35. D.N. Page and C.D.Geilker, *Phys. Rev. Lett.* **47**, 979 (1981).

36. L. de Broglie, in *Scienziati e Tecnologi Contemporanei*, E. Macorini, ed. (Mondadori, Milano, 1974).

PART 5

LOGIC AND PROBABILITY IN QUANTUM MECHANICS

FOUNDATIONS OF QUANTUM MECHANICS :
A QUANTUM PROBABILISTIC APPROACH

Luigi Accardi
Princeton University
Department of Statistics
Princeton, New Jersey 08544

ABSTRACT

We investigate the following topics :

1   Four basic questions in the foundations of quantum theory.
2   The reality issue in quantum theory.
3   The "proof" that the notion of reality is incompatible with quantum theory.
4   Collapses and paradoxes.
5   The quantum probabilistic solution : the statistical invariants.
6   An example of statistical invariants.
7   Deduction of the quantum formalism from Heisenberg principle.
8   Model independent formulation of some basic facts of quantum theory.
9   Quantum probabilistic analysis of the measurement process.
10   Statistics of first kind measurements, Zeno paradox and collimators.
11   Quantum probabilistic analysis of the measurement process : discrete case.
12   Deduction of von Neumann's measurement postulate.
13   Quantum probabilistic analysis of the measurement process : continuous case.
14   Deduction of the Luders-Zumino postulate.
15   The operational approach.
16   Iterated measurements, joint probabilities and Wigner's formula.
17   Ludwig's approach.
18   Classical operator valued processes and the Barchielli,
     Lanz, Lupieri,Prosperi construction.
19   Quantum Markov processes.
20   Quantum Markov theory and the quantum measurement process.
21   Quantum stochastic processes.
22   The maximal and the minimal hidden variables program.
23   von Neumann's theorems on hidden variables.
24   Gleason's theorem and some of its corollaries.
25   The equivalence between the conditions in the theorems of von Neumann
     and of Bell-Kochen-Specker.
26   A trivial construction, its complication and a known example.
27   Dropping down the one to one correspondence between quantum observables
     and hermitean operators.
28   Some more examples of hidden variables theories.
29   Some misunderstandings about locality and Bell's inequality.

G. Tarozzi and A. van der Merwe (eds.), The Nature of Quantum Paradoxes, 257–323.
© 1988 by Kluwer Academic Publishers.

1.   FOUR BASIC QUESTIONS IN THE FOUNDATIONS OF QUANTUM THEORY

Quantum theory represents nowadays the deepest level of our knowledge of nature. Its successes in explaining known phenomena and in predicting new ones are beyond any reasonable doubt. However, since its very beginning this theory was accompanied by a heated debate on its foundations which is going on still today. The debate on the foundations of quantum theory has nowadays definitively gained a place among the most famous conceptual debates in the history of science. With very few exceptions all the great physicists in the last sixty years and several first rank mathematicians, logicians and philosophers, brought their contributions to this debate.

The intimate manner in which mathematics, physics, epistemology, philosophy and history of science intertwine themselves in this debate is worth the best traditions of natural philosophy.

Conceptual problems motivated deep mathematical theorems, new physical experiments, an analysis of the conceptual structure of physical theories which, even if far from being complete, constitutes an important contribution to the theory of knowledge.

Usually scientists are not very sympathetic with philosophical problems, and in general our contemporary world seems much more inclined to use the results of science rather than to meditate on their mutual coherence and their conceptual foundations. Therefore it is quite natural to ask oneself why the purely conceptual problem of the foundations of quantum theory could attract, in the last sixty years, the attention of so many distinguished scientists.

I believe that the reason for this should be looked for in the four sources of dissatisfaction about the present status of the interpretation of quantum theory listed below :

1. Dissatisfaction with the idea, emerging from Heisenberg' s principle, that our deepest level of knowledge of the natural phenomena should be of statistical nature. Does this originate from a limitation on human knowledge or does it reflect an intrinsic indeterminism of nature ?

2. Dissatisfaction with the commonly accepted statement that quantum theory compels us to abandon the notion of objective, observer independent, reality (what precisely is meant with this statement is explained in Sec. (2)). This statement puts the interpretation of quantum theory in contradiction with some basic principles of relativity theory such as locality or causality. Is this contradiction merely a matter of interpretation , or is it something experimentally observable , which tells us something about the real world ?

3. Dissatisfaction with the "non intuitive" aspects of the mathematical description of quantum theory. Why the states of a physical system should be described by rays in some Hilbert space ? and the observables by self-adjoint operators ? We know that this description works remarkably well , but somebody would also like to know why does it work.

4. Dissatisfaction with the usual theory of quantum measurement.

The problems here are both of a principle and of a technical nature. The main technical problems are : (i) One would like to have a formula to compute the probabilities of measurements concerning observables with continuous spectrum. (ii) One would like to have a mathematical model for those measurements which cannot be schematized as ideal instantaneous measurements i.e., those in which the finite time interval in which the system interacts with the measuring apparatus cannot be ideally approximated by a single instant.

The main principle problem is that one would like to reconcile, at least approximatively, the reversible, continuous Schroedinger evolution with the irreversible, discontinuous evolution usually postulated in the quantum theory of measurement.

In the last ten years a new branch of probability theory, quantum probability, was developed partly motivated by problems in the foundations of quantum theory, partly by some more specific problems in mathematics and in quantum physics.

From this theory a new point of view emerged on the foundational problems of quantum theory and, as is the case for several mutually contradictory theories on this problem, also quantum probability claims it provides a solution to the main interpretational problems of quantum theory.

In the first part of the present paper (Sects. (2) to (5)) I shall outline the quantum probabilistic analysis of the reality issue (Problem (2) of the list above). In Sect. (7) I mention how the problem of deducing the quantum formalism from physically meaningful assumptions found a natural solution within the framework of quantum probability (Problem (3) of the above list). Sections (8) to (21) are devoted to a quantum probabilistic analysis of the measurement process (Problem (4) of the above list). All the known formulas, which are usually postulated in a rather arbitrary manner, are deduced in such a way that their precise range of applicability emerges clearly. Some new formulas are obtained and it is shown how these problems are naturally embedded in the general theory of quantum stochastic processes. Finally, in Sects. from (22) to (29), the problem of hidden variables is discussed, which is related to the Problems (2) and (1) of the above list .

2.   THE REALITY ISSUE IN QUANTUM MECHANICS

It is usually said that quantum theory forces us to abandon the notion of an objective, observer-independent , reality. However, the fraction of the huge literature developed in the last sixty years on the foundational problems of quantum theory that is expressly devoted to explain why must we do so and what precisely this means is remarkably small.

The problems connected with other physical notions such as locality, separability, action at distance, etc. are all derived from the problem of reality in quantum theory. And, since in the last years the debate on these questions has acquired a new life, both at a theoretical level (with the flood of papers on Bell' s inequalities) and at an experimental level (with the Aspect-Rapisarda type of experiments), it might not be completely useless to spend a few words in order to explain, with the least ambiguity possible, what are we talking about. This is our goal in the present section.

Let S denote a physical system, B an observable physical magnitude of S , and t any instant of time. Consider the statements :

-1. At time t the observable B assumes one and only one of its values.

The statement (1) is experimentally true, in the sense that, whenever one performs a measurement of B on the system S ,one finds that (1) is verified. This fact induced some physicists, among which Einstein, to believe that the statement (1) expresses an intrinsic physical property of the system S, an objective reality , independent of any observer. In other words, these physicists believe that the statement (1) is equivalent to the statement :

-2. At time t the observable B assumes one and only one of its values whether somebody is looking at it at that moment or not.

Some other physicists, however, claim that the statement (2) is in general wrong and that the correct statement is :

-3a. At each time the observable B assumes one and only one of its values if a measurement of B is actually performed on S at that time.

-3b. If at time t no measurement of B is actually performed on S, then if one supposes that the statement (1) is true , one will arrive necessarily at a contradiction with the experiments .

In colloquial language, the statements (3a) and (3b) can, respectively, be translated as :

-4a. An observable looked at assumes one and only one of its values .

-4b. An observable not looked at cannot assume one and only one of its values.

B.Russel gave another colloquial translation of (3b) which approximately reads : "It cannot rain in a country where there are no eyes to check whether it rains or not".

Assertion (2) is quite familiar in classical physics : it simply states that the natural phenomena have an objective reality, independent of any eventual observer. Assertion (3a) [or (4a)] is nothing more than the registration of some experimental results , the experimental errors in the determination of these quantities being of a different order of magnitude with respect to the effects we are discussing. The really characteristic feature in the orthodox interpretation of quantum theory is thus contained in assertion (3b) or (4b).

A physical system (necessarily not looked at) for which an observable B cannot assume one and only one of its values is said to be in a state of physical superposition . This notion has to be distinguished from the notion of mathematical superposition, which is a perfectly well defined formula.

There are no doubts that assertion(3b) spells out the more or less explicit attitude of the majority of the physicists concerning what a superposition state is. However, also in this case, the explicit literature on this point is surprisingly poor. I hope that the following citations will convince the reader that the term "orthodox interpretation" even if not referring to a systematically codified or axiomatized body of knowledge, i.e. to a formal theory, nevertheless singles out a well defined intersection among the attitudes of several scientists concerning the notion of physical superposition.

Of an atom in a box, divided in two by a wall with an hole in it, Heisenberg says *(Physics and Philosophy)* : "The atom can find itself, according to classical logic, either in the left half of the box or in the right one. There is no third possibility, ' tertium non datur' . In quantum theory however, we must admit, supposing we want to use the terms "atom" and "box" - that there exist other possibilities which are strange mixtures of the first two possibilities" and later on he adds : "the question whether the atom is in the left half of the box or in the right is not decided. But the term 'not decided' is in no way equivalent to on the left, only that we do not know where it is. But 'undecided' means a different situation, expressible only with a complementary statement"

Of an electron which can reach a screen only via two holes in another screen (cf. Sect. (5) in the following) , Feynman says : "... to conclude that the electron passes through one or the other slit when one doesn' t look, is equivalent to do an error in the prediction" (cf. the following Sect. 3 for a more precise statement).

Similar ideas are expressed by Regge in his popular version of the EPR paradox : "Suppose to put in an urn a white ball and a black ball. If I put one of these in one of my pockets without looking at it; a friend of mine makes the same thing with the other one; we then take planes which fly in opposite directions and, once separated, we observe the color of the ball. If I find a black ball I am sure that the other one has a white ball and conversely".

"... intuition says that in my pocket there is always a ball of a well determined color and that the color 'exists' independently on my observation.

In quantum mechanics however one must renounce this determinism.

The black and the white are the result of a measurement and became real only when the measurement is performed.

" ... if I pick out the ball and measure it black , I should say that I have created the black by measuring it, that the black became real only in the instant in which I observed it. It follows that also the white in the pocket of my friend is now real, as a consequence of my measure.

Therefore I have created a reality at distance instantaneously through the simple measurement of a local attribute."

The words of Dyson (in: his book *Disturbing the Universe*) synthesize the conclusion of many physicists on the situations described above : "It is a field [microscopic physics] in which the dogma of Monod : 'the cornerstone of the scientific method is the postulate that nature is objective' turns out not to be true...." and, more specifically concerning the meaning of the EPR paradox : " .... there is a famous experiment, originally suggested by Einstein, Podolsky and Rosen in 1935 as an ideal experiment to illustrate the difficulties of quantum theory, which proves that the idea that an electron exists in an objective state, independent on the observer, is untenable."

In more technical terms the same idea is expressed by Prosperi [53] : "...the appearance of the so called interference terms in the expression of the transition probability prevents one from ascribing to a certain quantity a value independent on actual observation of it ".

These citations (and many more could be added - cf. for example [27]) have purposefully been chosen from several distinguished practicing physicists most of whom hardly ever published a paper entirely devoted to foundational issues; they have been taken from different sources, ranging from highly specialized scientific journals to technical monographs or proceedings of conferences, to textbooks for students, to books or articles of philosophy of science,...... . If these citations came only from scientific papers, one could always say : "The author was not really interested in this pseudo-philosophical question; he inserted them 'en passant', without paying much attention to them, since the real core of the paper is elsewhere ". If these citations came only from popularization or philosophical publications, one could always say : "Of course, the author, in order to make the ideas of science accessible to a non specialized public, had to use simple images and words, thus he.necessarily altered the precise physical ideas he had in mind ."

But when all the indications from several different types of scientific literature, different authors, different nationalities, different and generations, point towards the same conclusion; when no published evidence can be produced against the claim that these statements correctly express the commonly accepted point of view ; when the few published arguments explicitly dealing with the problem are weaker than the arguments leading to the paradoxical conclusion (cf. the critique to Popper' s point of view in Sect.(5) in the following); then one has the right to say that this conclusion really represents the point of view of a community . To this point of view, which essentially consists in accepting the validity of statement (3b) or (4b) we shall refer as to " the orthodox interpretation." In this statement the roots of all the interpretational problems of quantum theory should be recognized because once you have accepted it, it is a relatively easy game to produce several paradoxes and strange situations.

The main critique one could address to most of the literature on the interpretational problems of quantum theory is that it takes this statement as point of its departure rather than of its arrival. It takes for granted that statement (4b) or some more implicit formulation of it, is a truth of nature, proved by the quantum physicists, and either ignores its consequences or goes on elaborating on them without taking into account their root. What is stranger is that even those who have strongly opposed this statement in name of "realism" have not gone into a theoretical analysis of the empirical and theoretical bases of statements like (3b) above, and in some cases the validity of such statements was identified with the validity of quantum theory, in the sense that an experiment confirming some elementary predictions of quantum theory was interpreted as confirming the nonlocal character of the physical laws (cf. the discussion of Bell' s inequality and of the related experiments in Sect.(29) below).

But in order to provide a satisfactory answer to the four problems listed at the beginning of this paper and explain what have been called "the mysteries of quantum theory " , one must go back to the origins of the problem and analyze in detail the reasoning which lead physicists to assert that observables not looked at have no values. As repeatedly stated, the literature on this topic is very poor, but a remarkable exception is constituted by Feynmann' s papers [35a] , [35b], [35c] , where previous analyses by Heisenberg and other authors are expressed in a synthetic and clear manner . For this reason we will assume Feynmann' s analysis as the starting point of our investigation.

3.   THE PROOF THAT THE NOTION OF REALITY IS INCOMPATIBLE WITH QUANTUM THEORY

I have already discussed in detail Feynmann 's analysis of what he calls "the origin of all the mysteries of quantum theory" ,therefore only the main ideas of this analysis will be outlined here, and I refer to the papers [7], [9], [14],[19] for further information.

Let A,B,C denote three observables; a,b,c, their values; P(a,b), P(b,c), P(c,a) the probabilities of the transitions $a \to b$; $b \to c$; $c \to a$ respectively; and $\phi(a,b)$, $\phi(b,c)$, $\phi(c,a)$ the corresponding quantum mechanical amplitudes.

The transition probabilities P(x,y) (x,y = a,b,c) are experimentally measurable and according to the elementary rules of classical probability, one should have :

$$P(a,c) = \sum_b P(a,b) \cdot P(b,c) \tag{1}$$

the sum being taken over all the values of the observable B.

Some experiments show however that that the identity (1) is wrong : If one substitutes for the transition probabilities P(x,y) the approximate relative frequencies obtained in an experiment in which the relative frequency of the value x is measured in an ensemble of particles prepared so that the value y is realized with certainty, then one finds that the difference between the right and the left hand side of (1) is some orders of magnitude bigger than the experimental errors. The difference between the left and the right hand side of (1) is called, in the physical literature, " the interference terms ". According to the usual rules of quantum theory, these interference terms are calculated starting from the experimentally correct formula

$$\phi(a,c) = \sum_b \phi(a,b) \phi(b,c) \tag{2}$$

(again the sum is over all the b' s) and making use of the connection between amplitudes and probabilities :

$$| \phi(x,y)|^2 = P(x,y) \quad ; \quad x,y = a,b,c \tag{3}$$

The reality issue in quantum theory did not arise from a metaphysical whim of some physicists, but from the necessity of explaining the contradiction between an elementary consequence of some basic principles of classical probability – the identity (1) – and the results of some experiments.

The way to overcome this contradiction, deviced by the orthodox interpretation, is best explained through the words of Feynmann ([35b], pg.369):

5 "...Looking at probability from a frequency point of view (1) simply results from the statement that in each experiment giving a and c, B had some value. The only way (1) could be wrong is the statement, "B had some value", must sometimes be meaningless. Noting that (2) replaces (1) only under the circumstance that we make no attempt to measure B, we are led to say that the statement, "B had some value", may be meaningless whenever we make no attempt to measure B."

The term "meaningless" here does not mean at all that for us it is indifferent how an object behaves when it is not looked at. On the contrary, Feynmann's analysis leaves no doubts on the fact that the correct interpretation of this term must be : if, under the circumstance that we make no attempt to measure B, we say that the statement " B had some value " is true, then we are led to a contradiction with experiment.

Thus the results of the experiments tell us that an observable always assumes one and only one of its values but, according to the orthodox interpretation, if we assume that this fact is an intrinsic, objective property of the physical systems, independent on any attempt we make to measure B, then we run into a contradiction. If we call "objective reality " the results of the experiments, then the above statement means that the validity of quantum theory is incompatible with the notion of objective reality. This is what we called "the reality issue in quantum theory".

It is important to remark that the above analysis deals only with relative frequencies and elementary (classical) probability theory. This means that the contradiction between formula (1) and formula (2) will arise in any theory whose experimental predictions agree with those of quantum theory (in particular in any hidden variable theory which does not falsify quantum theory).

Feynmann' s analysis, which I have synthesized above, gives the clearest account of the reality issue in quantum theory I could find in the published literature. I did not find any formulation of this problem which could not be reduced to Feynmann' s analysis. I think that anybody who believes that there is a reality issue in quantum theory, and that it is different from the one discussed by Feynmann, should formulate the problem with the same clarity and coherence as Feynmann. While those who do not believe there is such an issue, should produce some evidence to show, either how the above contradiction was solved, or in which points the analyses of Heisenberg, Feynmann, Regge,... were not right. I insist that I found no published reference in this direction.

With these premises the first problem we have to attack if we want to find a satisfactory solution to the "mysteries of quantum theory" is the following : Is the conclusion of Feynmann's analysis correct ? is it true that "... looking at probability from a frequency point of view (1) simply results from the statement that in each experiment giving a and c, B had some value." ?

In the of papers mentioned at the beginning of this section (cf. also the following section (5)) I have shown that the identity (1) relies upon a rather subtle, and by no means obvious, mathematical postulate. This leads to the suggestion that, before declaring the notion of objective reality incompatible with quantum theory, one may try more modestly to understand the consequences of the fact that a certain mathematical axiom , implicitly underlying all the classical statistical theories, is not compatible with the results of some experiments suggested by quantum theory. Maybe the right conclusion we should draw from the contradiction between (1) and (2) is not that we must abandon the notion of objective reality, but only that we must abandon some axioms on which all the variants of the classical probability calculus were built. If we convince ourselves that there are good physical reasons to question a priori the universal validity of these axioms, if moreover we could find a rigorous way of discriminating between the mathematical models underlying, respectively, classical statistics and quantum statistics, then we will be fully justified in adopting towards probability, an attitude we have learned to adopt towards geometry. In particular we will accept the idea that different physical situations might be described by different probabilistic models ; that the classical one is just one among the infinitely many possibilities; that our usual intuitive notion of reality is perfectly compatible also with other probabilistic models (as it is with other geometric models). The standard way to distinguish between geometrical models is through "geometrical invariants" whose importance, from the mathematical and physical point of view is that they are intrinsic, i.e. they depend only on measurable characteristics (such as angles). One of the main results of quantum probability is to provide the analog of the geometrical invariants for probability theory : the "statistical invariants" (cf. Sect.(5)).

A last remark is that the choice on which of the two notion to drop --physical reality or the classical probabilistic model -- is not symmetric : the incompatibility of the axioms defining the classical probabilistic model with the experimental data arising from some quantum mechanical experiments is a fact you can prove -- not a matter of opinions. Once you have made this choice, whether to abandon or not the usual notion of reality becomes a matter of personal opinion.

By contrast, once you accept the orthodox interpretation and the consequent abandonment of the usual notion of reality, you are nevertheless obliged to abandon the classical probabilistic model, as long as you accept the experimental validity of quantum theory (which seems to be reasonably well established).

Thus, at least from the point of view of the economy of thought, the quantum probabilistic interpretation is definitively better than the orthodox interpretation.

4. COLLAPSES AND PARADOXES

In the previous section we have seen that the notion of "physical superposition" (as opposed to "mathematical superposition", which is a perfectly well defined way of computing probabilities) arose from the unjustified application of a hidden axiom of classical probability theory to the statistical data arising from some experiments. In this section we show how, once one accepts the idea that a physical superposition is a new kind of objectively existing physical state, all the so called paradoxes of quantum theory easily follow.

If these hypothetical physical superposition states exist in nature, and the act of measurement makes them to collapse into pure states, then some questions naturally arise : Is
this collapse a discontinuous, instantaneous process (as in von Neumann' s scheme of
measurement [60]) or is it a smooth proces which looks swift just because of the
differences in the characteristic times of the micro and the macroscopic objects (as in
the Daneri Loinger , Prosperi approach [33]) ? Is this physical collapse triggered by a
psychological fact (like Wigner' s friend' s conscience) ? To our limited minds the idea
of Schroedinger ' s cat, which when not looked at is in a superposition state of dead and
alive states and , when we look at it, suddenly turns into a definite state of life or
death, sounds somewhat strange. Are we really convinced , with Ludwig , that the only
way out of this odd situation is to draw a demarcation line between the macroscopic
and the microscopic world and to attribute physical reality only to the macroscopic
phenomena (to which our senses have direct access), considering the microscopic objects
only as "mediators of the interactions among the macroscopic objects " (cf. Sect.(17) in
the following) ? And, even if we accept this modern version of the old philosophycal
idea that " esse est percipi ", which will be the exact boundary between the microscopic
and the macroscopic world ? Why should the "argumentum acervi" of the old scholastic logic not apply to the macroscopic bodies (considered as acerva of molecules and
atoms) ? Moreover, and more important from a physical point of view, even if we
accept Ludwig' s statement (cf. [45], Sec. 4)that ; "... only trajectories of the macrosystem.... are registered. No microsystem appears in the result of the measurement. The
microsystem is a 'theoretical invention' to explain the experiment......", this will not
help us very much in overcoming the contradiction between theory and experiments,
described in Sect.(3) , from which the interpretational difficulties of quantum theory
started. In fact this contradiction involves only some probabilities , i.e. relative frequencies, which can be described purely in macroscopic terms (counting flashes, or clichs, or
spots on photographic plates, etc. . Instead of "electron", we can speak, as Ludwig suggests, of "interaction mediator" between the photographic plate and the apparatus emitting the x-rays (both macroscopic objects) . But, independently on the level of linguistic
complication that we introduce, the probabilities will remain the same, hence Feynmann' s analysis will apply as described in Sec.(3). Thus Ludwig' s approach, which
surely introduces some original and useful ideas concerning continuous measurements (
cf. Sec.(17) of the present paper), does not help to solve the contradiction pointed out in
Feynmann' s analysis.

If the collapse from superposition to pure states is a physical phenomenon, then it is
sufficient to consider spatially separated systems related by an exact conservation law
and in a superposition state to produce "collapses at distance". This is the situation
envisaged in the Einstein-Podolsky-Rosen paradox but, as remarked by Bergia, Cannata,
Russo and Savoia [29] , extending a previous remark made by Margenau and Park [47]
(cf. also [20], for a mathematical comment), this situation arises in fact in any pure
state of any composite system.

But if collapses are physical phenomena, then collapses at distance are physical actions
at distance − against the relativity principle.

These are the reknowned "nonlocal " effects in quantum theory: they only arise if we
accept that the "collapse of the wave packet " is a physical transition , and they should
be carefully distinguished from the non-local effects arising from the fact that one is
using a nonrelativistic form of the Schroedinger equation . The former are principal,
model independent effects characteristic of a particular interpretation of quantum

theory; the latter are model dependent effects arising from the fact that one is choosing a priori a nonrelativistic approximation to describe a physical system, and are quite familiar also in classical physics (think of the diffusion equations).

It does not help, in overcoming the above-mentioned contradiction with the relativity principle, to claim − as some authors did − that we cannot use these actions at a distance to send superluminal messages : The relativity principle has to do with all the interactions in nature and not only with those that men can use to exchange messages.

Also the relaxation time arguments, which try to smooth out in time the collapse process by taking into account the interaction with the measurement apparatus, are of no help at all against the EPR argument, since the space separation between the systems can be chosen so large that the "relaxation time" would not be sufficient for an interaction to propagate from one system to another at the velocity of light.

It must also be said that the arguments which led to the notion of physical superposition are scientifically more convincing than those which were advanced against it : For example Popper opposed the notion of physical superposition and insisted that (cf. [52] , p.20): "... this so-called 'collapse of the wave packet' is in reality something that can occur in every probabilistic theory ..." However, as shown by Feynmann' s analysis discussed in Sec.(3) , the physicists arrived at the notion of "physical superposition", and the consequent notion of collapse, just because when they tried to apply the usual rules which "...occur in every probabilistic theory..." to the results obtained in some experiments, they arrived at a contradiction. Thus a quantum physicist could have correctly answered Popper , that what he was proposing as a solution, had in fact been the source of all the problems which eventually led to the notion of "collapse of the wave packet".

In order to give a rational foundation to Popper's statement and to give a scientific ground to his , a posteriori justified, diffidence about the relevance for the foundations of quantum theory of such notions as "reality", "locality", "collapse of the wave function", and all that , one has to provide sound mathematical and physical arguments to show why it is not correct to apply the usual probabilistic rules to the statistical data obtained from some quantum mechanical experiments, and why this has nothing to do with observables not taking values when nobody is looking at them. This program was realized by quantum probability theory.

5.   THE QUANTUM PROBABILISTIC SOLUTION: THE STATISTICAL INVARIANTS

The quantum probabilistic point of view on the reality issue in quantum theory is based on some simple considerations that we illustrate with the following example . Consider the two problems :

(G) Given a set of geometrical data, say the triples of inner angles of a given family of triangles, construct a mathematical model for these data.

(P) Given a set of statistical data, say transition probabilities between pairs of values of different observables, construct a mathematical model for these data.

It is well known how to solve the geometrical problem : There are mathematical constraints on the geometrical data, called geometrical invariants, which provide necessary and sufficient conditions for the given set of data to be describable within a geometrical model. For example, if all the triples of inner angles add up to π , then we can describe them within the usual mathematical model of Euclidean space. If their sum is less than π we try hyperbolic geometry model; if greater —an elliptic model.

We are now familiar with the idea that different physical situations (e.g. high concentrations of masses in the region where the geometrical measurements are done) might require different geometrical models, and nobody would say that if the rules of Euclidean geometry fail to account for some set of geometrical data, then the objects to which these data refer do not exist when nobody looks at them.

The starting point of quantum probability is to look for something analogous to the geometrical invariants for the statistical data. This naturally leads to the notion of statistical invariant.

By analogy with the geometrical invariants, we call a "statistical invariant" -- for a given set of statistical data, with respect to a given probabilistic model -- a mathematical constraint on the given set of data, whose realization is a necessary and sufficient condition for this set of data to admit a description within the given probabilistic model.

In order to make this intuitive idea mathematically rigorous, one has to give a precise definition of what is meant for a given set of statistical data to admit a given statistical model. Rather than give such a general definition (for this we refer to [16]), let us sketch this idea on the simple and important example of the two-slit experiment (a full discussion of this experiment from the point of view of quantum probability can be found in [19], [14]).

In this experiment a source emits particles (neutrons in the recent experiments made by Rauch [54]) which can pass a first screen through two slits, labeled 1 and 2 , and are collected on a second screen. In this experiment the set of statistical data is given by the probabilities :

$P(X)$ = probability of hitting the region X of the screen

$P(X|1)$ = probability of hitting X when the slit 2 is closed

$P(X|2)$ = probability of hitting X when the slit 1 is closed

We say that the above data admit a Kolmogorovian model if there exists a probability space (Ω, F, μ) and events in Ω, still denoted X, 1 and 2 , such that :

$$P(X) = \mu(X) \tag{1}$$

$$P(X \mid 1) = \frac{\mu(X \cap 1)}{\mu(1)} \tag{2}$$

$$P(X \mid 2) = \frac{\mu(X \cap 2)}{\mu(2)} \tag{3}$$

This means simply that in the classical, Kolmogorovian model the conditional probabilities $P(X|1)$ and $P(X|2)$ must be given by the usual Bayes' formula.

But it is easy to verify (cf. [14], [19]) that this can be the case if and only if the given probabilities satisfy the constraints

$$0 < \frac{P(X) - P(X \mid 2)}{P(X \mid 1) - P(X \mid 2)} < 1 \tag{4}$$

If one changes the statistical data in (4) , i.e. the probabilities, so that (4) is still fulfilled, then the property of admitting a Kolmogorovian model will remain unaltered.

In this sense we say that the relation (4) is a statistical invariant for the set of statistical data $P(X)$, $P(X|1)$, $P(X|2)$ with respect to the kolmogorovian model.

We can see that condition (4) is by no means a tautology on the relative frequencies. By no means the failure of (4) suggests that the neutron should have not passed through one and only one of the two slits.

It is also easy to show that condition (4) is equivalent to the existence of two positive numbers $\mu(1)$ and $\mu(2)$ satisfying :

$$\mu(1) + \mu(2) = 1 \tag{5}$$

$$\mu(1){\cdot}P(X \mid 1) + \mu(2){\cdot}P(X \mid 2) = P(X) \tag{6}$$

and (6) (with $\mu(1) = \mu(2) = 1/2$) is just the relation which, according to Feynmann's analysis , should hold if the neutrons had passed through one and only one slit even in absence of measurement.

But in the deduction of (6) the assumption that (4) is satisfied is equally important as Feÿnmann's assumption that the events 1 and 2 are disjoint (when no measurement to decide between them is made). Moreover, while Feynmann's assumption is in principle uncheckable, condition (4) is expressed uniquely in terms of measurable quantities (it is a statistical invariant) . It is an experimental fact that the probabilities obtained in the two-slit experiment do not satisfy (4). Therefore the application of the usual probabilistic manipulations to these data is not allowed. This is the mathematical error in Feynmann's deduction of formula (6). It is a subtle error, since for more than three hundred years probabilities have been manipulated according to the classical rules and this has built up the implicit conviction that these rules could be universally applied to all families of statistical data, irrespective of the fact that these data might come from completely different experimental data. The same error was repeated, about twenty years later , by Bell in the deduction of his reknowned inequality (cf. the discussion in Sec.(29) below).

Thus the only proof of the necessity to attribute physical reality to the superposition states is shown to be based on a mathematical error, and the consequences of this statement , such as the reality issue, the notion of collapse of the wave packet , the locality issue, etc. loose their theoretical foundations.

It is important to remark, in connection with the hidden-variable issue, to be discussed in Secs.(23)-(28), that quantum probability is not incompatible with the existence of an hypothetical quantum mechanical state space. The only thing it proves is that on this hypothetical space there cannot exist a single probability measure expressing certain sets of experimentally given statistical data by means of the usual formulas of classical probability theory. It also shows that the above one is the only conclusion that one can draw from a correct analysis of the two-slit experiment as well as of the Bell's inequality.

6. AN EXAMPLE OF STATISTICAL INVARIANTS

In the previous section we have shown that there are constraints, called statistical invariants, that a given set of statistical data has to satisfy in order to admit a kolmogorovian model.

In this section we study the same question with respect to the usual Hilbert space model of quantum mechanics. That is, we consider a set of discrete observables and the family of all the transition probabilities among their values ; we want to compute the mathematical constraints which these transition probabilities have to satisfy in order to guarantee the existence of a single Hilbert space model in which all the given observables are maximal and in which the given transition probabilities can be expressed as the square moduli of the scalar products between the eigenvectors corresponding to the given values. We give the complete solution of the problem for an arbitrary family of two valued observables; for two three valued observables the solution is known; for observables with more than four values (even in the case of only two observables), only necessary conditions are known.

The result proved in this section was originally proved in [11] in the case of three observables. The proof was then considerably simplified by Gudder and Zanghi [37a], and here we use their ideas to obtain the general solution. It is interesting to remark that in the case of more than three observables, a new statistical invariant ,of cohomological nature, arises.

This result shows the power of the method of the statistical invariants by solving an old open problem in the mathematical foundations of quantum theory (cf. [11] and the bibliography therein), i.e., to explain why just complex, and not, say, real or quaternionic, Hilbert spaces arise in quantum theory. The answer is that the complex numbers are written into the, experimentally measurable, transition probabilities.

A last remark: it is easy for the inexperienced reader to become confused between the results obtained with the method of the statistical invariants and the old von Neumann theorem on the impossibility of defining joint probabilities for noncommuting operators (cf. Santos' review [55] as an example of such confusion). Of course, the crucial difference is that von Neumann's argument is a model dependent statement, while the statistical invariants theory is model independent. The conclusion of von Neumann's theorem can only be: if you want joint probabilities for non compatible observables, then you must look for a mathematical model different from the usual Hilbert space model of quantum theory. The theory of statistical invariants shows that, for a given set of experimentally measurable probabilities, whatever mathematical model you choose, the joint probabilities do not exist. Moreover, since the von Neumann theorem is stated inside the Hilbert space model, the problem solved in this section (i.e. the computation of the statistical invariants of the Hilbert space model itself) cannot even be formulated in that language.

The mathematical formulation of the problem is the following :

Let T be a set and let $\{ P(i,j) : i,j \in T \}$ be a family of bi-stochastic matrices of order 2 , i.e. for each i,j in T

$$P(i,j) = \begin{bmatrix} p(i,j) & 1 - p(i,j) \\ 1 - p(i,j) & p(i,j) \end{bmatrix} = \begin{bmatrix} p_{1,1}(i,j) & p_{1,2}(i,j) \\ p_{2,1}(i,j) & p_{2,2}(i,j) \end{bmatrix}$$

<u>Remark</u> In our notation, $p(i,j)$ is the (1,1)-element of the matrix $P(i,j)$ and not the (i,j)-th element of some matrix.

Our problem is to find the conditions on the $p(i,j)$'s that guarantee that, for each j in the set T, there exists an orthonormal basis $\{ \phi_1(j) , \phi_2(j) \}$ of C^2 such that :

$$| <\phi_\alpha(j) , \phi_\beta(k) > |^2 = p_{\alpha,\beta}(i,j) \qquad (2)$$

for each j,k in T and for each $\alpha , \beta = 1,2.$

In the 2x2 case this is equivalent to find, for each j in T , a unit vector $\phi (j)$ in C^2 such that :

$$| <\phi(j) , \phi(k) > |^2 = p(j,k) \qquad (3)$$

for each j, k in T such that $j \neq k$. In this case, in fact, one can choose

$$\phi_1(j) = \phi(j) \; ; \; \phi_2(j) = \phi'(j) \tag{4}$$

where $\phi'(j)$ is any unit vector orthogonal to $\phi(j)$.

In the following, unless the contrary is explicitly stated, C^2 will be referred to the basis

$$\begin{pmatrix} 1 \\ 0 \end{pmatrix} ; \begin{pmatrix} 0 \\ 1 \end{pmatrix}$$

and we will assume that

$$0 < p(i,j) < 1 \; ; \; p(i,j) = p(j,i) \tag{5}$$

Theorem(1) The problem stated above has a solution if and only if , using the notations

$$p'(i,j) = 1 - p(i,j) \tag{6}$$

$$c_0(j,k) = \frac{p(j,k) - p(j,o)p(o,k) - p'(j,o)p'(o,k)}{2\sqrt{p(j,o)p(o,k)p'(j,o)p'(o,k)}} \tag{7}$$

one has

$$| c_0(i,j) | \leqslant 1 \tag{8}$$

$$[c_0(j,k)c_0(k,h) - c_0(j,h)]^2 = [1 - c_0(j,k)^2][1 - c_0(k,h)^2] \tag{9}$$

for each pair in the set of mutually different indexes $\{o,h,j,k\}$ in T. Moreover condition (8) is equivalent to

$$| p(j,k) + p(j,o) + p(o,k) - 1 | \leqslant 2\sqrt{p(j,k)p(j,o)p(o,k)} \tag{10}$$

Remark If T contains only three elements, then condition (9) looses meaning and (10) is the necessary and sufficient condition for the solution.

Proof Fix o in T . We can assume that

$$\phi(o) = \begin{pmatrix} 1 \\ 0 \end{pmatrix}$$

since for any solution $\{ \phi(j); j \in T \}$ and for any unitary U in C^2 also the family $\{ U\phi(j); j \in T \}$ is a solution. Moreover, since each $\phi(j)$ is defined up to a scalar of modulus 1, we can assume that each $\phi(j)$ has the form

$$\phi(j) = \begin{pmatrix} x_j \\ y_j \exp i \, \theta_j \end{pmatrix} \tag{11}$$

where x_j , y_j are real numbers (in the sense tha the problem has a solution if and only if it has a solution of that form). Now (3) implies that

$$x_j = \sqrt{p(o,j)} \; ; \; y_j = \sqrt{p'(o,j)} \tag{12}$$

Moreover, given (11) then (3) can be written :

$$p(j,k) = | x_j x_k + y_j y_k \exp i(\theta_j - \theta_k) |^2 + y_j^2 y_k^2 \sin^2(\theta_j - \theta_k)$$
$$= (x_j x_k)^2 + (y_j y_k)^2 + 2(x_j x_k y_j y_k)\cos(\theta_j - \theta_k) \tag{13}$$

for each j,k in T with $j \neq k$. In the following we also assume j,k $\neq$ o . From (13) we see that the problem has a solution if and only if :

$$\cos(\theta_j - \theta_k) = \frac{p(j,k) - p(j,o)p(o,k) - p'(j,o)p'(o,k)}{2\sqrt{p(j,o)p(o,k)p'(j,o)p'(o,k)}}$$

And we recognize that condition (8) is the necessary and sufficient condition for the above equation to admit a solution in the difference $\theta_j - \theta_k$.

Now assume that a solution of the problem exists, then we have

$$\cos(\theta_j - \theta_h) = c_0(j,k)$$

and, due to the identity

$$\cos(\theta_j - \theta_k) = \cos(\theta_j - \theta_k)\cos(\theta_k - \theta_h) - \sin(\theta_j - \theta_k)\sin(\theta_k - \theta_h)$$

one deduces that, for j,h,k in T , different from o and different among themselves

$$c_0(j,k) = c_0(j,k)c_0(k,h) - \sqrt{[1 - c_0(j,k)^2][1 - c_0(k,h)^2]}$$

which is equivalent to (9).
Conversely, if the functions $c_0(j,k)$ satisfy (8), then we can define, modulo 2π, an angle $\theta_{j,k}$ by

$$\cos(\theta_{j,k}) = c_0(j,k)$$

If moreover, also (9) is satisfied, then we also have

$$\theta_{j,k} + \theta_{k,h} = \theta_{j,k} \ (mod \ 2\pi)$$

and therefore there exists a family (θ_j) of real numbers, defined modulo 2π such that

$$\theta_{j,k} = \theta_j - \theta_k \ (mod \ 2\pi)$$

and this allows one to define the $\phi(j)$'s .

Finally , condition (8) is equivalent to

$$| p(j,k) - (x_j x_k)^2 - (y_j y_k)^2 | \leqslant 2(x_j x_k)(y_j y_k) \tag{14}$$

and with the notation

$$p = p(j,k) ; x = x_j x_k ; y = y_j y_k \tag{15}$$

(14) becomes

$$| p - x^2 - y^2 | \leqslant 2xy$$

or

$$p^2 + x^4 + y^4 - 2px^2 - 2py^2 - 2x^2y^2 = (p + x^2 - y^2)^2 - 4px^2 \leqslant 0$$

Substituting for p,x,y the values given by (12), (15) one finds (10) .

Note that it is easy to construct mathematical examples of triples (i.e. the set T has three elements) of 2x2 bi-stochastic matrices that do not admit any complex Hilbert space model (hence a fortiori no Kolmogorovian model − cf. [11]). For example, denoting p,q,r the three parameters defining the three bistochastic matrices (p = p(1,1) ; .....) , it is easy to see that, if q + r = 1 and p > 4q(1 - q), then condition (8) cannot be satisfied. Much more difficult is the problem of constructing some physically meaningful example exhibiting this behavior ; i.e. an example of a physical system whose statistical behavior cannot be described either by a Kolmogorovian or by a usual quantum model.

Recently a remarkable example of this kind has been constructed by Aerts [21]. Aerts ' example shows that the appearance of non-Kolmogorovian models in probability theory is not a peculiarity of quantum theory and suggests that the potentialities of these models should be more thoroughly investigated.

7. DEDUCTION OF THE QUANTUM FORMALISM FROM HEISENBERG'S PRINCIPLE

We barely mention here the strategy of the quantum probabilistic approach to the problem of deducing the main mathematical features of the quantum theoretical formalism from some basic physical principles. For a discussion as well as for the proofs of the statements, we refer to [9], [16].

First of all here the radical program of quantum probability, of deducing all the information from the transition probabilities, cannot be completely carried out. In fact, as shown in [6], the Heisenberg indeterminacy principle cannot be itself read into the transition probabilities: it contains physical information which transcends pure statistics.

This fact led , on the one hand to the model dependent analysis of the Hisenberg principle and to the mathematical definition of the notion of "complementarity" proposed in [16] ; and on the other hand it gave rise to the conviction that the Heisenberg principle should be taken itself as a postulate (cf. sec.(8) for a model-independent formulation of that principle).

The main result of [16] is that, if we introduce the model-independent formulation of Heisenberg' s principle , mentioned above, into the useful framework of the "algebras of measurements", introduced by Schwinger, then the resulting mathematical structure is not too wide, and we can obtain a complete classification theorem for the resulting finite dimensional Heisenberg algebras (which correspond to maximal observables assuming only a finite number of values).

The general case is dealt with through a limiting procedure, but a precise mathematical treatment is still lacking in the literature.

The classification theorem in [16] also suggests some highly nontrivial geometrical generalizations of the usual quantum formalism which are strictly related to gauge theories and to a geometric generalization of the imprimitivity systems.

8. MODEL INDEPENDENT FORMULATION OF SOME BASIC FACTS OF QUANTUM THEORY

The main emphasis of the quantum probabilistic approach to the foundational problems of quantum theory is on model independent analysis. Its attitude can be summarized in Hilbert' s words : "... one should formulate the physical requirements so completely that the mathematical model is uniquely determined by them..." In the previous sections we mentioned how, with a quantum probabilistic model independent, analysis it was possible to deduce the main structures of the quantum model, including the differentiation between the real and the complex numbers, and to overcome the interpretational problems connected with the reality issue.

Starting from this section, we begin a quantum probabilistic analysis of the quantum measurement process, as a result of which all the known formulas which are usually introduced as postulates, as well as some new formulas ,will emerge.

As a first step we shall formulate in the remain of this section the problem in a model-independent way (cf. [7],[8], [19] for a more detailed analysis).

In the following the notions of physical system,independent copies of a physical system,observable, measurement of an observable at a given time, etc. will be assumed as primitive notions. Observables will be denoted by capital letters A,B,C...., and their values (which, without loss of generality, can be assumed to be real numbers) with the corresponding small letters a,b,c,... . If the set of values of an observable is made up of isolated points, then we say that the observable is discrete, otherwise we call it continuous. The event that at a given time t the value of the observable A is a, will be denoted by

$$A_t = a \; ; \; or \; A(t) = a$$

Sometimes in the following we shall use symbols like A,B,C,... to denote a set (finite or infinite) of compatible (cf. further) observables $A = (A_1, A_2, \dots, A_n, \dots)$; correspondingly, the symbol a will denote the (ordered) set $(a_1, \cdots, a_n)$, and we will speak of the observable A to mean the set of observables $(A_1, A_2, \dots)$. the idea of measurement we have in mind can be schematized by the diagram :

$$----\rightarrow \quad C_{s]} \quad ----\rightarrow \quad A_t \quad ----\rightarrow \quad A_t \in I \qquad (1)$$

where s < t are two different instants of time and $C_{s]}$ denotes a set of physical operations, performed before time s (not necessarily at one and the same time), as a result of which a certain number $N(C_s)$ of copies of a given physical system $N(C_s)$ of copies of a given physical system is obtained, and on each of these systems a certain set of physical properties coincide (for example, they all have the same energy, or the same spin in a given direction,.... .

This class of measurements, even not exhaustive with respect to the effectively realizable physical procedures is nevertheless strictly larger than the class of the so called first-kind measurements , since no assumption is made about the values of the observable before or after the measurement (cf. Sec.(10) below). To begin with we stick to the idea of instantaneous measurement thus the A_t block in (1) means that one measures the observable A at time t on each of these systems, and the apparatus will allow to pass only those systems for which the value of A lies in the interval I of the real line.

We do not exclude a priori measurements as a result of which the system is destroyed ;however,whenever speaking of repeated measurements, these measurements will be tacitly excluded.

The following "definitions" should not be meant as formal mathematical statements, but rather as a compromise between the arid and sometimes inexpressive language of purely formalized theories and the totally sloppy language that one sometimes meets when reading about these topics.

Definition(1. By a preparation before time t we mean a family of physical operations the result of which is a family of $N(C_s)$ copies of a given system such that:

    i) For any time t > s , for any observable A , and for any interval I of the reals, the relative frequencies of the event [$A_t \in I$] tend to stabilize around a fixed number, called the probability of the event [$A_t \in I$] given the preparation C_s. In symbols

$$\frac{N(a_t \mid C_s)}{N(C_s)} \approx P(A_t \in I \mid C_s) \qquad (2)$$

ii) The family $C_{s]}$ can be reproduced at different places, at different times, with preservation of property (i).

This is of course not a formal definition, but I hope it gives a sufficiently precise idea of this important physical notion. The intuitive idea is that a preparation is a set of reproducible physical operations, the result of which is an ensemble of copies of a given physical system in which any observable has a well defined probability distribution. How to associate a definite probability with a, usually fluctuating, relative frequency is the problem of statistics and will not be discussed here. A related notion worth mentioning is that of "partial preparation", i.e., a set of physical operations $C_{s]}$ such that the conditional probabilities $P(A_t \in I \mid C_{s]})$ are well defined for some observables but not for all. For example, if P,Q denote the usual quantum mechanical momentum and position operators, then the formula proved in [16], Sec.(5) suggests that the events $[\, Q \in I\,]$ are partial preparations for P , and the events $[\, P \in J\,]$ are partial preparations for Q (if I,J are bounded Borel sets); however it is clear that these events cannot be preparations for all the observables, since the corresponding projections are not trace class .
In probabilistic terms, this means that using the above mentioned formula it makes sense to speak of the "relative conditional probability" that the momentum of a particle is in a certain region J given that its position is in a region I , but the fact that the position of the particle is in I cannot be a "good conditioning" for any other observable (in the sense that the corresponding probability may not be defined).
Note that the probabilistic counterpart of the physical notion of preparation is the notion of conditioning. In the following we shall use the two terms interchangeably.

Definition (2. Two preparations $C_{s]}$ and $C'_{s]}$ are called equivalent if for any observable A, for any time t > s , and for any interval I ,one has :

$$P(\, A_t \in I \mid C_{s]}) = P(\, A_t \in I \mid C'_{s]})$$

i.e., if they give rise to the same probability distributions for any observable at any time.

Definition(3. A state is an equivalence class of preparations (for the equivalence relation just defined).

Definition(4. Two observables are called compatible if, for each pair of values a of A and b of B , and for each time t , there exists a preparation $C_{t]}$ such that, for small dt , the conditional probability distributions of A_{t+dt} and B_{t+dt} are both delta-like and centered around a and b respectively.

In the situation described by Definition (4) we say that the observables A and B have sharp values in the preparation $C_{t]}$. This is a probabilistic formulation of the well known notion that two observables are compatible if they can be simultaneously measured with arbitrary precision. Definition (4) is extended in an obvious way to any finite number of observables.

Definition(5. A family $\{\, A_\alpha\, \}$ of observables is called compatible, if each finite sub-family of it is compatible.

Definition(6. An observable B is called a function of the family of compatible observables $\{\, A_\alpha\, \}$ if there exists a function F , of parameters $\{\, \lambda_\alpha\, \}$ such that the range of F coincides with the set of all values of B and

$$B = F(\, \lambda_\alpha\,)$$

whenever $A_\alpha = \lambda_\alpha$ for each index α.

Definition(7. A family of compatible observables is called maximal , or complete, if, for any preparation C in which all the A_α have sharp (i.e. exact) values, an observable B can have sharp values if and only if it is a function of the A_α's.

If $\{ A_\alpha \}$ is a maximal family of compatible observables, the symbol

$$P(B_t \epsilon I \mid A_\alpha(s) = a_\alpha ; \text{for each } \alpha)$$ (4)

will denote the probability

$$P(B_t \epsilon I \mid C_{s]})$$

where $C_{s]}$ is any preparation in which A_α has the sharp value a_α for each α.
Note that, for the moment , the notation (4) has been defined only in the case when the family $\{ A_\alpha \}$ is maximal.

Definition(8. A preparation $C_{s]}$ is called pure if there exists a maximal set of compatible observables $\{ A_\alpha \}$ such that , for any observable B and any t $>$ s

$$P(B_t \epsilon I \mid C_s) = P(B_t \epsilon I \mid A_\alpha(s) = a_\alpha ; \text{for each } \alpha)$$ (5)

Usually a pure preparation corresponds in the mathematical formalism to a single unit vector (up to a phase). However, given such a unit vector, the corresponding maximal family of compatible observables is not always physically well defined: for example, a coherent state of the free electromagnetic field is certainly a pure state, but rather than to a complete set of compatible observables, the physical preparation of such a state is realized by producing some stationary distributions of some given observables (typically the photon number).

Definition(9. Two families $\{ A_\alpha \}$ and $\{ B_\beta \}$ of compatible observables are called equivalent if each one is function of the other one, i.e. if there exist functions F_α and G_β such that :

$$A_\alpha = F_\alpha(B_\beta) ; B_\beta = G_\beta(A_\alpha) ; \text{for each } \alpha , \beta.$$ (6)

The following postulate underlies any classical physical theory:

PostulateC. Any two maximal families of observables are equivalent.

A corollary of Postulate (C) is the existence of the "state space" associated with the system in question. More precisely :

Corollary(10) Let be given a system satisfying Postulate (C) above. Then there exists a set S and a one-to-one correspondence between observables and functions $f : S \dashrightarrow R$ such that, if the observables A,B correspond respectively to the functions f,g , and if A = F(B) for some function F, then f = F(g).

Proof Let $\{ C_\alpha \}$ be a maximal family of compatible observables of the system. For each α use the notation

$$S_\alpha = \text{the set of all values of } C_\alpha$$

$$S = \Pi_\alpha S_\alpha$$

By maximality, for each observable A of the system there exists a function $F_A : S \to R$ such that

$$A = F_A(C_\alpha)$$

and it is easy to check that the correspondence $A \longrightarrow F_A$ has the required property.

It will help, in the discussion of hidden variables and of Bell's inequality (Secs.(22), (29) below), to keep in mind that the postulate of classical physics only guarantees the existence of the "state space" S, without any constraint on the probabilities that might be obtained from a set of different experiments.
The following postulate abstracts that feature of Heisenberg indeterminacy principle that constitutes the main breaking point between classical and quantum physics.

Postulate(Q. There exist at least two inequivalent maximal families of compatible observables.

A system for which Postulate (Q) is true will be called a quantum system. Anticipating the discussion in Sec.(22), let us remark that for such a system no state space in the sense of classical physics can exist. In fact, if such a space S existed, we could always identify it to a subset of a product of a suitable number of real lines; hence the coordinate functions would be a maximal set of compatible observables and any other maximal set of compatible observables should be able to express all the coordinate functions (by definition of maximality) and therefore all the functions on S .
Moreover, a theory accepting Postulate (Q) must necessarily be a statistical theory. In fact, having fixed the exact value a of a maximal set A of compatible observables, the best we can hope for those observables which are not functions of the set A, is that the conditional probabilities

$$P(B \epsilon I \mid A = a) \tag{10}$$

are physically well defined, i.e. the corresponding relative frequencies do not fluctuate too wildly. This is a natural generalization of the notion of "B being a function of A " to which it reduces in the case when the conditional probabilities (10) are delta-functions for each value a of A

It is intuitively clear that , by adjoining sufficiently many observables to any set of compatible observables one will obtain eventually a maximal set of compatible observables. We formalize this requirement in the following postulate :

Postulate. Any set of compatible observables is contained in at least one maximal set of compatible observables.

As a consequence of this postulate, any preparation B_s at time s can be realized by choosing a maximal set of compatible observables B and a set I of its values so that the preparation B_s is equivalent to the preparation defined by the event $[B(s) \epsilon I]$.
If the set I is reduced to a single value of B, then we obtain a pure preparation; otherwise it will be called impure.

9.   QUANTUM PROBABILISTIC ANALYSIS OF THE MEASUREMENT PROCESS

In this section we begin the investigation of the following problem :

Let $C_{r]}$ be a preparation before time r, and let $P(A_t \epsilon I \mid C_{r]})$ be the probability distribution of some observable A at time t $> r$,given the preparation $C_{r]}$. Assume that at time s, with r $< s < t$, we acquire some information on the system. How will the probability distribution of A vary as a consequence of this acquired information ?

We want to find a formula for this probability, and we want to prove that this formula is compatible with the basic principles of quantum theory. To solve this problem and the variants obtained from it by considering several (possibly infinite) measurements at different intermediate times, is the main goal of the quantum theory of measurement.

We shall consider in this paper only the first part of the problem, which is preliminary to the second one . In fact we shall find several different formulas corresponding to different physical situations, and therefore the coherence problem should be studied for all these formulas, and not only, as it has been until now, for one of them (the original von Neumann one). The problem of the asymptotic coherence of von Neumann's formula (cf. section (12)) with the basic principles of quantum theory has been investigated in a series of well known papers by Daneri, Loinger and Prosperi (cf.[33] and references therein). The same problem for the more general formulas discussed in the following, is strictly related to the theory of open quantum systems and to the theory of quantum stochastic processes (cf. [34] , [35d]).

From the probabilistic point of view, the above problem is the problem of conditioning, which in classical probability theory was solved by Bayes ' analysis . In the following we will extend this analysis by bringing in it the new elements due to the Heisenberg principle. The conceptual subtleties arising in this context were discussed in [19] ; here we shall limit ourselves to the deduction of the formulas which solve the problem in different situations, and to the description of the possible experiments which should allow us to discriminate among them.

Our attempt will be to deduce the answer to the above problem uniquely from considerations on relative frequencies (cf. the Secs.(9)-(13)). This however will not always be possible and we shall be obliged to introduce some model dependent assumptions (cf. Secs. (14)-(21)). In these cases I tried to make as clear as possible the discrimination between the technical assumptions and the physically observable properties, as well as the distinction between what is postulated and what is deduced (a distinction which I wish were more widespread in the literature on the topic).

Relative frequency considerations have been used by Ludwig [45] and his school in order to justify from a physical point of view the general notions of " operation " and "effect" (cf. [12] for a particularly lucid exposition of this point of view). Our level of analysis is finer, since it is not confined to the general features of the formalism but corresponds, in Ludwig's language, to the deduction of the specific form of the operation associated with a given experiment uniquely from relative frequency considerations. I have also avoided introducing the delicate notion of "state" of a physical system (cf. [8] for a detailed analysis of this notion) and tried to develop the whole analysis uniquely in terms of directly observable entities such as "the values of an observable", "the relative frequencies of these values",etc. .

Given two instants of time $r < s$ and two preparations $C_{r]}$,and B_s we denote by

$$B_s \cap C_{r]} \tag{1}$$

the preparation corresponding first to the performance of $C_{r]}$ and then of B_s (we will assume that this is still a preparation).

Experimental Fact Let $r < s < t$; let $C_{r]}$ be any preparation before r ; and let B_s be a pure preparation at time s . Then for any observable A and for any interval I of the real numbers, one has

$$P(A_t \in I \mid B_s; C_{r]}) = P(A_t \in I \mid B_s) \tag{2}$$

In terms of relative frequencies this means that

$$\frac{N(A_t \in I \mid B_s)}{N(B_s)} \approx \frac{N(A_t \in I \mid B_s ; C_{r_1})}{N(B_s ; C_{r_1})} \tag{3}$$

where here and in the following the symbol $\approx$ will mean "approximately equal", without further specification. The identity (2) or (3) is model independent in the sense that it is a statement on relative frequencies, independent of the particular mathematical model used to describe the probabilities. It amounts to saying that the probability distributions of A , obtained from the two different experimental situations schematized by the diagrams

$$--\to C_{r_1} --\to B_s ----\to A_t --\to A_t \in I \tag{4}$$

$$----\to B_s ----\to A_t --\to A_t \in I \tag{5}$$

coincide. It is an experimental fact that the identity (2) is generally false if B_s is not a pure preparation. Our next goal is to study what happens in this case.

Because of the discussion at the end of Sec.(8) any impure preparation can be realized by prescribing that a maximal family of compatible observables B takes values in a given set I , which we will take to be an interval of the real line. Therefore, the problem we are interested in can be described as follows : Let r,s, and C_{r_1} be as above; let B be any maximal observable; let A be any observable ; and let I,J be any intervals of real numbers. We want to find a formula for the probabilities

$$P(A_t \in J \mid B_s \in I ; C_r) \approx \frac{N(A_t \in J \mid B_s \in I ; C_{r_1})}{N(B_s I ; C_{r_1})} \tag{6}$$

The measurement corresponding to the event $[\, B_s \in I \,]$ will be called an incomplete (if I contains more than one value of B) selective measurement.

10. STATISTICS OF FIRST KIND MEASUREMENTS, ZENO PARADOX, AND COLLIMATORS

If one confuses between multitime conditional probabilities and multitime joint probabilities, paradoxes might arise. In this section we illustrate this statement with an analysis of first kind measurements . We shall come back to the problem of joint probabilities in Sec.(16).

Definition1. Let A be a discrete maximal observable. A measurement of A at time t is called a first kind measurement if

$$P(A_{t+dt} \neq a \mid A_t = a) = o(dt) \tag{1}$$

where, as usual we write that a quantity is o(dt) if o(dt)/dt tends to zero as dt tends to zero.

Remark1. believe that the above definition is an acceptable mathematical formulation of the characteristic property of first kind measurements , which is colloquially expressed by saying that the value of A "immediately after" a first kind measurement is the same as the result of the measurement.

Remark2. The difficulty of extending Definition (1) above to continuous maximal observables is that , contrary to what happens in the discrete case, the result of a measurement of a continuous observable at a time t cannot be considered as a preparation for another measurement.

Example. Let A be a maximal observable of a given physical system; denote with the same letter A the operator corresponding to A ; and denote with H the hamiltonian of the system. Let ψ denote the eigenvector of A corresponding to the eigenvalue a, and assume that ψ is in the domain of H^2. Then an elementary computation shows that

$$| <\psi, \exp idt H \; \psi >|^2 = 1 - dt^2 \Delta_\psi(H) + o(dt^2) = 1 - o(dt)$$

where $\Delta_\psi(H)$ is the dispersion of H in the state ψ (cf. Sec.(23) for the definition). This means that the measurement of the value a of A is a first kind measurement in the sense of Definition (1) and implies that such measurements exist in nature.

An interesting physical consequence of the property (1) is the possibility of doing "collimators" i.e. apparatus which "force" an observable of a physical system to keep the same value in a given fixed interval of time. The idea of the construction is as follows: consider the multi-time joint probabilities defined by

$$P(A_{t_n}; A_{t_{n-1}}; \ldots A_{t_1}; C_0) \approx \frac{N(A_{t_n}; A_{t_{n-1}}; \ldots; A_{t_1}; C_0)}{N(C_0)} \quad\quad (2)$$

Note that the choice of the normalization factor in (2) is essential in order to interpret the left hand side as a joint, and not a conditional, probability. Writing the right hand side of (2) in the form

$$\prod_{j=0}^{N} \frac{N(A_{t_j}; A_{t_{j-1}}; \ldots C_0)}{N(A_{t_{j-1}}; A_{t_{j-2}}; \ldots; A_{t_1}; C_0)} \quad\quad (3)$$

Now, since the measurements take place without destruction :

$$N(A_{t_j} | A_{t_{j-1}}; \ldots; C_0) = N(A_{t_j}; A_{t_{j-1}} \ldots; C_0)$$

therefore the expression (3) is approximately equal to :

$$\prod_j P(A_{t_j} = a | A_{t_{j-1}} = a ; \ldots; C_0)$$

which, because of Dirac's jump assumption is equal to :

$$\prod_j P(A_{t_j} = a | A_{t_{j-1}} = a) \quad\quad (4)$$

Thus, using (1) with $t_j - t_{j-1} = t/n$ one obtains

$$P(A_t = a_j ; A_{t_n} = a ; A_{t_{n-1}} = a ; \ldots; C_0) = (1 - o(t/n))^n \approx 1 \quad\quad (5)$$

This relation is experimentally confirmed by the existence of "collimators" of beams of particles.
The the multitime joint probabilities (2) of first kind measurements can never be obtained from a classical stochastic process (which should be necessarily markovian because of Dirac's jump assumption). In fact, given a natural integer N ; N different values of A, $a_1, \ldots, a_N$; we can choose $\epsilon > 0$ such that $N(1-\epsilon) > 1$, and n so large that the left hand side of (5) is greater than $1 - \epsilon$ for each $a = a_1, \ldots, a_N$. Correspondingly we will have that

$$\sum_{j=1}^{N} P(A_{t_n} = a_j ; \ldots; A_{t_1} = a_j | C_0) > N(1 - \epsilon) > 1$$

which is impossible, since we are summing over disjoint events. This fact is the mathematical content of the so called " Zeno paradox". The reason why it is physically unjustified to interpret the expressions (7) below as joint probabilities of a single classical stochastic process has been discussed in [14].

11. QUANTUM PROBABILISTIC ANALYSIS OF THE MEASUREMENT PROCESS; DISCRETE CASE

Let us first discuss the problem stated in the previous section in the case when the set I --corresponding to the incomplete measurement of the observable B — is discrete, i.e., made up of isolated points :

$$I = (b_1, b_2, ...) \tag{1}$$

In the measurement process defined by the diagram

$$-\rightarrow C_{r1} \;---\rightarrow\; B_s \;--\; B_s \in I \;---\rightarrow\; A_t \in J \;---\rightarrow \tag{2}$$

denote by

$$N(A_t \in J \mid [B_s \in I] \cap C_{r1}) \tag{3}$$

the number of systems of the ensemble which are allowed to pass through the apparatus. Since a system is not filtered away in the intermediate step if and only if $B_s = b_j$ for some $b_j \in I$, one has

$$N(A_t \in J \mid [B_s \in I] \cap C_{r1}) = \sum_{b_j \in I} N_I(A_t \in J \mid [B_s = b_j] \cap C_{r1}) \tag{4}$$

where the subscript I in the number $N_I(A_t \in J \mid [B_s = b_j] \cap C_{r1})$ means that this number is measured for a situation in which all the values $b_1, b_2, ...$ are allowed to pass, and a priori should be distinguished from the number $N(A_t \in I \mid [B_s = b_j] \cap C_{r1})$ which is measured in a situation in which only the value b_j is allowed. Using (4) we find with obvious steps :

$$P(A_t \in J \mid [B_s \in I]; C_{r1}) \approx \frac{N(A_t \in J \mid [B_s \in I]; C_{r1})}{N([B_s I]; C_{r1})} = \tag{5}$$

$$\sum_{b_j \in I} \frac{N_I(A_t \in J \mid [B_s = b_j]; C_{r1})}{N(B_s = b_j \mid C_{r1})} \cdot \frac{N(B_s = b_j \mid C_{r1})}{N(B_s \in I \mid C_{r1})} =$$

$$\approx \sum_{b_j \in I} \frac{N_I(A_t \in J \mid [B_s = b_j]; C_{r1})}{N(} \; B_s = b_j \mid C_{r1}) \frac{P(B_s = b_j \mid C_{r1})}{P(B_s \in I \mid C_{r1})}$$

To go further we introduce the following : *Postulate (indifference postulate).*

$$\frac{N_I(A_t \in J \mid [B_s = b_j] \cap C_{r1})}{N(B_s = b_j \mid C_{r1})} \approx \frac{N(A_t \in J \mid [B_s = b_j] \cap C_{r1})}{N(B_s = b_j \mid C_{r1})} \tag{(}$$

We call this -the "indifference postulate" since it means that, for the computation of the probability

$$P(A_t \in J \mid [B_s = b_j] \cap C_{r1}) \tag{7}$$

it is indifferent whether the relative frequencies are measured for a situation in which all the $b_1, b_2, ...$ are allowed or for a situation in which the single value b_j is allowed. Since the two sides of the (approximate) identity (6) are obtained under different experimental conditions, the validity of (6) should be checked experimentally case by case. It would be interesting to realize some experiments in which the identity (6) is not satisfied. One should also , a priori distinguish in the computation of the numbers $N_I(\,.\,)$, the case for which the values in the interval I are filtered by an exact measurement of each of them , from the case ın which they are filtered using, for example, a threshold mechanism. Also in this case the equality of the two numbers cannot be postulated a priori.

On the other side, Feynman' s analysis of the two-slit experiment [35a] provides at least one example of an experimentally realizable situation in which the identity (6) is verified.

Assuming that the indifference postulate holds, then we can substitute the probability $P(\,A_t \epsilon J \mid [\,B_s = b_J\,]; C_{r\rfloor})$ for either side of (6) and therefore (5) becomes :

$$P(\,A_t \epsilon J \mid [\,B_s \epsilon I\,] \bigcap C_{r\rfloor}) = \tag{8}$$

$$= \sum_{b_j \epsilon I} P(\,A_t \epsilon J \mid [B_s = b_j] \bigcap C_{r\rfloor}) \cdot \frac{P(\,B_s = b_j \mid C_{r\rfloor})}{P(B_s \epsilon I \mid C_{r\rfloor})} =$$

$$= \sum_{b_j \epsilon I} P(\,A_t \epsilon J \mid B_s = b_j\,) \cdot \frac{P(B_s = b_j \mid C_{r\rfloor})}{P(\,B_s \epsilon I \mid C_{r\rfloor})}$$

where in the last step we have used Dirac' s jump assumption (cf. Sec.(9)). The probabilities in the two sides of (8) are measured in different physical situations; therefore, as anticipated in Sec.(5) , the identity of the two sides of (8) is by no means a tautology on relative frequencies.

12.   DEDUCTION OF VON NFUMANN' S MEASUREMENT POSTULATE

In the previous section we have shown that, under the assumptions :

$-$ *probabilities are identified with relative frequencies.*   (1)

$-$ *Dirac' s jump assumption hold.*   (2)

$-$ *the indifference postulate holds.*   (3)

one can deduce the validity of the ıdentity

$$P(\,A_t \epsilon J \mid [\,B_s \epsilon I\,] \bigcap C_{r\rfloor}) = \sum_{b_j \epsilon I} P(\,A_t \epsilon J \mid B_s = b_j\,) \cdot \frac{P(\,B_s = b_j \mid C_{r\rfloor})}{P(B_s \epsilon I \mid C_{r\rfloor})} \tag{4}$$

Now the probabilities on the right hand side of (4) can be expressed in the usual quantum mechanical formalism through the substitutions :

$C_{r\rfloor} \longrightarrow W_{r\rfloor}$ density matrix

$[\,A_t \epsilon J\,] \longrightarrow E^{A_t}(J)$ *spectral projection of* A_t

$[\,B_s = b_j\,] \longrightarrow E^{B_s}(b_j)$ *rank one spectral projection of* B_s

$P(\,B_s \epsilon I \mid C_{r\rfloor} = Tr(\,E^{B_s}(I) W_{r\rfloor})\,; P(\,A_t \epsilon J \mid B_s = b_j\,) = Tr(E^{A_t}(J) E^{B_s}(b_j\,))$

where all the operators act on some Hilbert space H and Tr denotes the trace on H . Using these correspondences, one easily checks that (4) becomes equivalent to

$$P(\,A_t \epsilon J \mid [\,B_s \epsilon I\,] \bigcap C_{r\rfloor}) = Tr(E^{A_t}(J) E(\,B_s \epsilon I\,; W_{r\rfloor})) \tag{5}$$

where

$$E(B_s \in I \; ; W_{r]}) = \sum_{b_j \in I} \frac{E^{B_s}(b_j) W_{r]} E^{B_s}(b_j)}{Tr(E^{B_s}(I) W_{r]})} \tag{6}$$

In particular, if we choose :

$$W_{r]} = |\psi_r> <\psi_r| \; ; \psi \; in \; H : ||\psi_r|| = 1 \tag{I(7)}$$

$$I = \text{the set of all possible values of } B$$

and if we denote $\phi_j(s)$ the eigenvector of B_s corresponding to the eigenvalue b_j , then the right hand side of (6) becomes

$$\sum_{all \; b_j} | <\phi_j(s), \psi_r > |^2 \cdot | \phi_j(s) > <\phi_j(s)| \tag{8}$$

which is the formula proposed by von Neumann to describe the density operator at time s of a system which at time r < s was in the quantum state ψ_r and on which at time s an incomplete measurement of the maximal observable B has been performed (without filtration).

13. QUANTUM PROBABILISTIC ANALYSIS OF THE MEASUREMENT PROCESS: CONTINUOUS CASE

In this and the following sections the assumption that the observable B, corresponding to an intermediate measurement, is discrete will be dropped. The scheme of the measurement is given by the diagram :

$$---\rightarrow \; C_{r]} \; ---\rightarrow \; B_s \in I \; ---\rightarrow \; A_t \in J$$

where, to fix the ideas, the sets I,J will be thought to be intervals of the real line. In this case the analysis of Sec.(11) carries over, with the only difference that the identity (4) of that section should be substituted for by the identity

$$N(A_t \in J \; | \; B_s \in I \; ; C_{r]}) = \sum_k N_I(A_t \; | \; B_s \in I_k \; ; C_{r]}) \tag{1}$$

where $I_1, I_2, \ldots$ is any (finite or countable) partition of the interval I in disjoint sets.
As before, the index I on the right hand side of (1) denotes that the number $N_I(A_t \in J \; | \; [B_s \in I_k] \bigcap C_{r]})$ is referred to an experimental situation in which all the values of B_s in the interval I are allowed, and not only those in the interval I_k . Thus in this case the indifference principle takes the form

$$N_I(A_t \in J \; | \; B_s \in I_k \; ; C_{r]}) = N(A_t \in J \; | \; B_s \in I_k \; ; C_{r]}) \tag{2}$$

for any interval I_k contained in I , where the right hand side refers to a situation in which only the values of B_s belonging to the interval I_k are allowed .
Assuming the validity of the indifference principle, we drop the subscript I on the right hand side of (1). Under this assumption we can extend the arguments used in the discrete case by means of the following heuristic considerations : Assume that

$$P(B_s \in I \; | \; C_{r]}) = 0 \tag{3}$$

then ,since

$$P(B_s \epsilon I \mid C_r) \approx \frac{N(B_s \epsilon I \mid C_r)}{N(C_r)} \tag{4}$$

it follows that

$$N(B_s \epsilon I \mid C_r) \approx 0 \tag{5}$$

hence a fortiori

$$N(A_t \epsilon J \mid [B_s I]; C_r) \approx 0 \tag{6}$$

Thus the positive (not necessarily probability) measure

$$I \subseteq IR \longrightarrow \frac{N(A_t \epsilon J \mid B_s \epsilon I ; C_r)}{N(C_r)} \tag{7}$$

is absolutely continuous with respect to the measure

$$IR \longrightarrow P(B_s \mid C_r) \tag{8}$$

and we can denote by

$$P(A_t \epsilon \mid B_s = x ; C_r) \tag{9}$$

the Radon-Nykodim derivative of the measure (7) with respect to the measure (8), which is defined by the equality

$$\frac{N(A_t \epsilon J \mid B_s \epsilon I ; C_r)}{N(C_r)} = \int_I P(A_t \epsilon J \mid B_s = x ; C_r) \cdot P(B_s \epsilon dx \mid C_r) \tag{10}$$

Now, introducing a partition of the interval I whose generic element we denote by dx, we obtain :

$$P(A_t \epsilon J \mid B_s \epsilon I ; C_r) \approx \frac{N(A_t \epsilon J \mid B_s \epsilon I ; C_r)}{N(B_s \epsilon I \mid C_r)} \tag{11}$$

$$= \sum_{dx} \frac{N(A_t \epsilon J \mid B_s \epsilon dx ; C_r)}{N(B_s \epsilon dx \mid C_r)} \cdot \frac{N(B_s \epsilon dx \mid C_r)}{N(B_s \epsilon I \mid C_r)}$$

and, introducing the approximations

$$\frac{N(B_s \epsilon dx \mid C_r)}{N(B_s \epsilon I \mid C_r)} \approx \frac{P(B_s \epsilon dx \mid C_r)}{P(B_s \epsilon I \mid C_r)} \tag{12}$$

$$\frac{N(A_t \epsilon J \mid B_s \epsilon dx ; C_r)}{N(B_s \epsilon dx \mid C_r)} \approx P(A_t \epsilon J \mid B_s = x ; C_r) \tag{13}$$

we find :

$$P(A_t \epsilon J \mid B_s \epsilon I ; C_r) \approx \int_I P(A_t \epsilon J \mid B_s = x ; C_r); \frac{P(B_s \epsilon dx \mid C_r)}{P(B_s \epsilon I \mid C_r)} \tag{14}$$

One could now postulate that a generalized form of Dirac' s jump assumption holds also for continuous observables, which means that the probability density

$$P(A_t \epsilon J \mid B_s = x ; C_r) \tag{15}$$

depends only on A_t , B_s , J , x but not on C_r. In this case there should exist a function

$$P(A_t \epsilon J \mid B_s = x) \tag{16}$$

such that

$$\int_I P(A_t \epsilon J \mid B_s = x) \frac{P(B_s \epsilon dx \mid C_r)}{P(B_s \epsilon I \mid C_r)} \tag{17}$$

To complete our analysis, we should then find a formula for the function (16) above. A natural candidate, by analogy with the discrete case, might be the choice :

$$P(A_t \epsilon J \mid B_s = x) = E^{A_t}(J)(x^{B_s}, x^{B_s}) \tag{18}$$

where $E^{A_t}(.)$ is the spectral measure of A_t and

$$E^{A_t}(J)(x^{B_s}, y^{B_s})$$

is the kernel of the projection $E^{A_t}(J)$ in a representation in which B_s is diagonal . Such a choice would lead to the formula :

$$P(A_t \epsilon J \mid B_s \epsilon I ; C_r) = \int_I E^{A_t}(J)(x^{B_s}, x^{B_s}) \cdot \frac{Tr(E^{B_s}(dx)W_r)}{Tr(E^{B_s}(I)W_r)} \tag{19}$$

It is an interesting analytical problem to understand precisely for which observables A,B and which hamiltonian evolutions the formula (19) has a precise mathematical meaning, i.e. either the kernels of the projections involved are true functions or they are distributions, but the spectral measures of B_s are so smooth that the right hand side of (19) makes sense.

It is an interesting experimental problem to check the empirical validity of the formula (19) for some cases in which it makes sense (examples are easily constructed).

Formula (19) above has been deduced on the assumption that the probability $P(B_s \epsilon I \mid C_r)$ is expressed by the usual quantum formula. However formula (17) is model independent and one can substitute for the above probability with the formulas suggested by some generalization of the usual quantum mechanical formalism, for example the operational formalism of Ludwig [45] and Davies and Lewis [35]. Doing that, we obtain a formula similar to the one recently proposed by Ozawa [50] in the framework of the operational formalism -- the difference being that in Ozawa's paper the measure appearing in the integral on the right hand side of (17) comes from an operation valued measure, and the integrand has not the form suggested in (19) (I am grateful to Alberto Barchielli for indicating this reference to me).

14.   DEDUCTION OF THE LUDERS-ZUMINO FORMULA

In the previous sections we were lead to guess a general formula for the probabilities

$$P(A_t \epsilon J \mid [B_s \epsilon I] \bigcap C_r) \tag{1}$$

- corresponding to an incomplete measurement of the observable B — through a model independent analysis, i.e. without introducing any assumption on the mathematical models for the observables or the states.

In this section we shall introduce some model dependent assumptions in our analysis, i.e., we postulate that the observables of the theory are in one-to-one correspondence with the self-adjoint operators on some Hilbert space H , and let the events [$A \epsilon J$] correspond to the spectral projections $E^A(J)$. For simplicity we assume the observables to be real valued and denote the observables with the same letter as the corresponding operator.

Under these assumptions, for each fixed $C_{r]}$, each interval I, and instants $r < s < t$, and for any observables A, B (B -maximal), the maps

$$I \subseteq \mid R \ ----\rightarrow \ P(B_s \in I \mid C_{r]}) \tag{2}$$

$$J \subseteq \mid R \ ----\rightarrow \ P(A_t \in J \mid [B_s \in I] \bigcap C_{r]}) \tag{3}$$

induce probability measures on the projections of B(H) -the algebra of all bounded operators on H . Therefore, by Gleason' s theorem (cf. Sec.(24) in the following), if dimH > 2 , there exist density matrices $W_{r]}$ and $E_t (B_s \in I \ ; W_{r]})$ satisfying

$$P(B_s \in I \mid C_{r]}) = Tr(E^{B_s}(I) W_{r]}) \tag{4}$$

$$P(A_t \in J \mid B_s \in I \ ; C_{r]}) = Tr(E_t (B_s \in I \ ; W_{r]}) E^{A_t}(J)) \tag{5}$$

Introducing relative frequencies, we obtain :

$$Tr(E_t (B_s \in I \ ; W_{r]}) E^{A_t}(J)) = P(A_t \in J \mid B_s \in I \ ; C_{r]}) \approx \tag{6}$$

$$\approx \frac{N(A_t \in J \mid B_s \in I \ ; C_{r]})}{N(B_s \in I \mid C_{r]})} \approx \frac{N(A_t \in J \mid B_s \in I \ ; C_{r]})}{N(C_{r]})} \cdot \frac{1}{P(B_s \in I \mid C_{r]})}$$

This suggests the introduction of the operator

$$L_t (B_s \in I \ ; W_{r]}) = P(B_s \in I \mid C_{r]}) E_t (B_s \in I \ ; W_{r]}) \tag{7}$$

which has the property

$$Tr(L_t (B_s \in I \ ; W_{r]}) E^{A_t}(J)) \approx \frac{N(A_t \in J \mid B_s \in I \ ; C_{r]})}{N(C_{r]})} \tag{8}$$

Since systems are not destroyed after the measurements, one must have

$$N(A_t \in \mid R \mid B_s \in I \ ; C_{r]}) = N(B_s \in I \mid C_{r]}) \tag{9}$$

Thus, in particular

$$Tr(L_t (B_s \in I \ ; W_{r]})) \approx \frac{N(A_t \in \mid R \mid B_s \in I \ ; C_{r]})}{N(C_{r]})} = \tag{10}$$

$$= \frac{N(B_s \in I \mid C_{r]})}{N(C_{r]})} \approx P(B_s \in I \mid C_{r]})$$

Thus, using (4), we obtain

$$Tr(L_t (B_s \in I \ ; W_{r]})) = Tr(W_{r]} E^{B_s}(I)) \tag{11}$$

which implies in particular that the map

$$W_{r]} \ ----\rightarrow \ Tr(L_t (B_s \in I \ ; W_{r]}) \tag{12}$$

from density matrices on H to positive real numbers, is linear (more precisely —affine).

A stronger property of the map $L_t(\ . \ )$ can be obtained if we introduce , in analogy with what we did in Sec.(11), an "indifference principle" also for the condition $C_{r]}$. More precisely assume that the preparation condition $C_{r]}$ is a mixture of two preparations $C_{r]}'$ and $C_{r]}''$, in the sense that , among the $N(C_{r]})$ copies of the system we are considering, exactly $N(C_{r]}'$ have been prepared with the procedure $C_{r]}'$ and $N(C_{r]}''$

with the procedure $C_r]^"$, and that , furthermore,

$$N(C_r]) = N(C_r]) + N(C_r]) \tag{13}$$

Now clearly

$$N(A_t \epsilon J \mid B_s \epsilon I ; C_r])$$

$$= N_{C_r]}(A_t \epsilon J \mid B_s \epsilon I ; C_r]) + N_{C_r]}(A_t \epsilon J \mid B_s \epsilon I ; C^*_r]) \tag{14}$$

where the suffix $C_r]$ on the right hand side of (16) means that these numbers are measured for an experimental situation in which both types of conditions $C_r]'$ and $C_r]^"$ are simultaneously present.

The indifference postulate consists, for this case, in the statement that

$$N_{C_r]}(A_t \epsilon J \mid B_s \epsilon I ; C_r]) = N(A_t \epsilon J \mid B_s \epsilon I ; C_r]) \tag{15}$$

and the same for $C^*_r]$.

Using this and dividing both sides of (16) by $N(C_r])$, we obtain :

$$\frac{N(A_t \epsilon J \mid B_s \epsilon I ; C_r])}{N(C_r])} = p \cdot \frac{N(A_t \epsilon J \mid B_s \epsilon I ; C_r])}{N(C_r])} + \tag{I(16)}$$

$$(1 - p) \cdot \frac{N(A_t \epsilon J \mid B_s \epsilon I ; C^*_r])}{N(C^*_r])}$$

with

$$p = \frac{N(C_r])}{N(C_r])} \tag{17}$$

So, if W' and W" denote the density operators corresponding, respectively, to the preparations C', C" (from now on, for notational simplicity, we omit the index r]), then from (8) and (18) we deduce that

$$Tr(L_t(B_s \epsilon I ; pW' + (1 - p)W")) \cdot E^{A_t}(J))$$

$$= pTr(L_t(B_s \epsilon I ; W')) + (1 - p)Tr(L_t(B_s \epsilon I ; W")) \cdot E^{A_t}(J)) \tag{18}$$

Since, A and I are arbitrary, (20) is equivalent to the statement that the map

$$W \dashrightarrow L_t(B_s \epsilon I ; W) \tag{19}$$

is a linear (affine) map from any space of mutually commuting density operators in H to the positive trace class operators in H.

Under some additional mathematical assumptions (and again using Gleason' s theorem) we can deduce the linearity (affinity) of the map (19) from the space of all the density operators on H to the positive trace class operators on H. In the following we shall assume this linearity property, in analogy with what is done in the approaches of Ludwig [45], Davies and Lewis [35] , Holevo [38], Helstrom [38] ,and others.

Note that, from a physical point of view , the linearity of the map (19) is a far from obvious requirement, since it involves the behavior of the probabilities (16) under mixtures of mutually incompatible preparations.

The linearity of the map (19) is crucial in order to prove the following result, which is a corollary of a theorem due to B.Davies ([34] , Chap.2).

<u>Theorem 1.</u> Assume that the map (19), defined above, has the following properties :

i) is linear
ii) maps pure states into pure states
Then it must have one of the following forms :

$$L_t(B_s \epsilon I ; W) = Tr(W E^{B_s}(I)) |\psi> <\psi| \tag{20}$$

for some unit vector ψ in H , not depending on W but , a priori, depending on everything else (I, r, s, t, B).

$$L_t(B_s \epsilon I ; W) = U E^{B_s}(I) W \cdot E^{B_s}(I) U^* \tag{21}$$

where U is a linear or conjugate linear partial isometry on H satisfying

$$U^* U \geqslant E^{B_s}(I) \tag{22}$$

and not depending on W , but a priori depending on everything else (I, r, s, t, B).

<u>Proof</u> By the above mentioned theorem of Davies, the map $L_t(.)$ must have one of the forms :

$$L_t(B_s \epsilon I ; W) = Tr(W A) |\psi> <\psi| \tag{23}$$
$$L_t(B_s \epsilon I ; W) = A W A^* \tag{24}$$

where in (23) A is a bounded linear operator and ψ is a unit vector in H , while in (24) A is a bounded linear or conjugate linear operator. In the case (23), the identity (11) yields

$$Tr(WA) = Tr(W E^{B_s}(I)) \tag{25}$$

or equivalently, since W is arbitrary :

$$A = E^{B_s}(I) \tag{26}$$

whence (20) follows. In the case (24), the identity (11) yields :

$$Tr(A^* A W) = Tr(E^{B_s}(I) W) \tag{27}$$

and again the arbitrariness of W implies that

$$A^* A = E^{B_s}(I) \tag{28}$$

so that the polar decomposition of A will be :

$$A = U \cdot E^{B_s}(I) \tag{29}$$

U -being a partial isometry , linear or conjugate linear according to what A is, satisfying (22). And this proves the theorem.

Using the definition of $L_t(.;.)$ given by the identity (7), we find that the form of $E_t(.;.)$ corresponding to the choice (20) of $L_t(.;.)$ is :

$$E_t(B_s \epsilon I ; W) = |\psi> <\psi| \tag{29}$$

corresponding to a complete pure measurement. In other words, this formula means that the measurement corresponding to the filtration of the values of the maximal observable B that are in the interval I is equivalent to a complete pure measurement putting the system in the state ψ independently on what the state W was. According to the usual quantum theory, this can be the case if and only if the only point in the spectrum of B, lying in the interval I, is a simple eigenvalue corresponding to the eigenvector ψ. In this case, formula (29) is nothing but a mathematical formulation of Dirac's jump assumption (cf. Sec.(9)).

The choice of L_t $(. ; .)$, corresponding to formula (21), will yield instead

$$E_t(B_s \in I \; ; W) = \frac{U_{s,t} E^{B_s}(I) W \cdot E^{B_s}(I) U_{s,t}^*}{Tr(E^{B_s}(I) W)} \tag{30}$$

where the partial isometry or anti-isometry $U_{s,t}$ can depend a priori also on s, I, B.
In the particular case when $U_{s,t}$ is independent on s, I, and B, and it is a unitary evolution i.e.

$$U_{s,t} U_{r,s} = U_{r,t} \; ; r < s < t \tag{31}$$

then we can interpret it as the Schroedinger evolution of the system which, according to the usual quantum mechanics, should be applied when the system can be considered to be isolated. In this case we have, assuming no interactions in the interval [r , s]

$$E^{B_s}(I) = U_{r,s}^* \cdot E^{B_s}(I) U_{r,s} \tag{32}$$

and defining

$$W(r \, , s) = U_{r,s} \cdot W_{r]} U_{r,s}^* \tag{33}$$

the identity (30) becomes

$$E_t(B_s \in I \; ; W_{r]}) = \frac{U_{s,t} E^B(I) W(r \, , s) \cdot E^B(I) U_{s,t}^*}{Tr(E^B(I) W(r \, , s))} \tag{34}$$

in which we can distinguish three parts :

i)   an initial Schrodinger evolution :

$$W_{r]} ----\rightarrow W(r \, , s) = U_{r,s} \cdot W_{r]} U_{r,s}^* \tag{35}$$

ii) an instantaneous change due to the measurement at time t :

$$W(r \, , s) ----\rightarrow \frac{E^B(I) \cdot W(r \, , s) \cdot E^B(I)}{Tr(E^B(I) \cdot W(r \, , s))} \tag{36}$$

iii) again a Schrodinger evolution of the right hand side of (40) under the same evolution $U_{s,t}$ yielding the final result (38).

In probabilistic terms, the formula (34) means that, under the assumptions of Theorem (1), for any time t $> s$, any observable A, and any interval J in $\mathbb{R}$, the conditional probability

$$P(A_t \in J \; | \; [B_s \in I] \bigcap C_{r]}) \tag{37}$$

has the form :

$$\frac{Tr(U_{s,t}\cdot E^{B}(I)W(r,s)E^{B}(I)U^{*}_{s,t}\cdot E^{A}(J))}{Tr(W(r,s)E^{B}(I))} \tag{38}$$

In the language of usual quantum theory, the transition (ii) above is conventionally assumed to describe the instantaneous change at time s of the state of a system which has undergone in that moment a selective instantaneous measurement with filtration of those values of B_s which lie in the numerical interval I (cf.for example [34] Chap.2).

The above analysis shows however that this formula hides some very stringent, both physically and mathematically , assumptions. For example, the requirement that an incomplete selective measurement maps a pure state into a pure state is in contradiction with the generalized von Neumann formula which we derived in Sec.(12) for a discrete observable (and the analysis in this section didn't introduce any assumption on the observable B, so it applies to discrete observables as well). But since we know that at least in some cases von Neumann' s formula yields the right experimental results, this means that at least in these cases the Zumino-Luders formula is experimentally false. Therefore the applicability of this formula should be checked case by case.

I would be rather surprised if there existed in nature incomplete selective measurements mapping pure states into pure states ; but of course the last word must be to the experiments, and I would urge the experimentalists to clarify this point in a definitive manner.

Finally, let us remark that the incompatibility with von Neumann' s formula is a consequence of a more general property of the Zumino-Luders formula, namely its incompatibility with what we called the "indifference postulate". In fact, this postulate requires the validity of the identities (1) and (2) of Sec.(13), and because of the identity (8) of this section, these imply that the map

$$I \quad |R \longrightarrow \frac{N(A_t \in J \mid [B_s \in I] \bigcap C_{r})}{N(C_r)} = Tr(L_t(B_s \in I ; W_r)E^{A_t}(J)) \quad (39)$$

is a positive measure. Because of the arbitrariness of A and J, (43) is equivalent to the statement that the map

$$I \quad |R \longrightarrow L_t(B_s \in I ; W) \tag{40}$$

is, for any fixed W, a measure with values in the positive trace class operators on H . But this is incompatible with the Zumino-Luders formula (since W is arbitrary and since we are not allowing superselection rules), and it is compatible with the other formula of Theorem (1) if and only if the state ψ ,appearing in this formula, is independent of I . In this case we have already seen that the operator $E_t(. ; W)$ is independent of W and describes a complete pure measurement of B.

15. THE OPERATIONAL APPROACH

In the previous section we have shown that the requirement that the map

$$W \quad \longrightarrow \quad L_t(B_s \in I \ ; W) \tag{1}$$

(affine) and maps pure states into pure states, determines in an essentially unique way the Luders-Zumino formula (29) of Sec.(14) .

This requirement of preservation of purity, which seems to have been considered rather uncritically in the literature, is however too stringent and not plausible in most physically interesting cases. Dropping this requirement, and requiring only that the map (1) be a positive linear map of the set of the trace class operators into itself, leads naturally to the notion of operation. So, by definition, an operation is a positive linear map L from the space of trace class operators on some Hilbert space into itself. By duality an operation induces a positive linear map

$$L': B(H) \longrightarrow B(H) \tag{2}$$

characterized by the property

$$Tr(L(W)a) = Tr(W \cdot L'(a)) \tag{3}$$

for each density operator W and each positive operator a in B(H).

If L' is completely positive, then the operation L will be called completely positive (cf. [42], [44] for a definition of complete positivity and for the physical motivations of this property. The mathematical deduction of this property in the framework of the general theory of quantum Markov processes was obtained in [1], [2] , [3]).

According to Kraus' representation Lemma (cf. [42], Sec.3), a completely positive operation L must have the form

$$L(W) = \sum_k A_k W A_k^* \tag{4}$$

Thus, assuming that the map (1) above is completely positive and using the identity (13) in section (14) , one deduces that there must exist operations

$$A_k(I) \ ; \ k \in IN$$

such that

$$L_t(B_s \in I \ ; W) = \sum_k A_k(I)W A_k^*(I) \tag{5}$$

$$\sum_k A_k^*(I)A_k(I) = E^{B_s}(I) \tag{6}$$

It is an open problem to determine the most general family of operators { $A_k(I): k \in IN$ } satisfying (6). It is however easy to solve the problem in the particular case when one assumes that the $A_k(I)$'s have disjoint supports. In this case, there must exist a partition (I_k) of the interval I such that, for each k, $A_k(I)$ has the form

$$A_k(I) = U_k E^{B_s}(I_k) \tag{7}$$

where U_k is a partial isometry satisfying

$$U_k^* U_k \ >= \ E^{B_s}(I_k)$$

Correspondingly the formula for L becomes

$$L_t(B_s \in I \ ; W) = \sum_k U_k E^{B_s}(I_k)W E^{B_s}(I_k)U_k^* \tag{8}$$

which is a natural generalization of the Luders- Zumino formula, but which depends on the choice of the partition (I_k). One could, of course, justify such a choice with arguments on the limits of precision of the instruments. However a martingale theorem should be proved in order to prevent a bad behavior with respect to finer and finer partitions. It is known that there are mathematical obstructions to the existence of such theorems, at least if one insists on properties such as normality and faithfulness [58]. However a detailed analysis, combining these mathematical results with what one should expect physically , seems to be lacking.

16. ITERATED MEASUREMENTS, JOINT PROBABILITIES AND WIGNER'S FORMULA

In the previous sections we deduced several formulas for the probability

$$P(A_t \in J \mid [B_s \in I]; C_r)$$ (1)

which correspond to different physical realizations of the measurements involved. In this section we want to study in more detail the case in which the preparation C_r consists itself of several measurements performed in a sequence of different times. That is, we will consider conditional probabilities of the form

$$P(A_t \in J \mid B_{t_n}^{(n)} \in I_n ; \dots B_{t_1}^{(1)} \in I_1 ; C_{t_0})$$ (2)

where $t_0 < t_1 < \dots < t_n$ are instants of time; $I_1, \dots, I_n$, J are intervals in $\mathbb{R}$; $B^{(1)}, \dots, B^{(n)}$ are maximal observables ; A is any observable ; and C_{t_0} is a preparation before time t_0 . In the following, for notational convenience, we will choose at each time the same observable, however it will be clear from the discussion that the results hold in the general case.

The same application of Gleason's theorem as in the beginning of Sec.(14) shows that there must exist a density matrix, denoted by

$$E_{t,t_n \cdots t_0}(B_{t_n} \in I_n ; \dots; B_{t_1} \in I_1 ; W_{t_0})$$ (3)

such that the probability (2) has the form

$$Tr(E_{t,t_n \cdots t_0}(B_{t_n} \in I_n ; \dots, B_{t_1} \in I_1 ; W_{t_0}) E^{A_t}(J))$$ (4)

where the operator W_{t_0} denotes the density operator corresponding to the preparation C_{t_0} .

Introducing the preparation condition

$$C_{t_{n-1}} = [B_{t_{n-1}} \in I_{n-1}] ; [B_{t_{n-2}} \in I_{n-2}]; \dots; [B_{t_1} \in I_1]; C_{t_0}$$ (5)

the probability (2) can be written as

$$P(A_t \in J \mid [B_{t_n} \in I_n]; C_{t_{n-1}})$$ (6)

By the arguments of Sec.(14) , this must have the form

$$Tr(E_t(B_{t_n} \in I_n ; W_{t_{n-1}}) E^{A_t}(J)))$$ (7)

for some density operator $E_t(B_{t_n} \in I_n ; W_{t_{n-1}})$, where $W_{t_{n-1}}$ denotes the density operator corresponding to the preparation $C_{t_{n-1}}$, defined by (5) .

But from the preceding considerations we know that

$$W_{t_{n-1}} = E_{t_n, \ldots, t_0}(B_{t_{n-1}} \epsilon I_{n-1} ; \ldots, B_{t_1} ; W_{t_0}) \tag{8}$$

hence, equating (4) and (7) and keeping into account (8), we obtain

$$E_{t, t_n, \ldots, t_0}(B_{t_n} \epsilon I_n ; \ldots, B_{t_1} \epsilon I_1 ; W_{t_0})$$
$$= E_t(B_{t_n} \epsilon I_n ; E_{t_n}, \ldots, t_0(B_{t_{n-1}} \epsilon I_{n-1} ; \ldots ; B_{t_1} ; W_{t_0})) \tag{9}$$

By iteration of (9) one finds eventually :

$$E_{t, t_n \to 0}(B_{t_n} \epsilon I_n ; \ldots ; B_{t_1} \epsilon I_1 ; W_{t_0})$$
$$= E_t(B_{t_n} \epsilon I_n ; E_{t_n}(B_{t_{n-1}} \epsilon I_{n-1} ; \ldots (E_{t_2}(B_{t_1} \epsilon I_1 ; W_{t_0})))) \tag{10}$$

that is, in conclusion :

$$P(A_t \epsilon J \mid B_{t_n} \epsilon I_n ; \ldots ; B_{t_1} \epsilon I_1 ; C_{t_0}) =$$
$$= Tr(E_t(B_{t_n} \epsilon I_n ; E_{t_n}(B_{t_{n-1}} \epsilon I_{n-1} ; \ldots ; E_{t_2}(B_{t_1} \epsilon I_1 ; W_{t_0}) \ldots)) \cdot E^{A_t}(J)) \tag{11}$$

where the density operators $E_t(B_s \epsilon I ; W_r)$ are characterized by the property

$$P(A_t \epsilon J \mid B_s \epsilon I ; C_r) = Tr(E_t(B_s \epsilon I ; W_r) \cdot E^{A_t}(J)) \tag{12}$$

for any observable A and any interval J in $\mathbb{R}$.
We want to underline that the left hand side in the identity (11) is a conditional probability, not a joint probability, and that this conditional probability is experimentally measurable for every choice of the observables, the times, and the intervals appearing in the formula.
A source of considerable confusion in the literature on quantum measurement was due to the (implicit) introduction of the following postulate :

Postulate (K) The family of conditional probabilities :

$$P(A_t \epsilon I \mid B_{t_n} \epsilon I_n ; \ldots ; B_{t_1} \epsilon I_1 ; C_{t_0}) \tag{13}$$

for a fixed choice of the condition C_{t_0} and of the observables A,B but for the intervals I_j 's and the times t_j 's varying arbitrarily, admit a single Kolmogorovian model.

Postulate (K) above is equivalent to postulating the existence of a family of joint probabilities

$$P(A_t \epsilon J ; B_{t_n} \epsilon I_n ; \ldots ; B_{t_1} \epsilon I_1 \mid C_{t_0}) \tag{14}$$

which are related to the experimentally measurable probabilities (13) by the formula

$$P(A_t \epsilon J \mid B_{t_n} \epsilon I_n ; \ldots ; B_{t_1} \epsilon I_1 ; C_{t_0}) = \frac{P(A_t \epsilon J ; B_{t_n} \epsilon I_n ; \ldots ; B_{t_1} \epsilon I_1 \mid C_{t_0})}{P(B_{t_n} \epsilon I_n ; \ldots ; B_{t_1} \epsilon I_1 \mid C_{t_0})} \tag{15}$$

If such a formula were true, then one could determine inductively the joint probabilities (14) using (15) and (11). The result of the induction would then be :

$$P(A_t \epsilon J ; B_{t_n} \epsilon I_n ; \ldots ; B_{t_1} \epsilon I_1 \mid C_{t_0})$$

$$= TrL_t[B_{t_n} \epsilon I_n \; ; L_{t_n}(B_{t_{n-1}} \epsilon I_{n-1} ; \ldots ; L_{t_2}(B_{t_1} \epsilon I_1 ; W_{t_0}) \ldots))\}E^{A_t(J)} \quad (16)$$

where the operators L_t $(.,.)$, already introduced in Sec.(14), are given by the formula

$$L_t(B_s \epsilon I \; ; W_r) = E_t(B_s \epsilon I \; ; W_r) \cdot P(B_s \epsilon I \; ; C_r) \quad (17)$$

The particular case of formula (16), corresponding to the choice of the Luders-Zumino operator for L_t $(.,.)$,i.e. (cf. formula (25) of Sec.(14))

$$L_t(B_s \epsilon I \; ; W_r) = U_{r,s} E^{B_s}(I)W_r E^{B_s}(I)U_{r,s}^* \quad (18)$$

is known as " Wigner's formula for the joint probabilities".

However the term " joint probability " is misleading, since Wigner's formula is not even additive on disjoint intervals I_k 's and therefore it cannot represent a joint probability in any reasonable sense of the word.

Finally, let us conclude with the remark that the fact that the right hand side of (16) is not a probability in the usual sense of the word (let alone a joint probability , i.e. a projective family) is by no means limited to the specific form of the Luders- Zumino operator, in fact we know from Sec.(15) that the most general completely positive operator L_t has the form

$$L_t(B_s \epsilon I \; ; W) = \sum_k A_k(I)WA_k^*(I) \quad (19)$$

which only in very special cases can be additive in the interval I (relatively trivial cases in absence of superselection rules cf. [57]).

17. LUDWIG' S APPROACH

In the previous section we have seen that, in general, the experimentally measurable conditional probabilities

$$P(A_t \epsilon J \; |B_{t_n} \epsilon I_n \; ; \ldots ; B_{t_1} \epsilon I_1 ; C_{t_0}] \quad (1)$$

associated with a multitime quantum measurement , cannot be obtained from a family of joint probabilities

$$P(A_t \epsilon J \; ; B_{t_n} \epsilon I_n \; ; \ldots ; B_{t_1} \epsilon I_1 |C_{t_0}) \quad (2)$$

through the usual Bayes' formula. One could however invert the problem and ask oneself : Assuming that a well defined (in particular additive and projective) family of joint probabilities of the form (2) exist, what can be said about their mathematical structure ?

For notational convenience and without loss of generality we can assume that $A = B$. With the definition

$$S \; = \; all \; values \; of \; B \subseteq IR \quad (3)$$

the Daniell-Kolmogorov theorem tells us that to giving a projective family of joint probabilities of the form (2) is equivalent to give a measure on the space

$$\Omega = \prod_{IR} S = F(IR \; ; S) \quad (4)$$

of all functions from IR to S. Such a function will be called a trajectory of the system.

Therefore, denoting W_{t_0} the density operator corresponding to the preparation C_{t_0} , in our assumptions we have a probability measure

$$P(\,.\mid W_{t_0})\qquad\qquad\qquad(5)$$

on the space Ω of all the trajectories of the system.

<u>Theorem</u>1. Let H be a complex Hilbert space and let be given a correspondence

$\qquad$ $W \in$ *density operators on H* $\ -\!\!-\!\!\rightarrow\ P(\,.\mid W)$ *a probability measure on* Ω (6)

which is : (i) linear (ii) weakly continuous. Then there exists a positive operator valued measure

$$M : A\ \subseteq \Omega \text{---} \rightarrow\ M(A)\epsilon B(H)_+ \qquad\qquad(7)$$

such that, for any subset A of Ω and for any density operator W, one has

$$P(A\ \mid W) = Tr(W\cdot M(A))\qquad\qquad(8)$$

<u>Proof</u> One extends in a standard way the correspondence (6) to a linear functional on all the trace class operators. The continuity assumption and the fact that the dual of the trace class operators is $B(H)$ imply that the identity (8) holds for some $M(A)$ in $B(H)$. The positivity of $M(A)$ and the weak additivity of the map $A\qquad \Omega\text{---}\rightarrow M(A)$ are obvious.

$\qquad$ Theorem (1) above allows one to compare the previsions of such a theory with those of the usual quantum theory. In fact, writing :

$$X_t : \omega \in \Omega\text{---}\rightarrow X_t(\omega) = \omega(t)\epsilon S\qquad\qquad(9)$$

for the coordinate functions (X_t evaluates the trajectory ω at the point t), by construction for any I contained in S , the events

$$X_t^{-1}(I)\ ;\ [X_t\epsilon I]\ ;\ [B_t\epsilon I]$$

coincide. Therefore one should have :

$$Tr(W\cdot M(X_t^{-1}(I))) = P(X_t\epsilon I\ \mid W_{t_0}) = P(B_t\epsilon I\ \mid C_{t_0}) = Tr(W_{t_0}\cdot E^{B_t}(I))(10)$$

Thus, in order to have agreement with the usual quantum mechanics, one should have that the positive operator valued measures M(.) satisfy

$$M(\ X_t^{-1}(I)) = E^{B_t}(I)\qquad\qquad(11)$$

But the nontrivial positive operator valued measures satisfying a condition such as (11) are certainly very rare (in fact, even if I could not quote any precise theorem on this regard, under some natural additional regularity conditions, they should reduce to the usual projection valued measures).
This means that the formalism of the generalized positive operator valued measures cannot give the correct multi-time joint probabilities in all those cases in which the single time transition probabilities of the usual quantum theory are experimentally correct.
However, this does not mean that the formalism of positive operator valued measures is physically meaningless in fact the precise quantum theoretical statement concerning the transition probabilities $P(B_t\epsilon I\ \mid C_{t_0})$ is that they are correctly predicted by the right hand side of (10) only in the case when during the interval $[\,t_0\,,\,t]$, between the preparation and the measure, the system has undergone no interaction.
If in that time interval the system has been in continuous, non negligible, interaction with another system, say the measurement apparatus, then according to quantum

mechanics, one should write the Schroedinger equation of the coupled system, solve it (possibly) ,and then compute the probabilities $P(B_t \epsilon I \mid C_{t_0})$ by the usual quantum mechanical formalism, in which , however, the spectral projections of B at time t ,i.e.

$$E^{B_t}(I) = U_t E^B(I)U_t^* \qquad (12)$$

are obtained by letting the system evolve through the dynamics of the coupled system. But in case of a macroscopic system, such as a measurement apparatus, it is practically impossible to calculate this dynamics, and therefore in this case the prescriptions of the usual quantum theory are useless.

Ludwig's ansatz is that in such situations, or at least in some of them, it should be possible to find an operator valued measure M (.) such that the relevant probabilities, i.e. the probabilities (2), are correctly predicted by Eq.(8). The physical intuition underlying this ansatz is , very schematically, the following : Ludwig looks at micro-objects (like electrons, photons,etc.) as "theoretical inventions ", "mediators of interactions between macro-objects ". In his approach the space Ω represents the sample space of the macro-object (e.g. the measurement apparatus). According to Ludwig , who in this way theorizes and develops an idea already present in the late Bohr, there is a qualitative irreducible difference between the physics of the macro- and of the micro- world. The macroscopic objects should be described by classical physics and by the classical probabilistic model and the microscopic objects, not directly accessible to our experience, should be described by the usual quantum mechanical formalism. In particular to each state of a micro-object (a density operator in the model) should correspond a probability distribution of the values of some macroscopic apparatus (which is what we measure). But we have seen from Theorem (1) above that, under the additional assumption that the probability distribution depends linearly on the state of the micro-object, these assumptions lead naturally to the notion of positive operator valued measures. Of course, also in this case the considerations made in the remark before Theorem (1) in Sec.(14) above on the delicate assumptions hidden in the additivity requirement, apply. It might happen however that there are physically meaningful situations which cannot be described by the usual formalism of quantum mechanics, but which can by the formalism of positive operator valued measures. The program naturally emerging from Ludwig's analysis is thus to produce examples of situations of the above type. Recently considerable progress in the realization of this program has been obtained by Barchielli, Lanz, Prosperi,Lupieri,... (cf. [23] [24] ,and others).

The positive operator valued measures formalism and its applications is discussed in the monographs of Helstrom [38] , Holevo [39], and Kraus [42] . The question whether from the probabilistic point of view this formalism is really different from the Hilbert space formalism, i.e., if one can distinguish between this formalism and the usual quantum mechanical one by means of statistical invariants, similarly to what one did with the Kolmogorovian, the complex, and the real Hilbert space models, was posed in [16] , but at the moment the matter seems to be still open.

18. CLASSICAL OPERATOR VALUED PROCESSES AND THE BARCHIELLI-LANZ-LUPIERI-PROSPERI CONSTRUCTION

In this section we outline a theory of classical operator valued processes and apply to this theory the general quantum Feynmann-Kac perturbation technique developed in [5], [18] . We then show how the operator valued classical processes constructed by Barchielli,Lanz, Lupieri, and Prosperi [23], [24], and consequently also the classical operator valued processes studied by Davies under the name of "quantum stochastic processes", can be obtained as a particular case of our construction.

In the next sections it will be shown how this theory can be obtained by restricting to a particular abelian subalgebra the quantum markovian processes introduced in [1], [2], [3], [4].

In [13] the quantum Feynmann-Kac perturbation scheme was used to construct non-trivial (i.e. non gaussian) examples of quantum stochastic processes. Up to now all the known nontrivial examples of quantum stochastic processes have been constructed through a more or less implicit application of this scheme.

Let B denote the algebra of all bounded operators on some Hilbert space H (most of the following considerations apply however to more general algebras). Let (Ω, F, P) be a probability space and let :

$$X_t : \Omega \dashrightarrow B \tag{1}$$

denote a stochastic process with values in B. Let further $F_{(s,t)}, F_{s]}$ denote the σ-algebras generated respectively by the random variables X_r with r in (s,t) and $r \leqslant s$.

It is well known (cf. [55a], Theorem (1.22.13)) that the map

$$f \otimes b \in L^\infty(\Omega, F, P) \otimes B \rightarrow f(.)b \in L^\infty(\Omega, F, P; B) \tag{2}$$

defines a unique isomorphism of W^*- algebras. Moreover the maps :

$$f \otimes b \rightarrow P(f)b \; ; \; f \otimes b \rightarrow E_{s]}^P(f)b \tag{3}$$

where P(f) denotes the P-expectation of f and $E_{s]}^P$ denotes the P-conditional expectation with respect to the σ-algebra $F_{s]}$, extend to weakly continuous conditional expectations :

$$E : L^\infty(\Omega, F, P; B) \rightarrow B \approx B \otimes 1 \tag{4}$$

$$E_{s]}^\circ : L^\infty(\Omega, F, P; B) \rightarrow L^\infty(\Omega, F_{s]}, P; B) \tag{5}$$

Remark that the image under E of the algebra $L^\infty(\Omega, F, P) \otimes 1$ is the multiples of the identity of B, since it is in the center of B. Conversely, it is clear that any conditional expectation of the form (4) uniquely determines a B-valued classical stochastic process (in fact it is easy to construct, given such a conditional expectation, a family of compatible joint probabilities for a B-valued stochastic process).

Now consider a map E as in (4) which is positive and identity preserving, but not necessarily a conditional expectation. In this case it is no longer true that the image under E of $L^\infty(\Omega, F, P) \otimes 1$ is the multiples of the identity in B, and therefore we cannot associate with such a map a family of joint probabilities. We can do this however, if we also fix a state ϕ_0 on B. In such a case in fact, we can define a state ϕ on $L^\infty(\Omega, F, P; B)$ by

$$\phi(F) = \phi_0(E(F)) \; ; \; F \in L^\infty(\Omega, F, P; B) \tag{6}$$

and, given, the state ϕ the joint probabilities can be defined by

$$P(X_{t_1} \epsilon I_1; \; \cdots \; X_{t_n} \epsilon I_n) = \phi(\chi_{I_1}(X_{t_1}) \; \cdots \; \chi_{I_n}(X_{t_n})) \tag{7}$$

where $t_1 < \cdots < t_n$ are real numbers and $I_1, \cdots, I_n$ are subsets of B.

The probability measure , on the space of all B-valued trajectories of this stochastic process, obtained in this way, will be denoted $P(. \mid \phi_0)$ and it is easy to see that it depends linearly and weakly continuously on ϕ_0 . Note that in this case the unique B-valued positive operator measure M(.), on this trajectories space , characterized by Eq.(8) of Sec. (17), is nothing but

$$M(K) = E(\chi_K) \tag{8}$$

K being any sub set of the trajectory space.

A simple and general way of constructing maps E of the type described above, and therefore nontrivial examples of operator valued measures, is to perturb the family of conditional expectations defined by (5) by means of the following construction.

First note that if the process (X_t) we started with is an independent increment process, then :

$$E^\circ{}_{t|s}(L^\alpha(\Omega, F_{(s,t)}, P\ ; B)) \subseteq B \tag{9}$$

Consider now a multiplicative functional, i.e. a two parameter family $U_{t,s}$ of operator valued functions with the following properties :

$$U_{t,s} \in L^\alpha(\Omega, F_{(s,t)}, P\ , B) \tag{9}$$

$$\text{if } r < s < t \quad \text{then} \quad U_{s,r}U_{t,s} = U_{t,r} \tag{10}$$

A multiplicative functional with an additional covariance property with respect to the time shift was called in [5] a "markovian cocycle". The maps $E_{t|s|}$ defined by

$$E_{t|s}(x) = E^\circ{}_{t|s}(x) \tag{11}$$

are clearly (completely) positive even if in general they will not be conditional expectations.However they enjoy the properties :

$$E_{t|s}(L^\alpha(\Omega, F_t], P\ ; B)) \subseteq L^\alpha(\Omega, F_s], P\ ; B) \tag{12}$$

$$\text{if } r < s < t \quad \text{then} \quad E_{s|r}]E_{t|s}] = E_{t|r}] \tag{13}$$

$$E_{t|s}(f_s|a_t]) = f_s]E_{t|s}(a_t]) \tag{14}$$

for each $f_s \in L^\alpha(\Omega, F_s], P)$ and for each $a_t] \in L^\alpha(\Omega, F_t], P\ ; B)$. In the language introduced in [1] the properties (12) and (14) mean that $E_{t|s}]$ is a "quasi-conditional " expectation with respect to the triple

$$L^\alpha(\Omega, F_s], P)\otimes 1 \subseteq L^\alpha(\Omega, F_s], P\ ; B) \subseteq L^\alpha(\Omega, F\ , P\ ; B) \tag{15}$$

If they also enjoy the property

$$E_{t|s}(1) = 1 \tag{16}$$

we say that the multiplicative functional $U_{t,s}$ is normalized. Moreover, since the conditional expectation $E^\circ{}_{t|s}]$ is associated with an independent increment process, one also has

$$E_{t|s}(L^\alpha(\Omega, F_{(s,t)}, P\ ; B) \subseteq B \tag{17}$$

These properties allow us to establish the connection between perturbed families of conditional expectations, in the sense described above, and operator valued stochastic processes in the sense of Barchielli, Lanz, Prosperi [23].

In fact, assume that all the above conditions are realized (including the independent increment assumption) and define for each subset $M_{s,t}$ of the space Ω, measurable for the σ-algebra $F_{(s,t)}$, a linear operator on B , by

$$L(t,s\ ;M_{s,t})[b] = E_{t|s}(\chi_{M_{s,t}} \otimes b) \tag{18}$$

Then clearly the operator $L(t,s\ ;\ M_{s,t})$ is (completely) positive, identity preserving, additive on disjoint subsets of $F_{(s,t)}$ and satisfies

$$L(t,s;\Omega) = 1 \tag{19}$$

and, because of the " quasi-conditional expectation" properties (12) and (14), one has :

$$L(s,r\ ;N_{r,s})L(t,s\ ;M_{s,t})[b] = E_{s|r}(\chi_{N_{r,s}} \otimes L(t,s\ ;M_{s,t})[b]) = \tag{20}$$

$$= E_{s|r}(\chi_{N_{r,s}} \otimes L(t,s\ ;M_{s,t})[b]) =$$

$$= E_{s|r}(\chi_{N_{r,s}} E_{t|s}(\chi_{M_{s,t}} \otimes b)) =$$

$$= E_{s|r} E_{t|s}(\chi_{N_{r,s} \cap M_{s,t}} \otimes b)$$

Thus in conclusion

$$L(s,r;N_{r,s})L(t,s;M_{s,t})[b] = L(t,r\ ;N_{r,s} \cap M_{s,t})[b] \tag{21}$$

Conversely , given a family ($L(t,s: .)$) ($s \leqslant t$) of maps from the $F_{(s,t)}$ -measurable subsets of Ω which is additive on disjoint sets and satisfies (19) and (21) above, then one can use the identity (21) to define the family $E_{t|s]}$, and it is easily checked that this family satisfies the conditions (12)-(17).

For each subset N of Ω the dual of the map $L(t,s\ ;\ N)$, which will be denoted $E(t,s\ ;\ N)$, acts on the states on B or equivalently on the density operators. The family of maps { $E(t,s\ ;\ .):s\leqslant t$ } is an operator valued stochastic process in the sense of Barchielli, Lanz, Prosperi [23] . The problem of constructing examples of such processes is then reduced to the problem of constructing families of normalized multiplicative functionals, in the sense defined above, and the main techniques developed up to now to achieve this are either the solution of an ordinary differential equation or the solution of a quantum stochastic differential equation in the sense of Hudson and Parthasarathy [40a].

19. QUANTUM MARKOV PROCESSES

The characteristic feature of quantum probability is that one can "lump together " in a single mathematical structure , i.e. the algebra of quantum observables and a state on it , infinitely many classical probability spaces (i.e. the restriction of the quantum state to the abelian subalgebras of the algebra of the observables). Similarly, if one "lumps together" infinitely many classical operator valued processes, one is led to the notion of quantum stochastic process.

The following heuristic considerations show that there is a natural way to achieve this program.

Let us begin by considering only two instants of time $s \leqslant t$ and two observables B_s^1 ; B_t^2 and a preparation W referred to some interval of time before s. Denote by

$$P(B_t^1 \epsilon I_1\ ;\ B_s^2 \epsilon I_2 \mid W) \tag{1}$$

the joint distributions of these two observables and denote by $e_j(I,t)$ the spectral projection of the observable B_t^j relative to the numerical interval I (j = 1,2).

The joint probabilities (1) give a probability measure on the partition of the identity $e_1(I,t) \otimes e_2(J,s)$ on the space $H \otimes H$ where the observables are defined. Letting the observables B' s vary in all possible ways, one obtains a probability measure on all the partitions of the identity on $H \otimes H$ of the form $e_1(I) \otimes e_2(J)$. These are not all the possible partitions of the identity on the space $H \otimes H$, but assume that, possibly under additional natural conditions, they are sufficiently many to allow the application of an hypothetical "product version" of the Gleason theorem (cf. Sec. (24) and the general problem stated there) and to conclude that any such a family of probability measures is obtained by restricting on the abelian algebras of the form $A_1 \otimes A_2$ a weakly continuous state on $B(H \otimes H)$ which we denote by $\phi_{s,t}$. Assume moreover that the correspondence

$$W \longrightarrow \phi_{s,t} \tag{2}$$

is linear (affine) and weakly continuous. Then by the same duality arguments used in Secs.(17) and (18), we conclude that there exists a positive continuous, linear map

$$E_{s,t} : B(H \otimes H) \longrightarrow B(H) \tag{3}$$

with the property that, for every density matrix W on H and for every operator x in $B(H \otimes H)$, one has

$$\phi_{s,t}(x \mid W) = Tr(W E_{s,t}(x)) \tag{4}$$

Since W is arbitrary one also has

$$E_{s,t}(1 \otimes 1) = 1 \tag{5}$$

With the same physical justifications , mentioned in Sec.(15), we can assume that the map $E_{s,t}$ is completely positive.

A completely positive map of the form (4) with the property (5) has been called in [1] a transition expectation.

If the Hilbert space H is realized as the L^2 -space over some probability space (Ω, F, P) then we can let the elements of $L^\infty(\Omega, F, P)$ act by multiplication on H , and this gives a natural embedding

$$L^\infty(\Omega) \otimes B(H) \to B(H \otimes H) \tag{6}$$

In particular, denoting χ_N (N -a subset of Ω) the characteristic function of the set N, the map

$$b \in B(H) \to E_{s,t}(\chi_N \otimes b) \in B(H) \tag{7}$$

is an operation in the sense of Sec.(15), and the map

$$N \subseteq \Omega \to E_{s,t}(\chi_N \otimes .)$$

is an operation valued measure in the sense of Davies and Lewis [35]. Thus, just as by restricting a quantum state on an abelian algebra, one obtains a classical probability measure, so restricting a quantum transition expectation on an abelian algebra gives an operation valued measure.

The above argument can be extended without difficulty to the case of n-dimensional joint distributions leading, to the formula

$$P(B_{t_n}^n \in I_n ; \cdots ; B_{t_1}^1 \in I_1 \mid W) = \tag{8}$$

$$= Tr(W \; E_{t_o, t_1, \cdots, t_n}(e_n(I_n ; t_n) \otimes \ldots \otimes e_1(I_1 ; t_1)))$$

where $e_j(I, t)$ is the spectral projection of B_t^j and

$$E_{t_0, \ldots, _{t_n}} : B(\otimes^n H) \to B(H) \tag{9}$$

is a completely positive, identity preserving map whose structure is given by

$$E_{t_0, \ldots, _{t_n}}(b_1 \otimes b_2 \otimes \ldots \otimes b_n) = E_{t_1 t_0}(b_1 \otimes E_{t_2 t_1}(B_2 \otimes \ldots \otimes E_{t_n t_{n-1}}(b_{n-1} \otimes b_n) \ldots)) \tag{10}$$

where the maps $E_{t,s}$ are those defined by Eq.(4) above.

If we want that the joint probabilities given by the right hand side of (8) form a compatible (i.e. projective) family, then we have to impose the following compatibility conditions on the maps $E_{t,s}$:

$$E_{r,s}(a \otimes E_{s,t}(b \otimes 1)) = E_{r,s}(a \otimes b) \tag{11}$$

$$E_{r,s}(a \otimes E_{s,t}(1 \otimes b)) = E_{r,t}(a \otimes b) \tag{12}$$

$$E_{r,s}(1 \otimes E_{s,t}(a \otimes b)) = E_{s,t}(a \otimes b) \tag{13}$$

A family of transition maps satisfying the conditions (11),(12),(13) above, which are the quantum mechanical generalization of the well known Chapman-Kolmogorov equations of the classical theory of Markov processes, is called a markovian family of transition expectations [1], [3] . For such a family one can easily extend to the quantum case the Daniell-Kolmogorov construction. The result is a particular class of quantum Markov processes (called symmetric in [3]) with the property that the time ordered correlations uniquely determine the whole process . This class includes, as shown above, all the classical operator valued processes however, in the much wider class of quantum processes with continuous time parameter, the symmetric markovian processes are very few (there are however many examples with discrete time parameter [1],[6],[12]).

20. QUANTUM MARKOV CHAINS AND THE QUANTUM MEASUREMENT PROCESS

The simplest examples of quantum Markov chains can be obtained as follows: Take for each instant of time a copy of the algebra B(H) of all bounded operators on some Hilbert space H and form the infinite (C^*-) tensor product of this family :

$$\otimes_t B(H) = A \tag{1}$$

The symbol :

$$a_{t_1} \otimes a_{t_2} \otimes \cdots \otimes a_{t_n}$$

will denote the element of the tensor product (1) which has a_{t_j} at the t_j -th place (j = 1,...,n) and 1 elsewhere ; in particular, for n = 1 , we will write j_t (a) for a_t .

Now choose :

- a one parameter unitary group U_t in B(H)

- a normal state ϕ on B(H)

- an abelian sub-algebra C of B(H)

- a conditional expectation F : B(H) $\to$ C

and define, for $s < t$ and a,b in B(H) :

$$E_{t,s}(a \otimes b) = u_{-s}(F(a)F(u_t(b))) \tag{2}$$

where

$$u_t(a) = U_{-t}aU_t \tag{3}$$

It can be shown that the family ($E_{t,s}$) is a markovian family of transition expectations. Moreover the joint expectations

$$\phi(a_o \otimes a_{t_1} \otimes \cdots \otimes a_{t_n}) = \phi_o(F(A_o)F(u_{t_1}(a_1)) \cdots F(u_{t_n}(a_n))) \tag{4}$$

define a unique quantum Markov state on the algebra A .
If w_o is the density matrix of the system at time 0, we can define the density matrix at time t by the formula

$$Tr(w_t a) = \phi_0(F(u_t(a))) \tag{5}$$

Thus, if the algebra C is generated by the rank one projections

$$e_i = P(\psi_i) \tag{6}$$

corresponding to an orthonormal basis (ψ_i) of H, one finds that

$$w_t = \sum_j Tr(w_o e_j) u_t(e_j) \tag{7}$$

which is precisely von Neumann' s formula for the density operator at time t of a system which has undergone an incomplete measurement at time 0 of the observable corresponding to the atomic spectral measure (6). Thus, as a by-product of the theory of quantum Markov chains,we have proved that it is possible to construct the joint probabilities for incomplete measurements in a way which is compatible with von Neumann evolution even without abandoning the usual quantum mechanical formalism .

21. QUANTUM STOCHASTIC PROCESSES

The algebra (1) of the preceding section (the discrete tensor product)   is only the simplest example of algebra where on can construct quantum Markov processes. The next natural candidate is the continuous tensor product , which corresponds to the Fock space . More generally, the theory of local algebras provides a natural class of examples of quantum stochastic process . In the applications to the quantum theory of irreversible processes, it is quite natural to consider algebras which are generated by the sub algebras at fixed times . A quantum stochastic process on these algebras is uniquely determined by its "correlations kernels". The corresponding reconstruction theorem is given in [13]. In fact the class of stochastic processes defined in [13] includes all the examples constructed up to now.

22. THE MAXIMAL AND THE MINIMAL HIDDEN VARIABLE PROGRAM

The problem of hidden variables in quantum mechanics is explained by Bell as follows [25] : "...The question at issue is whether the quantum mechanical states can be regarded as ensembles of states further specified by additional variables, such that given values of these variables together with the state vector determine precisely the results of individual measurements. These hypothetical well-specified states are said to be 'dispersion free '...".
More explicitly one might say that one wants these additional variables to be experimentally measurable, and that the result of any other individual measurement is precisely determined by their values and the state vector. Under these circumstances, if such additional variables exist, then all the observables of the theory are compatible and the Heisenberg principle, even in the very weak formulation of section (8) is

experimentally falsifiable.

The unsuccesful attempts by Einstein to give a direct counterexample to the Heisenberg principle are well known and constitute the first phase of the famous Bohr-Einstein debate. It doesn' t seem that since Einstein' s times any new idea concerning the falsification of the Heisenberg principle has emerged. In fact, the scarcity of direct experimental verifications of the Heisenberg principle, underlined by Jammer in [40b] , might well be a symptom of scarsity of ideas in devicing an experiment suitable to challenge this principle. Thus the program naturally arising from Bell' s formulation of the hidden variables problem amounts to nothing less than the original Einstein program of the falsification of the Heisenberg principle. We will call this the "maximal hidden variable program".

After the passionate Bohr-Einstein debate, this maximal hidden variables program has not been widely pursued in the literature, while a remarkable amount of work has been produced since then on the the question of whether the existence of a hidden variable theory, in the sense of the maximal program, would be a priori contradictory with the established results of quantum theory or not. The supporters of the hidden-variables theories have tried to show that the basic principles of the theory are not a priori in contradiction with the assumption of an underlying exact theory of classical type, by constructing some mathematical models in which the observable predictions of quantum mechanics are expressed in terms of classical mean values over functions on a classical state space. The opponents of hidden variables have tried to rule out in principle such types of construction by showing that no classical theory can be equivalent to the usual quantum theory.

The reason why different people could arrive to different conclusions is that they used different notions of "equivalence" between a classical theory and the usual quantum theory. In the following I have tried to summarize some points of this debate . Even in the multiplicity of approaches, and of the corresponding conclusions, some general statements seem to emerge from this debate, now more than half a century old , namely :

i) If, in the notion of equivalence between a classical and a quantum theory, one requires a minimum of algebraic constraints, then the result is negative : No nontrivial hidden-variable theory satisfying these constraints can exist (cf. Secs.(23)-(27)).

ii) As soon as the notion of equivalence between the quantum and a classical theory is loosened enough, then immediately you have uncountably many hidden variables theories which fit into this notion of equivalence (cf. Secs. (26)-(28)).

This situation looks rather like a stale mate: the supporters of hidden-variable theories accuse the opponents of introducing unphysical algebraic constraints in their notion of equivalence (cf. for example Bell' s critique of the no-go theorems in [25]). The opponents of hidden variables accuse the supporters of being inconclusive : The fact that you can invent a notion of equivalence between a classical and a quantum theory, in such a way that there exist infinitely many classical theories equivalent in this sense to the usual quantum theory, does not increase very much our knowledge of the natural world until one has pointed out which of these theories is chosen and why.

There are a number of hidden-variables models which reproduce various aspects of quantum theory. The one which covers the widest class of quantum phenomena is undoubtly Nelson' s stochastic mechanics [48], [49] which in the last years, especially with the works of Guerra and his school (cf.[37b] and citations therein) , introduced some remarkable new ideas in classical probability theory. Stochastic mechanics is equivalent to quantum mechanics in the following sense : one fixes a (usually pure) quantum state and a quantum hamiltonian which is the generator of a positivity preserving semi-group. To these objects one associates a classical stochastic (in fact Markov) process for the position observable, with the property that at each fixed time the probability distribution of the random variable "position" coincides with the probability distribution of the quantum mechanical position operator in the given state. There are many classical processes with this property, but the one constructed by stochastic mechanics is characterized by some very natural mathematical constraints, the most relevant one being the stochastic generalization of Newton's equation of motion , due to Nelson. This theory does not contradict the results obtained using the statistical invariants, in fact in it both the initial condition (the pure state) and an "abelian section" of the full algebra of quantum observables (the position algebra) are fixed a priori, and only for these the kolmogorovian model is constructed. Moreover several phisically interesting hamiltonians are the generators of a positivity preserving semigroup, and therefore the above construction allows one to study the spectrum of such hamiltonians with the powerful tools of the theory of stochastic processes (this is also the mathematical essence of the euclidean technique to study quantum hamiltonians).

The main idea of quantum probability is that a quantum process is obtained by "lumping together " infinitely many classical processes in a nonclassical way. It might be that these infinitely many classical processes are those given by stochasic mechanics. In the following we shall limit our discussion to the case in which the algebra generated by the observables is the whole B(H), but many of the results have been extended to more general lattices (cf. [28] and [51] for a review of these extensions).

23. VON NEUMANN' S THEOREMS ON HIDDEN VARIABLES

For any *-algebra A (e.g., a matrix algebra) and for any state μ on A, the μ-dispersion of a ϵ A is defined by

$$\Delta_\mu(a) = \mu(|a|^2) - |\mu(a)|^2 \tag{1}$$

The state μ is called dispersion free if the μ-dispersion of any a ϵ A is zero.

Lemma 1. A state μ on A is dispersion free if and only if

$$\mu(a^* b) = \mu(a)^* \mu(b) \tag{2}$$

Proof Sufficiency : Letting b = a in (2) we find that the right hand side of (1) is equal to zero. Necessity : for any a,b in A , Schwartz' s inequality implies that

$$|\mu(a^* b) - \mu(a)^* \mu(b)| = |\mu(a^*[b - \mu(b)])| \leqslant \mu(|a|^2)^{1/2} \mu(|b - \mu(b)|^2)^{1/2}$$

Hence (2) follows if μ is dispersion free.

A corollary of the above lemma is that a dispersion free state can assume only the values 0 and 1 on the projections of A (i.e. the elements of A satisfying $a = a^* = a^2$). This fact will be frequently used in the following sections. The following theorem shows that a system can have sufficiently many dispersion free states if and only if it is a classical system, in the sense that all its observables are compatible.

Theorem 2. Let A be a *-algebra whose states separate the points (i.e. an element a of A is zero if and only if $\mu(a) = 0$ for each state μ on A).

If every state on A is a convex combination of dispersion free states, then A is commutative. If A is a C^*-algebra, then the converse statement is also true.

Proof. Necessity: Lemma (2) implies that for any dispersion free state μ on A one has :

$$\mu(ab - ba) = 0 \; ; \; a, b \in A$$

and, since the states on A separate the points, this implies that ab=ba for each a,b in A.

Sufficiency: If A is a commutative C^*-algebra, then it is isomorphic to the algebra C(S) of the continuous complex functions on some compact Hausdorff space S. By Baire's theorem, any state μ on C(S) has the form

$$\mu(a) = \int_S a(x) dm(x); a \in C(S)$$

for some probability measure m on S. Thus any state on A= C(S) is a convex combination of the dispersion free Dirac states.

If $a = a^* \in A$ is an observable and μ is a state on A, the number $\mu(a)$ is called the mean value of the observable a in the state μ . If μ is a dispersion free state, then any observable a in A has no fluctuations around its mean value, therefore one says that $\mu(a)$ is the exact value of a in the state μ. In a theory for which any state is a convex combination of dispersion free states, the only possible origin of the statistical fluctuations is a lack of knowledge about the dispersion free state in which the system is. The statistical component of such a theory is completely characterized by a probability measure on the family of dispersion free states. If we obtain precise information on the dispersion free state in which the system is, then all the values of all the physical observables of the system are uniquely determined. Therefore, if one could prove that the physical assumptions of the theory uniquely determine a mathematical model in which the observables are represented by elements of a non-commutative *-algebra, then Theorem (3) would force upon us the conclusion that our physical assumptions are such that the statistical elements in the description of the system cannot be traced back only to a lack of knowledge on the dispersion free states, but must have some other origin too. Von Neumann repeatedly tried to find such a proof , but without success. Nevertheless, he concluded ([60] , pg. 323) that "Hence, within the limits of our conditions , the decision is made , and is against causality; because all the ensembles have dispersions, even the homogeneous ..."

The reason why one cannot agree with this statement is that, in absence of such a proof, we can only conclude that the usual quantum model is not compatible with the assumption that the only origin of the statistical elements in the theory is a lack of knowledge about the dispersion free states. But this does not exclude the possibility that our non-commutative model is unnecessary and that all the results can be obtained using another model in which the only source of the statistical elements is lack of knowledge on the dispersion free state in which the system actually is. Theorem (3) shows that in such a model the family of physical observables should be represented by a commutative *-algebra, and since the standard model of such algebras is the family of all continuous complex valued functions on some compact space S, it follows that, if we want to look for such a model, then we have to look for:

- a measurable space S

- a correspondence between physical observables and functions on S

Concerning this alternative model, two attitudes are a priori possible:

i) The usual quantum theory is wrong, and the alternative model corresponds to a theory leading to different experimental predictions.

ii) The alternative model should lead to the same predictions as the usual quantum theory.

Up to now, most of the attempts to construct an alternative model for quantum theory have followed the latter attitude. Assuming this, any correspondence between physical observables and functions on a (state) space S will also induce a correspondence between self-adjoint operators on some Hilbert space H and functions on S. The conceptual scheme of von Neumann' s second hidden-variable theorem can be described as follows:

- One postulates a priori some mathematical properties of the correspondence between hermitean operators and functions on S (e.g. that the correspondence is one-to-one, that it is well behaved with respect to some algebraic properties,etc.)

- One then proves that no correspondence can exist with the nice mathematical properties required.

This scheme has been applied by a number of authors to generalize von Neumann' s theorem below ,to structures more general than the algebra of all operators on some Hilbert spaces (cf.[28]).

Theorem 2. Let H be an Hilbert space. Let S be a measurable space and let f: $a \rightarrow f_a$ be a map from the bounded self-adjoint operators on H to the measurable functions on S, such that :

$$f_{a+b} = f_a + f_b \tag{1}$$

$$f_{F(a)} = F(f_a) \tag{2}$$

for each continuous function $f : R \rightarrow R$. Then the map $a \rightarrow f_a$ is identically zero.

Proof. Clearly the map $a \rightarrow f_a$ is linear and, choosing in (2) $F(x) = x^2$ one sees that f_a is a positive function whenever a is a positive operator. Hence for each s in S the map $a \rightarrow f_a(s)$ defines a positive linear dispersion free functional on the algebra of all bounded operators on H. And only the zero functional can have these properties. (cf.[28]).

But again, as pointed out by Bell [25] von Neumann' s conclusion : " ... we need not to go any further into the mechanism of 'hidden parameters' since we now know that the established results of quantum mechanics can never be rederived with their help ...." . In fact, the algebraic constraints posed by von Neumann are by no means an obvious consequence of the assumption of the existence of some hidden-parameter model for quantum theory.

24. GLEASON' S THEOREM AND SOME OF ITS COROLLARIES

A deep theorem of Gleason states that any positive countably additive measure on the projections of some Hilbert space H is obtained by taking the restriction, on the lattice of the projections, of a weakly continuous state on the algebra of all operators on H. As a result of the contributions of many authors, Gleason's theorem has been extended to measures defined on the projections of arbitrary von Neumann algebras [51] (cf. also [43] for an outline of the main ideas of the proof) .

One of the key steps in the proof of Gleason theorem is the following technical lemma from which several theorems on hidden variables can be obtained as corollaries. Recently an elegant elementary proof of this lemma has been obtained by R. Cook, M. Keane and W. Moran [32].

Lemma 1. Let $S^{(2)}$ denote the unit sphere in R^3. Any bounded below function f : $S^{(2)} \to R$ such that for any orthonormal basis a,b,c of R^3

$$f(a) + f(b) + f(c) = 1 \tag{1}$$

is continuous.

Corollary 2. There exist no bounded below non-constant function $f : S^{(2)} \to R$ satisfying (1) and assuming only the values 0 and 1.

Proof. By Gleason's lemma such a function should be continuous, and, since $S^{(2)}$ is connected, it must be a constant if its range is the set {0,1}.

Corollary 3. There exist no non zero finitely additive probability measure on the projections of the 3x3 matrices, assuming only the values 0 and 1.

Proof. Let m be such a measure. With each unit vector a in R^3 we can associate a rank one projection P_a and the function

$$f(a) = m(P_a)$$

satisfies the conditions of the corollary above hence it must be zero.

The following consequence of Gleason theorem was noted by Bell [25] and subsequently by Kochen and Specker [41]:

Theorem 4. Let H be an Hilbert space of dimensions ≥ 3. There exist no measurable space (S,B) and nonzero map a $\to f_a$ from the bounded self-adjoint operators on H to the measurable real valued functions on S such that:

$$f_{F(a)} = F(f_a) \tag{3}$$

for each measurable function $F : R \to R$ and each hermitean operator a.

Proof. Equation (3) implies that, for each pair of bounded commuting self-adjoint operators a,b on H, one has :

$$f_{(a+b)} = f_a + f_b \tag{4}$$

$$f_{ab} = f_a \cdot f_b \tag{5}$$

in fact, if a and b commute, there exist an hermitean operator c and functions F and G such that

$$a = F(c) ; b = G(c)$$

thus, choosing H(x) = F(x) + G(x); K(x) = F(x).G(x) and using (3) one finds;

$$f_{a+b} = f_{H(c)} = H(f_c) = F(f_a) + G(f_c) = f_a + f_c$$

and similarly:

$$f_{ab} = f_{K(c)} = K(f_c) = F(f_c).G(f_c) = f_a \cdot f_b$$

From this one deduces that

$$f_0 = 0 ; f_{(-a)} = -f_a \tag{6}$$

$$f_{(\lambda a)} = \lambda f_a \tag{7}$$

for each rational number λ. Moreover, because of (4)

$$a \geqslant 0 \rightarrow f_a \geqslant 0 \tag{8}$$

hence, if a and b commute and a $\geqslant$ b, then $f_a \geqslant f_b$. Therefore, if λ is a real number, and λ',λ'' are rational numbers such that $\lambda' \geqslant \geqslant \lambda \geqslant \lambda''$ one has :

$$\lambda' f_a \geqslant f_{\lambda a} \geqslant \lambda'' f_a$$

whence for any real number λ

$$f_{\lambda a} = \lambda f_a.$$

These properties imply in particular that if f is identically zero on each rank one projector, then it will be identically zero for each operator a on H.

Assuming, by contradiction, that there exists a rank one projection b on which f is not identically zero then , possibly restricting oneself to a sub-space, one can assume that H has 3 dimensions over the reals. In this case, the mapping $a \rightarrow f_{a(s)}$ must be zero on each projection because of Corollary (3). Therefore, because of the above discussion, it must be zero on each operator, against our assumption.

It is well known that the case of a two dimensional (real or complex) Hilbert space is pathological for the purely algebraic aproaches of von Neumann, Kochen and Specker type. In fact in this case,for each vector a in the first quadrant of R^2 one can define in an arbitrary way a measurable function f_a from some measurable space with values in the set $\{0,1\}$. One then defines the function f_a on the vector a', in the second quadrant orthogonal to a, to be $1 - f_a$. This map is extended to self adjoint operators on R^2 ,defining

$$f_{(\alpha P_a + \beta P_{a'})} = \alpha f_a + \beta F_{a'}$$

This map is nonzero and satisfies the conditions of the Bell-Kochen-Specker theorem. The construction is trivially extended to the complex two-dimensional case.

The quantum-probabilistic approach, which does not introduce arbitrary algebraic conditions, but relies uniquely on the experimentally observable transition probabilities, removes this pathological behaiour of two-dimensional observables, whose physical meaning has never been clearly explained.

25. ON THE EQUIVALENCE BETWEEN THE CONDITIONS OF VON NEUMANN AND THOSE OF BELL-KOCHEN-SPECKER

The strategy of the proof in the von Neumann and in the Bell-Kochen-Specker theorems is similar and purely algebraic : In both cases one shows that there can exist no correspondence between quantum observables and measurable functions satisfying certain a priori given algebraic conditions.

The statistics of the values of the observables in question ,which is one of the main sources of comparison between quantum theory and experiments, doesn' t play any role in the proof of these theorems. But since, according to Bell [25] :...the function of the hypothetical dispersion free states is to reproduce the measurable peculiarities of quantum mechanics when averaged over..." one expects that the statistical properties of this correspondence should after all play a role.

In this section we show that, if in addition to the correspondence between quantum observables and measurable functions, we also require the existence of a correspondence between quantum states and probability measures on the dispersion free states which preserves the mean values, then the assumptions of the Bell-Kochen-Specker theorem become equivalent to the apparently stronger assumption in the von Neumann theorem.

Theorem 1. Let H be   an Hilbert space of dimensions greater than 3. Assume that there exist:

i) a measurable space S on which the probability measures separate the measurable functions

ii) a correspondence $a \to f_a$ between hermitean operators on H and measurable real valued functions on

iii) a correspondence $\phi \to \mu_\phi$ between (w-continuous) states on the algebra B(H) and probability measures on S

Suppose moreover that the following two conditions are satisfied:

iv) for any hermitean operator a

$$F(f_a) = f_{F(a)} \tag{1}$$

for any function $F : R \to R$

v) for any normal state ϕ on B(H) and for any a in B(H)

$$\phi(a) = \mu_\phi(f_a) \tag{2}$$

Then for any pair of hermitean operators a,b in B(H) one has :

$$f_{a+b} = f_a + f_b \tag{3}$$

Proof. Assume by contradiction that there exist two bounded self-adjoint operators a and b such that

$$f_{a+b} \neq f_a + f_b \tag{4}$$

Then, because of assumption (i) , there exists a probability measure ν on S such that

$$\nu(f_{(a+b)}) \neq \nu(f_a) + \nu(f_b) \tag{5}$$

Because of iv) the map

$$p \in Proj(H) \to \nu(f_p)$$

is a positive measure on the projections on H. Hence, by Gleason' s theorem , there exists a normal state ϕ_ν on B(H) such that, for any projection p on H

$$\phi_\nu(p) = \nu(f_p) \tag{6}$$

But condition (iv) implies that, if P(.) is the projection valued measure associated with some hermitean operator a, then, denoting χ_I the characteristic function of the interval I, one has

$$\chi_I(f_a) = f_{\chi_I(a)} = f_{P(I)} \tag{7}$$

From (7) it follows that if μ_ϕ is the probability measure associated with the normal state ϕ on B(H) , then the measure

$$I \subseteq R \to \mu_\phi(f_{P(I)})$$

is the distribution of the random variable f_a .

Applying this remark to ϕ_ν and using (6) , we deduce that for any hermitean operator x in B(H),

$$\phi_\nu(x) = \nu(f_x)$$

In particular

$$\nu(f_{a+b}) = \phi_\nu(a + b) = \phi_\nu(a) + \phi_\nu(b) = \nu(f_a) + \nu(f_b)$$

which contradicts (5). Hence, under our assumptions we must have that

$$f_{a+b} = f_a + f_b$$

and this proves the statement.

26. A TRIVIAL CONSTRUCTION, ITS COMPLICATION, AND A KNOWN EXAMPLE

In this section we discuss a trivial method for constructing hidden -variable theories (which from time to time has been noted by different authors). As a particular case of this construction we obtain a hidden variable theory constructed by Bell and discussed by Wightman in [61] .

Let H be any Hilbert space; denote by S the set of weakly continuous states on the algebra B(H) , considered as a measurable space with the Borel σ-algebra defined by the weak topology. For each hermitean operator a on H, define the function f_a on S by

$$f_a(\phi) = \phi(a); \phi - a \text{ state on } B(H) \tag{1}$$

For any (normal) state ϕ on B(H) , define the probability measure μ_ϕ on S to be the Dirac delta measure concentrated on the state ϕ on S .

It is easily checked that the correspondences

$$a - operator \rightarrow f_a - measurable \ function \ on \ S$$

$$\phi - state \rightarrow \mu_\phi - probability \ measure \ on \ S$$

enjoy the following properties:

$$f_{(a+b)} = f_a + f_b ; \ any \ a, b \ in \ B(H) \tag{2}$$

$$f_a \geqslant 0 ; if \ a \geqslant 0 \tag{3}$$

$$\phi(a) = \mu_\phi(f_a) \tag{4}$$

So this "hidden-variable theory" reproduces in a trivial way the predictions of quantum mechanics. Now let us complicate this construction as follows : Introduce a new, arbitrary , probability space and choose any function $F_a : SxT \rightarrow C$ subject to the only constraints of being measurable and

$$\int_T F_a(\phi, \lambda) d \nu(\lambda) = f_a(\phi) \tag{5}$$

With this construction we obtain two new correspondences:

$$a - operator \rightarrow F_a - function \ on \ SxT$$

$$\phi - state \rightarrow \rho_\phi = \mu_\phi \otimes \nu \ probability \ measure \ on \ SxT$$

still satisfying the condition

$$\phi(a) = \rho_\phi(F_a) \tag{6}$$

It is clear that one can choose in infinitely many ways some function F_a so that condition (5) is satisfied , and that to each choice there will correspond a "hidden variable " theory.

Now consider the particular case in which H is C^2 . Then B(H) is the algebra of 2x2 complex matrices and any expectation value $\phi(a)$ can be written in the form

$$\phi(a) = a_0 + a.\gamma \tag{7}$$

where $a_0 \in R$, $a \in R^3$ depend only on the matrix a and $\gamma \in R^3$ depends only on the state ϕ. In these notations, if we choose :

$$T = [-1/2 , 1/2]$$

and ν as the Lebesgue measure, and if we define:

$$F_a(\phi,\lambda) = a_0 + 1/2|| a ||sgn(\lambda||a||) + | a\gamma |sgn(a\gamma)$$

(sgn(x) = +1 for x > 0 ; = -1 for x < 0), then an elementary computation shows that the condition (4) is satisfied and that the resulting "hidden-variable " theory is the one constructed by Bell [25] and discussed in detail by Wightman in [61].

Let us conclude by showing that if we substitute the condition :

$$a \geqslant 0 \rightarrow f_a \geqslant 0$$

for the stronger condition :

$$F(f_a) = f_{F(a)}$$

appearing in the Bell-Kochen-Specker theorem, then their nonexistence result is turned into an existence, uniqueness (and triviality) result: The only possible correspondence $a \rightarrow f_a$ satisfying the above condition plus additivity on commuting observables is obtained from the trivial one, described at the beginning of this section, by possible restriction of the state space to a sub-set of all the normal states.

PROPOSITION 2. Let H be a Hilbert space of dimension $\geqslant 3$. Assume that there exists a measurable space S and a correspondence $p \rightarrow f_p$ between projections in B(H) and measurable functions on S such that , for any projection p on H and for any s in S,

$$f_{(p + q)} = f_p + f_q ; p ,q \text{ mutually orthogonal} \tag{8}$$

$$f_p \geqslant 0 ; \text{for any projection } q \tag{9}$$

Then there exists a map $s \rightarrow \phi_s$ from S to the states on B(H) such that

$$f_p(s) = \phi_s(p) \tag{10}$$

for every projection p and for every s in S.

PROOF We can restrict ourselves, without loss of generality, to the case in which H is finite dimensional . Conditions (8) and (9) imply that for every projection p on H :

$$f_1 \geqslant f_p$$

Thus, possibly disregarding a sub-set S_0 of S such that

$$f_p(s) = 0 ; \text{for each } s \text{ in } S_0 ; \text{each projection } p$$

we can normalize f so that

$$f_1(s) = 1 \text{; for each } s \text{ in } S$$

Accordingly, for each s in S, the map $p \to f_p(s)$ is a probability measure on the projections of B(H), and therefore, by Gleason's theorem, there exists a unique (normal) state ϕ_s on B(H) such that (10) holds. And this proves the statement.

27. DROPPING DOWN THE ONE-TO-ONE CORRESPONDENCE BETWEEN QUANTUM OBSERVABLES AND HERMITEAN OPERATORS

In order to apply Gleason's theorem to the correspondence $p \to f_p$, between projections and measurable functions, one needs this correspondence to be defined for all the projections on the Hilbert space in question. To some authors (cf. for example the recent book by Holevo [40]) this restriction appears to be too stringent, therefore it is desirable to get rid of it.

In their paper, Kochen and Specker produce an example of a finite lattice of projections to which the result of their theorem applies. A general method to construct examples of this kind was proposed by Bell [25]. The method is based on the following technical lemma:

LEMMA 1. Let ψ, ψ', ψ'' be an orthonormal basis of C^3, and let ϕ be a unit vector in the plane generated by ψ and ψ'. A necessary and sufficient condition for the existence of twelve real constants

$$\gamma \; ; \delta \; ; \epsilon \; ; \eta \; ; \theta \; ; \iota$$

(here and in the following, x will mean either x_+ or x_-, and when more dots enter the same expression, they are either all + or all -) such that

$$\gamma\phi + \delta\psi'' \text{ is orthogonal } \epsilon\psi' + \eta\psi'' \tag{1}$$

$$\theta(\gamma\phi + \delta\psi'') + \iota(\epsilon\psi' + \eta\psi'') = \psi + \psi'' \tag{2}$$

is that

$$| <\phi, |\psi>|^2 \geqslant 4/5 \tag{3}$$

PROOF Let α, β be complex numbers such that

$$\phi = \alpha\psi + \beta\psi'$$

With no loss of generality, one can assume that α is real. Thus

$$\alpha = <\phi, \psi >; |\beta| = ||\phi - \alpha\psi||$$

Hence, omitting everywhere the subscript ".", one finds

$$<\gamma\phi + \delta\psi'', \epsilon\psi' + \eta\psi'' > = \gamma\epsilon\beta + \delta\theta$$

Now, if $\beta = 0$, then it is easy to verify that the Eqs.(1) and (2) always admit a solution. Therefore, in the following we shall limit ourselves to the case $\beta > 0$. In this case one easily shows that also ϵ must be $\neq 0$. Therefore (1) is equivalent to

$$\gamma = -\delta\eta / \beta\epsilon$$

and, after simple manipulations, the left hand side of (2) becomes

$$(-\theta\delta\eta\alpha / \beta\epsilon)\psi - ((\theta\delta\eta / \epsilon) - \iota\epsilon)\psi' + (\theta\delta + \iota\eta)\psi''$$

Hence a necessary condition for (2) is that

$$\iota = \theta \delta \eta / \epsilon^2$$

With this choice of ι the left hand side of (2) becomes

$$(- \theta \delta \eta \alpha / \beta \epsilon)[\psi - \frac{\beta}{\alpha}(\frac{\epsilon}{\eta} + \frac{\eta}{\epsilon})\psi'']$$

Therefore our problem has a solution if and only if the equations

$$(\alpha / \beta)(\lambda + \frac{1}{\lambda}) = +1 \ or \ -1$$

have a solution . This is the case if and only if

$$|\beta / \alpha| = ||\phi - \alpha\psi|| / |\alpha| \leqslant 1/2$$

or equivalently, since α is real

$$1 - \alpha^2 \leqslant \alpha^2 / 4 \longleftrightarrow 4/5 \leqslant \alpha^2 = | <\phi, \psi >|^2$$

which proves the thesis.

COROLLARY 2. Let L be a set of projections on a Hilbert space H and let m : L → R be a map with the following properties:

(A)   For any mutually orthogonal vectors ϕ , ψ in H whose associated rank one projections (always denoted P(ϕ) , P(ψ) in the following) belong to S, one has

$$m(P(\phi)) = 1 \rightarrow m(P(\psi)) = 0$$

(B) If ϕ , ψ are as in (A) and :

$$m(P(\phi)) = m(P(\psi)) = 0$$

then for any pair of complex numbers α , β such that $P(\alpha\phi + \beta\psi))$ is in L,

$$P(\alpha\phi + \beta\psi)) = 0$$

Assume moreover that there exist three orthonormal vectors ψ , ψ' , ψ'' in H and a vector ϕ in the plane generated by ψ and ψ' such that

$$| <\phi, \psi >|^2 \geqslant 4/5$$

and L contains the rank one projections corresponding to the ten vectors

$$\psi , \psi' , \psi'' , \phi , \psi + \psi' , \psi - \psi'' , \gamma\phi + \delta\psi'' , \epsilon\psi' + \eta\psi''$$

where γ , δ , ... are eight constants satisfying the conditions of Lemma (1).

Then, if m(P(ψ)) = 1 , one must have m(P(ϕ)) $\neq 0$.

PROOF Property (A) implies that

$$m(P(\phi)) = m(\psi'')) = 0$$

Assume by contradiction that

m(P(ϕ)) = 0 Then, by property (B)

$$m(P(\gamma\phi + \phi\psi'')) = m(P(\epsilon\psi' + \eta\psi'')) = 0$$

Hence, using (2) and again property (B)

$$m(P(\psi + \psi'')) = m(P(\psi - \psi'')) = 0$$

which, by property (B) implies that

$$m(P(\psi)) = 0$$

against the assumption. Thus m(P(ϕ)) cannot be zero, and this proves the thesis.

COROLLARY 3. Let L_0 be any lattice of projections on H containing a set L with the properties described in Corollary (2). Then :

(i) Any probability measure on L_0 satisfies conditions (A) , (B) of Corollary (2) .

(ii) If m is a dispersion free measure on L_0, then necessarily

$$m(P(\phi)) = m(P(\psi))$$

for ϕ , ψ as in Corollary (2.).

(iii) For any measurable space S and for any map

$$p \in L_0 \rightarrow f_p \ -measurable \ function \ on \ S$$

such that

$$f_p^2 = f_p$$

$$f_{p+q} = f_p + f_q \ ; \ q,p \ -mutually \ orthogonal \ in \ L_0$$

one must have

$$f_{P(\phi)} = f_{P(\psi)}$$

for any ϕ , ψ as in Corollary (2).

PROOF (i) By a probability measure on L_0 we mean a finitely additive positive measure m on L_0 such that

$$sup[\ m(P) : P \ in \ L_0] = 1$$

Such a measure is monotone ($p \leqslant q \rightarrow m(p) \leqslant m(q)$). Thus property (A) follows from

$$1 \geqslant m(p + q) = m(p) + m(q)$$

for mutually orthogonal p and q. Property (B) follows from monotonicity and

$$P(\alpha\phi + \beta\psi) \leqslant P(\phi) + P(\psi)$$

for mutually orthogonal ϕ , ψ.

(ii) If m is dispersion free, then it takes only the values 0 and 1. So if ϕ , ψ are as in Corollary (2) and m(P(ψ)) = 1, then one must also have, m(P(f)) = 1.

iii) If $f \rightarrow f_p$ is a correspondence as described in point (iii) of the assumptions, then for each s in S the map $p \rightarrow f_p$ is a dispersion free measure on L_0. Hence the thesis follows from point (ii).

In [25] Bell produces some arguments against the physical necessity of condition (B) in Corollary (2).

28. SOME MORE EXAMPLES OF HIDDEN-VARIABLE THEORIES

The main goal of a hidden variable theory is to construct a correspondence between quantum observables and (measurable) functions on some state space S which in some sense preserves the statistical relations of quantum theory.

Denoting f_a the function on S corresponding to the quantum observable a, all the no-go theorems discussed up to now (with the exception of the one in section (25)) made use of the condition

range of f_a = { set of all values of a } .

From the point of view of the supporter of hidden-variable theories, the relation (1) has no particularly compelling physical meaning. It might well be in fact that the presently observed values of the quantum observables appear only as average values, over some hidden parameters, of some classical observable f_a . For example, one might try to associate with the projection P_a , corresponding to the eigenvalue a of the observable A, a gaussian random variable $f_{A,a}$, strongly peaked around its mean value a and such that, if $f_{B,b}$ is a similarly defined gaussian random variable, corresponding to a quantum observable B and its eigenvalue b, then

$$E([f_{A,a} - a\mathbf{I}[f_{B,b} - b]) = |\langle \psi_a^A, \psi_b^B \rangle|^2$$

where E(.) denotes gaussian expectation; ψ_a^A is the eigenvector of A corresponding to the eigenvalue a (assumed to be simple); and similarly for ψ_b^B .

In this way the quantum mechanical transition probabilities arise as the correlation functions of a classical stochastic process. The approach is physically plausible and preserves the symmetry between the two pure vectors defining the transition probability (while in the usual hidden-variable theories, this symmetry is broken, since one of the two state vectors is made to correspond to a function and another one to a probability measure). The following propositions show that "hidden-variable theories " of the type described above can always be realized.

PROPOSITION 1. Let T be any set (e.g. the time set) and let { A_t : t in T } be a family of self-adjoint operators on some separable Hilbert space, each with discrete simple spectrum. There exists a unique gaussian process { $f_{t,n}$: t in T ; n in N } such that

$$E([f_{t,n} - \lambda_{t,n}\mathbf{I}[f_{s,n} - \lambda_{s,n}]) = |\langle \psi_{t,n}, \psi_{s,m} \rangle|^2$$

E(.)-denoting gaussian expectation.

PROOF The function from $(TxN)^2$ to the reals, defined by

$$(t,n),(s,n) \rightarrow |\langle \psi_{t,n}, \psi_{s,n} \rangle|^2$$

is positive definite. Therefore there exists a unique gaussian process $f_{t,n}^0$ indexed by TxN, of mean zero and covariance given by the above scalar product. By construction, the process

$$f_{t,n} = f_{t,n}^0 + \lambda_{t,n}$$

is gaussian, has mean $\lambda_{t,n}$, and covariance given by the scalar product above.

In the case of a continuous spectrum, a modification of the above construction can be applied, yielding a gaussian process which depends not only on the given family of observables, as the process described above, but also on the choice of an initial state.

PROPOSITION 2. Let T be any set and let { A_t : t in T } be any family of self-adjoint operators on some Hilbert space H. Denote, for each t in T, by e_t (.) the spectral projection of A_t. For each unit vector ψ in H , there exists a unique gaussian process

$$f_{t,I} : t \text{ in } T \text{ ;} I \text{ } -a Borel \text{ set in } R$$

such that, for each s,t in R and I,J -Borel sets in R, the mean of the random variable $f_{t,I}$ is

$$\langle \psi, e_t(I)\psi \rangle$$

and the covariance is

$$\langle \psi, e_t(I)e_s(J)\psi \rangle$$

PROOF. Obvious analogue of the proof of Proposition (2).

29. SOME MISUNDERSTANDINGS ABOUT LOCALITY AND BELL'S INEQUALITY

In recent years several authors have claimed that Bell's inequalities can provide a test for the locality principle of relativity theory. Some experiments have been done [22] others are planned. A huge amount of papers have been written in which it was taken more or less for granted that the experimental validity of the old quantum mechanical singlet law (and the consequent failure of Bell's inequalities) was equivalent to an experimental falsification of the locality principle.

Again, as in the reality issue, discussed in section (2), before declaring that such an important principle is experimentally false, one would like to understand better the theoretical reasoning which led to this conclusion. It might be in fact that we use the wrong conceptual tool to interpret the experimental data, and that the contradictions we find are not between the locality principle and the experiments, but between the experiments and some mathematical model into which we wrongly pretend to fit some experimental data. It is completely useless to prove rigorous mathematical result with an asserted physical meaning if the applications of these results to the explanation of natural phenomena is based on vague and obscure considerations. Thus, since Bell's inequality is a mathematical result whose asserted consequence is the mutual contradiction between quantum theory and the locality principle, we should examine rigorously not only the chain of steps in the proof, but also the chain of steps which from the proof leads to the contradiction stated above.

To carry out this analysis is the goal of this section. I will first expose the qualitative arguments which point out the arbitrary steps done in connecting Bell's inequality with the locality problem. Then I will give a new proof of a generalized form of the Bell's inequalities in which the precise places where these arbitrary assumptions enter are identified very clearly.

The scheme of reasoning explained in Sec.(3) above, and leading to the conclusion that the notion of objective reality is incompatible with quantum theory, was the following:

(i) One implicitly postulates the applicability of the usual rules of classical probability theory to an arbitrary set of statistical data.

(ii) This implicit postulate leads to contradictions with the results of some experiments.

(iii) Not recognizing the role of the implicit probabilistic postulate, one is led to the conclusion that the only way to solve the contradiction is to postulate that quantum theory is incompatible with objective reality.

The same scheme of reasoning was applied by Bell almost forty years later and led him to the conclusion that quantum theory is incompatible with locality.

The reason why it took so long time to recognize the common mathematical assumption hidden in the prototypic two-slit experiment and in Bell' s argument, is due to some differences between the experimental situations in the two cases :

(1) in the two-slit experiment the statistical data are transition probabilities; in Bell' s example they are correlations.

(2) in the two-slit experiment one considers ensembles of individual particles (electrons, neutrons, etc.); In Bell' s example one considers pairs of particles which are approximately independent as far as their spatial degrees of freedom are concerned, but are strongly correlated in their spin (or polarization) degrees of freedom.

The conclusion of Bell' s analysis is that either :

(I) Quantum theory is wrong (in the sense that the results predicted by it in that experiment are not verified)

or :

(II) The laws of nature are nonlocal.

To prove the validity of this alternative, Bell considers the correlations of two spin-1/2 particles in singlet state and proves that there cannot exist six random variables (corresponding to the spin of each particle in three different space directions) defined on a single probability space and reproducing the correlations predicted by quantum theory.

Bell was motivated by considerations of locality, and in his proof he used a beautiful relativistic formulation of the locality property in any statistical theory (a mathematical formulation of this property was given independently in [15] and [49]). This part of Bell' s argument is model independent, and I think that any future formulation of quantum field theory should include it as an essential part even if with some modifications in order to keep into account the counterexample discussed at the end of this section.

However, the fact that Bell used the locality property in the derivation of his inequality, does not mean that this property is necessary to derive the inequality.

Therefore, if one shows that the same inequality can be derived without any locality assumption, that should prove that the real contradiction is not between the predictions of quantum theory and locality, but between these predictions and the assumptions made in the derivation of this inequality.

In particular, if one can deduce the Bell inequality under the only assumption that the (experimentally measurable) correlations involved admit a Kolmogorovian model, then one must admit that the only contradiction arising from the comparison of Bell' s inequalities with the experimental data is between the predictions of quantum theory and the existence of a Kolmogorovian model for a set of statistical data, that is the same type of contradiction we found in the two-slit experiment.

I shall now give a proof Bell' s inequality using only elementary inequalities among real numbers and (in a crucial way) an elementary property of the Kolmogorovian probabilistic model. In the case (almost uniquely considered in the huge literature on the argument) in which the observables assume only the values +1 or -1 (this is also the case which applies to the performed experiments) , the proof is simpler, and I gave it in

[19]. But, since in the original paper Bell considered the more general case of observables with values in the interval $[-1,+1]$, we will deal here with this case.

First let us establish the following elementary inequality (the proof given below was kindly communicated to me by G.Watson)

Lemma 1. Let a,b,c, be numbers in the interval $[-1, +1]$. then

$$ab - bc + ac \leqslant 1 \qquad (1)$$

Proof Consider the parabola

$$x^2 - x(b + c) + bc = q(x)$$

Its zeros are at b and c and the minimum is negative. Therefore, assuming (without loss of generality by symmetry and because the case b = c is trivial) that $b < c$, we find that, for $-1 <= a <= b$ and for $c <= a <= 1$

$$1 \geqslant a^2 \geqslant a(b + c) - bc$$

If $b < a < c$, then if $b + c > 0$ one gets

$$a(b + c) - bc \leqslant cb + c^2 - bc = c^2 \leqslant 1$$

Similarly, if $b + c < 0$, then

$$a(b + c) - bc \leqslant b^2 + bc - bc = b^2 \leqslant 1$$

and this proves the lemma.

Corollary(2) For any given a,b,c,d in the interval $[0,1]$ one has :

$$\mid ab - bc \mid \leqslant 1 - ac \qquad (2)$$

$$\mid ab - ac \mid \leqslant 1 + ac \qquad (3)$$

$$\mid ab + bc \mid + \mid ad - dc \mid \leqslant 2 \qquad (4)$$

Proof If $ab - bc \geqslant 0$, then (2) follows form Lemma (1). If $ab - bc < 0$, then (2) is equivalent to

$$bc - ab + ac \leqslant 1$$

and, with the substitutions

$$a' = c \; ; b' = a \; ; c' = b$$

the above inequality becomes

$$a'b' - b'c' + a'c' \leqslant 1$$

which again follows from Lemma (1). Thus (2) is true. Since (3) is obtained from (2) exchanging b into -b and c into -c , it follows that also (3) is true. Finally, changing c into -c in (2) we get

$$\mid ab + bc \mid \leqslant 1 + ac$$

and, adding this to (2) gives (4).

Theorem 3. Let (Ω, F , P) be a probability space , and let A,B,C,D be any four random variables defined on Ω and taking values in the interval $[-1,1]$. Then the following inequalities hold :

$$| E(AB) - E(BC) | \leqslant 1 - E(AC) \tag{5}$$

$$| E(AB) - E(BC) | \leqslant 1 + E(AC) \tag{6}$$

$$| E(AB) - E(BC) | + | E(AD) + E(DC) | \leqslant 2 \tag{7}$$

where E denotes expectation with respect to the P-measure, i.e.

$$E(F) = \int_{\Omega} F dP \; ; F : \Omega \to | R$$

Proof From Corollary (2) one finds

$$| AB - BC | \leqslant 1 - AC \; ; \; | AB - BC | \leqslant 1 + AC \tag{8}$$

Thus (5) and (6) follow by taking expectations of both sides of (8) and using the inequality $| E(F) | \leqslant E(| F |)$ one finds (5) and (6). To obtain (7) from (5) one argues as in the proof of Corollary (2).

These are the well known Bell's inequalities. As everybody can see, they follow only from elementary inequalities on real numbers plus the assumption that the four random variables are described by a single Kolmogorovian model . In fact, if this were not true, then one could no longer deduce (5) and (6) from (8), since in this case, the probability measure with respect to which to is to be taken the expectation value, would depend on the pairs of observables AB, BC, AC in such a way that the the three corresponding probability measures cannot be comprised in a single one, no matter how one enlarges the probability space a typical quantum probabilistic phenomenon.

It is well known that a corollary of these inequalities is that there cannot exist six random variables (corresponding in the EPR experiment to the spin of two spatially separated particles in three different directions) which are defined on a single probability space and which reproduce the quantum mechanical singlet correlations.

But again the implicit postulate of the existence of a single Kolmogorovian model is physically unjustified, since it is equivalent to postulating the existence of the joint probabilities for all the six observables while experimentally only a pair of them at a time can be measured.

Bell derived the above inequalities under an additional condition of (conditional) statistical independence which he called "locality". As mentioned above, Bell' s formulation of the locality property is beautiful and deep, but the theorem above show that it is completely irrelevant for the proof of his inequality.

Some authors have claimed that Bell's inequality contradicts some property different from what Bell calls locality. For example, the property that here (and in [6] , [14]) was expressed by saying that there exists a single probability space in which the observables A,B,C are represented by random variables and their correlations by the usual Kolmogorovian formulas, is called by Holevo "separability"(cf.[40]). What however should be made clear by Holevo (and in [40] is not) is whether he adopts the point of view of quantum probability or he believes , with Bell and several others ,that this property has something to do with the crucial issue of the compatibility between quantum theory and locality.

In the latter case he should carefully explain why since his separability notion has nothing to do with Bell's locality, and therefore the physical arguments of Bell do not apply to it.

Gradually the point of view of quantum probability is gaining consensus among scientist who probed for many years this aspect of the locality problem. For example, in the recent preprint [36], Garuccio and Gutkowski introduce a new notion of locality different from Bell's and which is satisfied both by quantum theory and by the Kolmogorovian model , and explicitly acknowledge (cf. section IV of [36]) that, if the experimental frequencies satisfy their notion of locality (as they seem to do, from the recent experiments), then the Bell inequality should be interpreted as discriminating between the quantum and the Kolmogorovian probabilistic model (as in the case of the two slit experiment).

In a recent paper [49a] Nelson has shown that Bell's notion of locality, which he calls "passive locality" ,is too strong a requirement, and has introduced the notion of "active locality", which is not in contradiction either with quantum theory or with stochastic mechanics. He also proves an inequality of Bell's type, in which he weakens Bell's (implicit) assumption of the uniqueness of the Kolmogorovian model. I am very grateful to E. Nelson for showing to me the manuscript of [49a] before publication, and I want to conclude this paper with some comments on Nelson's analysis.

Nelson considers six +1 or -1 valued random variables

$$S_a^1 ; S_b^1 ; S_c^1 ; S_a^2 ; S_b^2 ; S_c^2$$

corresponding respectively to spin (or polarization) measurements in the directions a,b,c , of two particles 1 and 2. He considers the transition probabilities

$$p_{a,c}(S_a^1 = +1 ; S_b^2 = -1 \mid X) \tag{9}$$

(and similar for $a \to b \to c$) where X denotes the common preparation of the two particles before the experiment (e.g. the condition that the two particles are in singlet state). Instead of imposing the constraint of the uniqueness of the Kolmogorovian model, Nelson imposes two different conditions, namely he requires that, for x = a,b,c

$$P_{x,x}(S_x^1 = +1 ; S_x^2 = -1 \mid X) + P_{x,x}(S_x^1 = -1 ; S_x^2 = +1 \mid X) = 1 \tag{10}$$

which is the usual singlet condition, and, for x ≠ y

$$P_{x,y}(S_x^1 = +1 ; S_y^2 = -1 \mid X) = P_{x,x}(S_x^1 \mid X) P_{y,y}(S_y^2 = 1 \mid X) \tag{11}$$

which is Bell's locality condition (passive locality in Nelson's terminology) expressing the idea that :

> given the knowledge of the common past (in this case, the fact that the two particles were prepared in the singlet state) the results of the measurements on particle 1 should be statistically independent on those on particle 2, since the two particles are spatially separated.

We will call this property "Bell's locality" [26].

Using (10) and (11), and denoting $p_x = P_{x,x}(S_x^1 \mid X)$ one finds

$$P_{x,y}(S_x^1 = +1 ; S_y^2 = -1 \mid X) + P_{x,y}(S_x^1 = -1 ; S_y^2 = +1 \mid X) = \tag{12}$$

$$p_x p_y + (1 - p_x)(1 - p_y) \geqslant 1/2$$

while according to the experimental results, for some choices of the directions a,b,c , the left hand side of (12) is less than 1/4.

The physical content of Nelson's argument can be stated as follows : if we require the validity of Bell's locality for all pairs of directions x and y with the exception of those pairs for which x = y , for which we require strong correlations, then we arrive to a contradiction.

In Nelson's proof, contrarily to Bell's one, there is no implicit mathematical assumption, and only experimentally measurable quantities come into each step. Does the contradiction between the singlet law (Eq.(10)) and Bell's statistical formulation of locality (Eq.(11)) mean that the natural laws are nonlocal, or that Bell's statistical formulation of the locality property is not completely satisfying from a physical point of view ?

To answer this question, consider the following situation : four beads of definite (constant in time) and mutually different colors, are in a box. Two experimenters extract randomly two beads each and go to two far away regions R_1 and R_2 . The probability that the experimenter in region R_1 , picking randomly one of his beads, will find the blue color , is 1/4. Thus, according to Bell' s locality one should have

$$P(blue \ in \ R_1 ; blue \ in \ R_2 \mid X) = P(blue \ in \ R_1 \mid X) \, P(blue \ in \ R_2 \mid X) = 1/8$$

but the left hand side is zero since the colors are conserved.

Since nothing in the above example contradicts the principles of relativity theory, we conclude that Bell's statistical formulation of locality sometimes is at odds with the most elementary classical situations .

Nothing in relativity theory excludes the possibility of a conservation law between systems which have interacted in the past and have then been separated.

Any such a conservation law will introduce strong correlations between spatially separated observables, in contradiction with Bell's locality which, by definition, excludes such situations. So it is not surprising that it leads to contradictions also with the simplest laws of quantum theory because this contradiction has been built into a physically unjustified statistical formulation of Einstein's locality principle.

NOTES 1.) Part of the results obtained in the present paper were included in my lectures at the "Advanced School on Quantum Probability and its Applications " (Leuven , 9-29 September 1985). Partial support from the European Community, the Direzione Generale Scambi Culturali, and the Office of Naval Research Contract #N00014-84-K-4421 is gratefully acknowledged.

BIBLIOGRAPHY

1. Accardi, L. : 1974, "Non-commutative Markov chains," in *Proceedings School of Mathematical Physics, Universita' di Camerino, Sept.-Oct. 1974.*

2. Accardi, L. : "On the noncommutative Markov property," *Funct. Anal. Appl. 9* (1975) 1.

3. Accardi, L. : "Nonrelativistic Quantum Mechanics as a noncommutative Markov process," *Adv. Math. 20* (1976) 329-366.

4. Accardi, L. : "Noncommutative Markov chains associated to a preassigned evolution. An application to the quantum theory of measurement," *Adv. Math. 29* (1976) 329-366.

5. Accardi, L. : "On the quantum Feynman-Kac formula," *Rend. Sem. Mat. Fis. Univ.*

Polit. Milano, 48(1978) 135-179.

6. Accardi, L. : "Topics in quantum probability," *Phys. Rep. 77* (1981) 169-192.

7. Accardi, L. : "Probabilita' e teoria quantistica," *Physis 23* (1981) 485-524.

8. Accardi, L. : "Stato fisico," *Enciclopedia Einaudi* Vol.XIII (1981) 514-548.

9. Accardi, L. : "Foundations of quantum Probability," in *Rend. Sem. Mat. Univ. Polit. Torino* (1982) 245-273.

10. Accardi, L., and C. Cecchini: "Conditional expecations in von Neumann algebras and a theorem of Takesaki," *J. Funct. Anal. 45* (1982) 245-273.

11. Accardi, L., and A. Fedullo: "On the statistical meaning of complex numbers in quantum theory," *ett.Nuovo Cim*

12. Accardi, L., and A. Frigerio: "Markovian cocycles," *Proc. Royal Irish Acad. 83A* (1983) 251-263.

13. Accardi, L., Frigerio, A., and J. Lewis: "Quantum stochastic processes," *Pub. R.I.M.S. Kyoto 18* (1982) 37-133.

14. Accardi, L. : "The probabilistic roots of the quantum mechanical paradoxes," in *The Wave Particle Dualism*, S. Diner, G. Lochak, and F. Selleri (eds.), Reidel, Dordrecht, 1983.

15. Accardi, L. : "Probabilita' classica," in *Dizionario delle Scienze Fisiche. Istituto dell' Enciclopedia Italiana*, Rome, 1983.

16. Accardi, L.: "Some trends and problems in quantum probability," in *Quantum Probability And Applications to the Quantum Theory of Irreversible Processes* Springer LNM 1055, 1-19.

17. Accardi, L., and A. Bach: "Quantum central limit theorems for strongly mixing random variables," *Z. Wahrsch. Verw. Geb. 68* (1985) 393-402.

18. Accardi, L. : "Quantum stochastic processes," in *Statistical Physics and dynamical systems: Rigorous Results*, Second Colloquim, Birkheuser, Boston, 1985.

19. Accardi, L. : "Non-Kolmogorovian probabilistic models and quantum theory," invited talk at the 45th ISI Conference, Amsterdam, August 1985, to appear.

20. Accardi, L. : "On the universality of the Einstein-Podolsky-Rosen phenomenon," to appear.

21. Aerts, D. : "A possible explanation for the probabilities of quantum mechanics," *J. Math. Phys.* to appear 1985.

22. Aspect, A., G. Dalibard, and G. Roger: "Experimental test of Bell's inequalities using time-varying analyzers," *Phys. Rev. Lett. 49* (1982) 1804-1807.

23. Barchielli, A., L. Lanz, and G. M. Prosperi: "A model for the macroscopic description of continual observables in quantum mechanics," *Nuovo Cim. 72B* (1982) 69.

24. Barchielli, A., and G. Lupieri: in *Quantum Probability and Applications II*, Springer *LNM* 1136 (1985) 57-66.

25. Bell, J. S. : "On the problem of hidden variables in quantum mechanics," *Rev. Mod. Phys. 38* (1966) 447-452.

26. Bell, J. S. : "The theory of local beables," *Epist. Lett. (1975).*

27. Bell, J. S. : "Bertlmann' s socks and the nature of reality," *J. de Physique Colloque C 2, Suppl. N3, 42* (1981) 41-62. *Epist. Lett.* (1976).

28. Beltrametti, E., and G. Cassinelli: *The Logic of Quantum Mechanics*, Addison-Wesley, New York, 1982.

29. Bergia S., F. Cannata, S. Russo, and M. Savoia: "Group theoretical interpretation of von Neumann' s theorem on composite systems. *Am. J. Phys. 47* (1979) 548-552.

30. Chentzov, N. N., and E. A. Morozova: "Noncommutative quantum logics," (in

Russian), Inst. Appl. Math. Keldish preprint No. 57 (1981).

31. Chentzov, N. N., and E. A. Morozova: "Probability distributions on noncommutative logics (finite dimensional theory)," (in Russian), Inst. Appl. Math. Keldish preprint No. 129 (1981).

32. Cook R., M. Keane, and W. Moran: "An elementary proof of Gleason's theorem," *Math. Proc. Camb. Phil. Soc. 98* (1985) 117-128.

33. Daneri, A., A. Loinger, and G. M. Prosperi: "Quantum theory of measurement and the ergodicity conditions," *Nucl. Phys. 33* (1962) 297-319.

34. Davies, E. B.: "Quantum Theory of Open Systems," Academic Press, New York, 1976.

35. Davies, E. B., and J. T. Lewis: "An operational approach to quantum probability," *Comm. Math. Phys. 17* (3) (1970) 239-260.

35a. Feynmann, R. P.: "The concept of probability in quantum mechanics," in Proc. 2nd Berkeley Symp. Math. Stat. Prob.; University of California Press, Berkeley, (1951) 533-541.

35b. Feynmann, R. P.: "Space-time approach to non-relativistic quantum mechanics," *Rev. of Mod. Phys. 20* (1948) 367-385.

35c. Feynmann, R. P.: *The Nature of Physical Law*, (Italian edn.) Boringhieri, Torino, Italy, 1971.

35d. Frigerio, A., and V. Gorini: *Dynamics of Nonrelativistic Quantum Systems*, to appear in Series Encyclopedia of Mathematics and its Applications.

36. Garuccio, A., and D. Gutkowski: "Comparison between local Kolmogorovian models and Bell's inequality," preprint, 1985.

37. Gleason, A. M.: "Measures on closed subspaces of Hilbert space," *J.Math.Mech.6* (1957) 885-893.

37a. Gudder, S., and N. Zanghi: "Probability models," *Nuovo Cim. B 79* (1984) 291.

37b. Guerra, F.: "Structural aspects of stochastic mechanics and stochastic field theory," *Phys. Rep. 77* (1981).

38. Helstrom, C. W.: *Statistical Theory of Signal Detection*, Pergamon Press, 1968, 2nd edn.

39. Holevo, A.: *Probabilistic and Statistical Aspects of Quantum Theory*, North Holland, Amsterdam, 1982.

40. Holevo, A.: *Statistical Structure of Quantum Mechanics and Hidden Parameters*, (in Russian) Znanie, Mosow, 1985.

40a. Hudson, R., and K. R. Parthasarathy: "Quantum Ito's formula and stochastic evolutions," *Comm. Math. Phys. 93* (1984) 301-323.

40b. Jammer, M.: *The Philosophy of Quantum Mechanics*, Wiley, New York, 1974.

41. Kochen, S., and E. P. Specker: "The problems of hidden variables in quantum mechanics," *J. Math. Mech. 17* (1967) 59-67.

42. Kraus, K.: *States, Effects and Operations*, Springer, LNP *190*, 1983.

43. Kruszinsky, P.: "Extensions of Gleason's," in *Quantum probability and applications*, Springer LNM 1055, 1984. 1984. 1984. 1984. 1984. 1984. 1984. 1984. 1984. 1984. 1984. 1984. 1984. 1984. 1984. 1984. 1984. 1984. 1984.

44. Lindblad, G.: "Entropy, information and quantum measurement," *Comm. Math. Phys. 33* (1973) 305-322.

45. Ludwig, G.: *Foundations of quantum mechanics*, Springer, New York, 1983.

46. Mackey, G. W.: *Mathematical Foundations of Quantum Mechanics*, Addison-Wesley, New York, 1976.

47. Margenau, H., and J. L. Park: "The logic of noncommutability of quantum mechanical operators and its empirical conseqences," in Perspectives in Quantum Theory, W. Yourgrau and A. van der Merwe (eds.), Dover, New York, 1971.
48. Nelson, E. : *Dynamical Theories of Brownian motion*, Princeton University Press, Princeton, NJ, 1967.
49. Nelson, E. : *Quantum fluctuations*. Princeton University Press, Princeton, NJ, 1985.
49. Nelson, E. : "The locality problem in stochastic mechanics," Presented to the Conference on New Techniques and Ideas in Quantum Measurement theory, New York Academy of Sciences, New York, January, 1986.
50. Ozawa, M. : "Quantum measuring process of continuous observables," *Journ. of Math. Phys. 25* (1985) 79-87.
51. Pashkievicz, A. : "Measures on projections of von Neumann algebras," *J. Funct. Anal. 62* (1985) 87-117.
51. Piron , C. : *Foundations of Quantum Physics*, Addison-Wesley, New York, 1976.
52. Popper, K. : "Realism in quantum mechanics and a new version of the EPR experiment," in Open questions in Quantum Physics, G.Tarozzi, A. van der Merwe (eds.) Reidel, Dordrecht, 1985.
53. Prosperi, G. M. : "Macroscopic physics and the problem of measurement in quantum mechanics," in *Foundations of Quantum Mechanics*, Varenna IL Corso 1971 Academic Press, New York.
54. Rauch, H. : "Tests of quantum mechanics by neutron interferometry," in *Open Questions in Quantum Physics*, G. Tarozzi, A. van der Merwe (eds.) Reidel, Dordrecht, 1985.
55. Santos, E. : "The wave particle dualism," (review) *Found. of Phys. 15* (1985) 229-231.
55a. Sakai, S. : C^*-*Algebras and* W^*-*Algebras*, Springer, New York, 1971.
56. Schwinger, J. : *Quantum Kinematics and Dynamics*, Academic Press, New York, 1970.
57. Srinivas, M. D. : "Collapse postulate for observables with continuous spectrum," *Comm. Math. Phys. 71* (1980) 131-158.
58. Stormer, E. : "On projection maps of von Neumann algebras," *Math. Scand. 30* (1972) 46-50.
59. Vogt, A. : "Position and momentum distributions do not determine the quantum mechanical state," in *Mathematical Foundations of Quantum Theory, A. R. Marlow (ed.)*, Academic Press, New York, 1978.
60. von Neumann, J. : *Mathematical Foundations of Quantum Mechanics*, Princeton University Press, Princeton, NJ, 1955.
61. Wightman, A. S. : "Hilbert' s sixth problem: mathematical treatment of the axioms of physics," in *Mathematical Developements Arising from Hilbert Problems, Proc. Symp. in Pure Math. 28,* (1976) AMS, Providence, 1976.

ON THE ADEQUACY OF A NONCLASSICAL LOGIC FOR QUANTUM THEORY

Rossella Lupacchini
Department of Philosophy
University of Bologna, Bologna, Italy

ABSTRACT. The aim of this paper is to assess whether and at which level
a nonclassical logic can be found that is satisfactory for a more adequate understanding of quantum theory. After analyzing the most meaningful proposals of quantum logic, we reach the following conclusions:
At the level of the theory it is hard to find arguments in favor of the
advisability of a nonclassical logic for a formal reconstruction of the
theory. If, on the other hand, the issue is treated at the metatheoretical level, then, in line with von Weizsaecker, it might be proper to restrict the nonclassical logic at the experimental propositions level and
to seek a consistent interpretation of it in a metalanguage conforming
to classical logic. From this perspective, the problem of finding a logic
adequate for quantum theory becomes eventually hardly distinguishable
from a problem of language.

1. INTRODUCTION

The questions and doubts which first accompanied the general acceptance
of quantum theory must be considered as yet unresolved. They involve
both the formal structure and the conceptual content of the theory which
consequently finds itself in a somewhat awkward situation.

The biggest difficulty can be identified in the lack of a language
capable of expressing effectively and without ambiguities the new physical picture emerging from quantum physics. The content of the theory is
mainly expressed by means of a "classical" language, due to certain analogies with the formal structure of classical mechanics. Clearly, however,
all the terms borrowed from classical physics must be interpreted within
the new quantum context according to the restrictions imposed upon their
usage by the postulates of the theory if one is to avoid any contradiction. Unfortunately this leads to representations which are only vaguely

325

G. Tarozzi and A. van der Merwe (eds.), The Nature of Quantum Paradoxes, 325–341.
© 1988 by Kluwer Academic Publishers.

connected with quantum phenomena.

The concept of "complementarity" introduced by Bohr, forces an am-
biguos usage of the classical concepts in accordance with the Heisenberg
uncertainty principle. Following Bohr, a "uniformly complete" interpretation of the formalism of quantum theory--i.e., an interpretation that
assigns to every term employed in the postulates a meaning which stays
the same in the various contexts of application of the postulates--must
attribute to quantum objects the characteristic traits of classical objects. This necessarily implies the attribution of a dual paradoxical
nature to quantum objects. A physical system can qualify as a model for
quantum theory provided it satisfies the uncertainty relations; however,
if the simultaneus values of position and momentum of a particle can only
be determined with limited precision, even in principle, then the term
"particle" must clearly have a meaning other than the classical one.
Which one, is to be established. An adequate description of quantum phenomena actually cannot be given by simply utilizing the language of classical physics: One should rather refer to the mathematical structure of
the theory, or perhaps one could attempt to define a more general formal
language which includes logical models of the theory conforming to the
mathematical scheme.

The problem concerning the *language* of quantum theory can thus be
approached along two different routes: The first consists of defining
the scope of the presently available language, seeking adequate definitions of the concepts of the theory in mathematical terms; the second
attempts to define a formal language adequate for the mathematical model
of the theory.

The latter alternative leads one to envisage a possible solution of
some of the problems encountered in the analysis of the theory that employs classical terms, via the construction of a system of nonclassical
logic: the "quantum logic."

In order to find a viable route through what van Fraassen calls
"the labyrinth of quantum logics,"[1] it may be convenient to group the
logical problems of quantum theory into two main categories. In the
first we may collect the *syntactical problems*, i.e., the problems internal to the formalism of quantum mechanics which are tightly connected
with the theory of measurement and with the probability function originated both from the state superposition principle and from the uncertainty principle. In the second we may collect the so-called *semantical
problems*, that is, the problems connected with the interpretation of
formalism, in particular with the interpretation of the Schroedinger
wave function, which lies at the basis of the wave-particle dualims.
From an analysis of syntactical problems there derive two fundamental
trends of quantum logic: non-distributive logic and three-valued logic;

on the other hand, von Weizsaecker, by introducing his "complementarity logic," aims at giving a possible solution for the involved problem of dualism.

Keeping well in mind such preliminary distinction, let us now attempt to reach into the labyrinth to try and clarify what exactly has been comprehended under the general denomination of "quantum logic."

2. NON-DISTRIBUTIVE LOGIC

Starting from an analysis of the predictability and measurability limits implicit in quantum theory, Birkhoff and von Neumann constructed a system of non-distributive logic which can be viewed as the first exhaustive proposal of a quantum logic. They demonstrate that the lattice structure of a physical theory is a mathematical model of the system of logic adequate for the theory and seek a formulation of quantum mechanics as a theory based upon a non-Boolean propositional structure—that is, an extension of classical propositional calculus which takes into account the restrictions imposed by Heisenberg's uncertainty principle on measurement.

In order to construct a lattice whose points correspond to the propositons of quantum mechanics, one must construct the lattice of vector spaces and of their subspaces; the logical implication will then be represented by the relation of inclusion between subspaces. The crucial point lies in the fact that the operation of union on the lattice of subspaces is not the operation of set union, but rather the subspace generated by the vectors of the two given subspaces; and, due to a characteristic propriety of such operation, the lattice of subspaces is non-distributive. Hence the logical structure associated with the lattice is not distributive.

The distinctive trait of quantum logic must be identified in the very nonvalidity of distributive laws. In this connection, one must emphasize how the validity of these laws in quantum mechanics is a logical consequence of the compatibility of observables, and that is why distributive laws cannot be retained in quantum logic. Here holds instead the so-called "modular identity"—if $a \subset c$, then $a \cup (b \cap c) = (a \cup b) \cap c$—which can be considered a logical consequence of the formal properties of probability.[2] The nonvalidity of distributive laws may also be interpreted as a logical consequence of the lack of symmetry between the quantum conjunction and disjunction.[3] Indeed, while the truth of a conjunction in quantum logic must invariably be equivalent to the truth of both members of the conjunction, the truth of a disjunction is not, generally speaking, equivalent to the truth of at least one member of the disjunction.

In quantum mechanics. the connection between the experimental propositions and the mathematical structure of the theory--the Hilbert space--
can be established by defining the set of mathematical representatives.
The mathematical representative of any experimental propositions is a
closed linear subspace of the Hilbert space and, furthermore, since all
operators of quantum mechanics are Hermitian, the mathematical representative of the *negative* of any experimental proposition is the *orthogonal
complement* of the mathematical representative of the proposition itself.
Therefore, for two experimental propositions A and B concerning a given
type of physical system, the following conditions are equivalent: (1)
The mathematical representative of A is a subset of the mathematical
representative of B; (2) A implies B--that is, whenever A can be predicted with certainty, so can B; (3) for any statistical set of system,
the probability of A is at most the probability of B. The set of mathematical representatives of the experimental propositions related to any
physical system represents mathematically the *propositional calculus* for
the given system.

Propositional calculus must be completed, in the case of quantum
mechanics, by means of the following postulate:

"The set-theoretical product of any two mathematical representatives of experimental propositions concerning a quantum-mechanical system, is itself the mathematical representative of an
experimental proposition."

When this postulate is added to the normal postulates of quantum theory,
the characteristic properties of vector subspaces allow one to deduce
that the set product, the closed linear sum, and the orthogonal complement of any two closed linear subspaces of the Hilbert space, which mathematically represents an experimental proposition relative to a given
quantum system, are mathematical representatives of an experimental proposition regarding that system as well. This brings to completion the construction of the calculus of experimental propositions concerning a given
quantum-mechanical system, as a calculus characterized by three oper--
ations and a relation of implication.

Henceforth the "quantum logic," interpreted as a "non-distributive"
logic, is the logic which is extracted from the mathematical formalism
of quantum theory and which can be described in axiomatic-algebraic
terms as a "complete orthomodular lattice."

The semantic interpretation of the logical structure of quantum
theory by means of a modular lattice can be viewed, in line with Birkhoff and von Neumann, as the analog of the interpretation of a formal
system by means of a particular geometry. Indeed, if the physically significant statements of classical mechanics constitute a Boolean algebra,

the propositional calculus of quantum mechanics show, for the physically significant statements of the theory, a structure which is actually more closely related to an abstract projective geometry of sorts. Moreover, it is important to emphasize that quantum theory implies irreducible propositional calculi of unbounded complexity, while in classical mechanics any given propositional calculus, involving more than two propositions, can be decomposed into independent constituents. From this stems the greater logical coherence of quantum mechanics when compared with classical mechanics, a conclusion corroborated by the impossibility of measuring different quantities independently.[5]

The interest of physicists for nondistributive logic, or indeed for quantum logic in general, is almost invariably associated with the search of a clarification of the controversy of "hidden variables." According to Bell,[6] the point to be stressed is that a knowledge of precise values of hypothetical hidden variables allows one to predict with certainty that the system will manifest properties P_a and P_b but not the property $P_{a \wedge b}$. This follows from the fact that the measurement of $P_{a \wedge b}$ involves an altogether different procedure than the measurement of P_a and P_b, hence the result of the measurement of $P_{a \wedge b}$ cannot be simply derived from a combination of the measurements of P_a and P_b. However the *problem of measurement* is quite separated from the *purely internal* problem concerning the possibility of representing the statistical states of the theory as probability measurements over a space of classical probability.[7]

A solution was proposed by Kochen and Specker.[8] Their analysis of the hidden-variable problem points out that the statistical states of a theory may be represented as measures on a classical probability space. Now, since this is always possible, in the case of quantum mechanics the problem does not concern the possibility of such representation, but rather the possibility of "preserving the structure of the set of physical magnitudes under such a representation."[9]

The relation of compatibility is the characteristic feature of quantum mechanics. The intransitivity of such a relation differentiates the algebraic structure of quantum-mechanical quantities from the commutative algebra of quantities belonging to classical mechanics. Therefore, the structure of quantum-mechanical quantities may be formalized as a partial algebra, and the problem of representing the statistical states of the quantum mechanics on a classical probability space may be reformulated as the problem of imbedding the partial algebra of quantum quantities into a commutative algebra. In this way, Kochen and Specker prove that the partial Boolean algebra of idempotent quantities on a Hilbert space of three (or more) dimensions cannot be imbedded into a Boolean algebra. This result entails the impossibility of representing the statistical states of quantum mechanics as probability measures on a classical prob-

ability space in such a way that the structure of the set of quantities
is preserved.[10]

3. THREE-VALUED LOGIC

The feasibility of a three-valued quantum logic must be considered with
reference to the proposals put forward by Paulette Février, Reichenbach,
and in particular by Suppes.

In her 1937 note,[11] Février stressed the advisability of interpreting Heisenberg uncertainty relations not as consequences of the mathematical formalism of quantum theory within the framework of classical logic,
but rather as fundamental laws of physical measure, starting from which
one can construct a logic adequate to the properties of quantum objects.
Quantum theory can thus be represented as a mathematical theory within
this logic.

The prerequisites for this kind of operation are, on the one hand,
an axiomatic system of principles as needed in view of experimental research, on the other, a set of propositions enunciated in agreement with
axiomatic conditions. With respect to the axiomatic principles established, these propositions can be: true, either necessarily or in a contingent way; false, in a contingent way; and, false, necessarily.

If the uncertainty relations are considered as parts of the axiomatic system, two propositions expressing respectively, the truth value
of the position coordinate and the truth value of momentum, can be separately true, but their conjunction cannot be true, since--in virtue of
Heisenberg's principle--their truth cannot be simultaneously assessed.
According to Février, the uncertainty relations are responsible for introducing, for given pairs of propositions called conjugate, bounds of
a nonclassical type which in turn give rise to special laws for the conjunction of this pairs of propositions. This suggest the notion of split-
ting the false of classical logic into two values: false "F," viewed as
"possible not realized," and absurd "A," viewed as "not realizable."
Whence derives the better adequacy of a three-valued logic for quantum
theory.

In disagreement with Février's approach, Reichenbach only takes
into consideration the adequacy of a three-valued logic with respect to
the mètalanguage of the theory.

In Reichenbach's view, a fundamental problem lies in the fact that,
according to an approach not much unlike the one presented by Février,
the statements regarding the values of non-observable quantities, "meaningless" according to the Copenhagen interpretation, must be included in
the language of quantum physics. If this undesirable consequence is to
be avoided, Reichenbach proposes to utilize "an interpretation which ex-

cludes such statements, not from the domain of *meaning*, but from the domain of *assertability*."[12]

With this objective in mind, a preliminary distinction between one
language for observation--the object language--and one *language for quantum theory*--the metalanguage--is introduced. The quantum language can
only be established by means of definitions: A proposition *A* of quantum
mechanics is defined in terms of the set of experimental propositions
which verify it.

An interpretation of the quantum language may well be "statistically
incomplete" with respect to a spacetime description, but will invariably
be "statistically complete" with respect to observation, inasmuch as, for
every possible state, the outcome of a measurement can be predicted with
a given probability. In order to avoid the ambiguities which could derive
from the usage of the word "complete," the interpretations of the quantum language are divided into "exhaustive" and "restrictive." The *restrictive interpretations* would allow us to get rid of the so-called
"causal anomalies"--those propositions which contradict the physical laws
for observables--and can be characterized in two ways: either as interpretations with "restrictive meaning," utilizing a definition of meaning
which allows one to exclude the "undesirable" statements from the language of quantum mechanics by making them *meaningless*; or as interpretations with "restrictive assertability," excluding the undesirable statements from assertions of quantum mechanics, rather than from the object
language.

The first kind of interpretation, with restrictive meaning, would
correspond to that developed by the Copenhagen School and based on a
semantic rule which fixes the expression "meaningless" into metalanguage:
In a physical state for which no measurement of a physical quantity has
been made, any statement relative to the value of such quantity is meaningless. According to Reichenbach, if this type of definition is utilized, one is forced to include in the quantum language the meaningless
statements.

The proposed alternative consist then in utilizing a three-valued
logic which allows for an intermediate truth value--the "indeterminate"--
to be attributed to the group of meaningless statements. It is not
necessary to exclude from the meaningful statements the statements relative to the value of a non-observable quantity, it is rather sufficient
to have a rule which allows us to consider them neither true nor false.
When the indeterminate is introduced as third truth value, it is important that the meaning of the term *indeterminate* is clearly distinguished
from the meaning of the term *unknown*. Hence, if there exists an intermediate value represented by the logical status "indeterminate," the
classical principle of "excluded third" does not hold any longer. As is

well known, this principle states that, even if the truth value of a
statement can be unknown, such a statement is nevertheless invariably
true or false.

The advantages of the introduction of the third truth value become
apparent in quantum mechanics when the results of the measurements of
conjugate quantities are considered. The measurement of the position
coordinate in any given physical state entails, as has already been observed, the impossibility of knowing what would have been the result in
case the measurement of the conjugate quantity momentum had been carried
out. On the other hand, trying to reproduce in another system the same
physical state, thereby to measure a second time the position coordinate,
would not be of much help, because the measurement outcome of position
can only be determined with a given probability, and there is no reason
to believe that a repeated measurement will recover the value obtained
in the first case.

Intuitively, the quantum logic proposed by Reichenbach seems to
aim at justifying the interpretation of the theory in terms of hidden
variables, an interpretation that views the uncertainty not so much as
limitation of principle inherent to quantum mechanics, but rather as a
"large-numer" effect.

Leaying aside a critical analysis of this kind of interpretation,
it will be pointed out that the main flaw of Reichenbach's proposal lies
in its failure to faithfully characterize the set of experimental propositions for which the quantum logic should be adequate. By adopting a
system of three-valued logic for the quantum metalanguage, Reichenbach
offers a solution to the problem of "causal anomalies," but his proposal appears unsatisfactory in relation to the analysis of quantum theory. The greatest problem lies in the difficulty of showing how this
system of logic can play a "functional role" in the development of the
theory. This is because the adequacy of the three-valued logic is restricted to the metalanguage of the theory, and thus the structure of
experimental propositions belonging to the object language is not constructed in such a way as to be truth-functional in this logic.

The non-adequacy of Reichenbach's system of logic is strongly upheld by Suppes. He outlines that

> "In physical or empirical contexts involving the application
> of probability theory as a mathematical discipline, the func-
> tional or working logic of importance is the logic of events
> or propositions to which probability is assigned, not the
> logic of quantitative or intuitive statements to be made about
> the mathematically formulated theory."[13]

In the logical discussion of quantum mechanics, a crucial aspect consists in the characterization of the set of propositions to which the logic examined is applied. Now, on one hand, quantum mechanics is presented as a physical theory which avails itself of the most advanced mathematical results, on the other hand, the discussions on logic involve the very principles of mathematics. If quantum theory utilizes a mathematical formalism in conformity with classical logic, the central problem concerns the possibility of a change of logic within the theory. Generally, the theory of probability based on the logic of events is regarded as implicit in the mathematical scheme of quantum mechanics. But then it must be emphasized that the structure of the algebra of events expresses exactly the logical structure of the theory. Consequently, if a truth-functional nonclassical logic is adequate for the assertions of quantum theory, then the logic of experimental propositions must also be nonclassical.

An alternative proposal by Suppes consists in adopting a truth-functional scheme of logic that assigns the usual truth values "true and false" to empirically meaningful propositions and the peculiar value "meaningless" to the rest. A three-valued logic, whose third value is "meaningless," allows one to easily "segregate" those propositions not having a definied empirical content and to call attention to "otiose features" of any model of the theory.[14]

In the case of the logic of experimental propositions, even if it is not particulary difficult to correlate the structure of subspaces arising in quantum mechanics with the orthocomplemented modular lattices as formally defined, one generally neglects to investigate in detail how this logic corresponds to the set of experimental propositions. Suppes suggests filling this gap by considering a "simpler and more classical" logic for the experimental propositions. If an "experimental event" is viewed as an event observable at least in principle by means of a certain configuration of experimental apparatus and if it is further assumed that all observable events can be expressed as functions of position and momentum, then the conclusion reached is that the logic of experimental propositions of the quantum mechanics of n particles, for which all observables are defined in terms of position and momentum, is constitued by a family of 2^{3n} Boolean algebras and by their subalgebras. It does appear advisable to adopt for quantum mechanics a more complicated set of values: Suppes' idea is to take as truth values not simply T, F, and M, but rather the ordered pairs (T,R), (F,R), and (M,R), where R is the set of coordinates of event corresponding to the proposition in question.[15]

The preferential system of comparison for the quantum logic of Suppes is definitively the one of non-distributive logic by Birkhoff

and von Neumann, which is considered of central importance in the mathematical formulation of quantum theory. However, the fact that the linearly closed subspaces of an Hilbert space form a "modular lattice" is
not sufficient, in Suppes' opinion, to maintain that this is the correct
expression of the logic for experimental propositions of the theory; indeed, for a given set of propositions, the distributive logic still holds.
The truth-functional system constructed by Suppes appears to be adequate
both for experimental propositions and for the mathematical model of the
theory. This is why Suppes' approach appears more convincing in connection with a quantum logic adequate for quantum theory in general.

    In the latter case, in fact, the algebraic structure isomorphous to
logical calculus cannot simply be a non-Boolean structure, as in the case
of nondistributive calculus, but it must be a family of Boolean algebras
and of their subalgebras. This type of algebraic structure emerges as
isomorphous to the complete mathematical model of the theory, hence also
to the set of experimental propositions. Consequently, nondistributive
logic may be interpreted in Suppes' system as the calculus isomorphous
to a specified subalgebra.

4. QUANTUM LOGIC AND CLASSICAL LOGIC

The problem concerning the significance of quantum logic for quantum theory, and the position of this logic with respect to classical logic, is
presented very clearly in the works of Maria Luisa Dalla Chiara and of
van Fraassen.[16] Without entering into a formal treatment of this subject, let us introduce a few intuitive considerations.

    As is well known, classical mechanics proves at once necessary and
inadequate for the formulation of quantum theory. This very fact can suggest, on the one hand, that logical constants of quantum theory must be
the constants of classical logic, on the other, that the language of
quantum theory should best be extended by addition of new quantum-logical
constants. Let L be the language of quantum theory, L' the extention of
L, and L* the smallest sublanguage of L' closed with respect to quantum-
logical constants.

    The description of language L* would thus allow us to extrapolate
a class of deterministic assertions from the properly indeterministic
context of quantum mechanics or, in other words, to interpret one kind
of deterministic sentences, which include no probabilistic variables,
in a constintent way. The language L* would express essentially one subtheory of quantum theory, and it is for this language that a logical
frame weaker than classical logic has to be valid.[17]

    The position of quantum logic in the context of weaker logics may be clarified by descriptions that utilize an algebraic semantics[18] or, alternatively, a Kripke semantics.[19] In this way, it has been shown that, from a *syntactical* point of view, quantum logic can be characterized as a subtheory of classical logic, while, from a *semantical* point of view, quantum logic turns out to be incompatible with classical logic. Such a conclusion could also be reached starting from the similar relationship connecting classical mechanics and quantum mechanics.

A "metalogical" property of quantum logic is detected in the non-validity of Lindenbaum's lemma; indeed, one result by Kochen and Specker shows that, in general, given a quantum-logic formal system, there does not exist a consistent and complete quantum-logical extension of the same. This purely logical result has a precise physical correspondence in von Neumann's demonstration--which actually cannot be considered conclusive-- that quantum theory in its present form cannot be transformed in a deter-ministic theory, since every modification in this direction would make the theory self-contradictory.

Coming now to the significance of quantum-logical constants, as com-pared to classical-logical constants, it may be interesting to consider whether they can be interpreted in a classical way. The semantic incom-patibility between classical logic and quantum logic does not exclude the possibility of a formal comparison between the two, which comparison is made feasible by the use of a modal logical frame that represents a certain extention of classical logic and not simply a classical logic.

It is proven that every algebraic realization of the language L* of quantum logic can be transformed into a modal realization of the lan-guage of modal logic. The physical meaning of a modal realization asso-ciated with an algebraic realization of L* can be explained considering that the "possible worlds" of modal realization can be interpreted as non-empty subspaces of Hilbert space, so that every non-empty subspace represents a *possible physical world* in modal realization. This result allows us to define a translation of the language of quantum logic into the language of modal logic: Every sentence of quantum logic is valid within quantum logic if and only if its modal translation is a valid sen-tence of modal logic.[20]

In this way the possibility of a "complete reduction" of quantum logic to classical logic is denied, but it is stated that the quantum-logical constants can be interpreted as a particular model of classical connectives.

The conclusions that can be drawn from the previous reasoning are the following: Quantum logic is the weaker logic when compared to clas-sical logic and proves adequate only for that particular mathematical model of quantum theory in which the set of "deterministic" propositions can be represented. Its role should be that of singling out, from the

set of propositions of the theory, those experimental propositions involving "predictable" measurement procedures. Such a role can evidently
be played in an equivalent way by a system of modal classical logic that
singles out propositions of the theory that are "necessarily" true. Therefore, it does not seem to be easy to find good reasons for upholding the
"alternative" character of this logic and, consequently, the need of its
utilization for an adequate comprehension of the theory.

However, it is possible still to attempt and consider whether a nonclassical logic may be adequate for quantum theory at the level of the
semantic interpretation of the theory.

5. THE LOGIC OF COMPLEMENTARITY

Actually the motivations for seeking an alternative system of logic for
quantum mechanics have originated at least initially in *semantic problems*. The difficulties involved in the analysis of metatheoretical concepts might have paved the way for their possible overcoming by means
of a "quantum-logical" interpretation of the theory. Among metatheoretical concepts, the one of "complementarity" has played a key role in the
Copenhagen interpretation, that is, in the commonly accepted interpretation of quantum theory.

By introducing the complementarity concept, Bohr aims at justifying the Heisenberg uncertainty principle by means of the wave-particle
dualism. In his view, the need to make use of the Einstein-de Broglie
relations in the derivation of the uncertainty principle is a clear manifestation of the wave-particle dualism or, in general, of the fact that
at the very foundation of quantum theory lie two mutually exclusive
"complementarity" descriptions of physical phenomena. Any contradiction
can however be avoided, provided one allows for a "complementarity principle" according to which *in no instance complementarity aspects can be
observed simultaneously.*

It must be pointed out that Heisenberg viewed the uncertainty relations as logical deductions from the mathematical formalism of the theory and, at least initially, he did not see dualism as a necessary assumption of the theory. In fact, the equivalence between Bohr's quantum
mechanics and Schroedinger's wave mechanics would prove, again according
to Heisenberg, that the particle representation and the wave representation are simply two different aspects of the same physical reality.

The irreducible dualism of the two alternative representations of
the same reality of phenomena does not present problems if one considers,
along with Heisenberg, that the mathematical formalism of the theory excludes contradiction. On the other hand, just because the formalism of
the theory expresses the dualism of the two representations, it must

necessarily be extremely flexible. The uncertainty deriving from the incompatibility between particle representation and wave representation,
which forms the fundamental principle of quantum theory, represents a
considerable weakening with respect to the principles of classical
physics.

Von Weizsaecker tackled the question from a logical point of view
and sought to clarify the terms of the problem by distinguishing between
the object language of the theory and the metalanguage which we utilize
to talk about it. The crucial problem of quantum theory appears then to
be the fusion of one part of the object language with the metalanguage,
and the solution proposed for it is to impose restrictive conditions on
the object language.

The original proposal of von Weizsaecker consists essentially in
the attempt to envisage complementarity as a fundamental concept of
logic.

Among the several attempts aimed at describing the formal structure
of the logic of quantum theory, the one due to Birkhoff and von Neumann
is considered the most rigorous insofar as "formal coherence" is concerned. These authors have shown, according to von Weizsaecker, that
the mathematics of quantum theory can be viewed as an extension of the
propositional calculus of classical logic. In what sense? The question
is posed in the following three partitioned formulations:
(a) Would it not be sufficient to talk about a change of physics rather
than about a change of logic?
(b) If indeed one has to talk about a change of logic, what concept does
such change first involve?
(c) What relationship exists between the logic employed by quantum theory in speaking about its objects and the logic we use in speaking about
theory? [21]

The complementarity concept, once transfered to logic, gives rise
to complementarity between sentences, while the concepts of position and
momentum indicate possible predicates of certain objects. The complementarity of position and momentum becomes then a complementarity of
two sentences, and the complementarity between two sentences is a sentence on their possible truth values.

Considers the two sentences U: "This particle has position r" and
V: "This particle has momentum r'." Once the truth of U is established,
in the case of V both truth and falsity can occur, but there is no way
of predicting with certainty which one will actually occur. According
to the Copenhagen interpretation, it is not even acceptable to regard V
as true or false *per se*, irrespective of our lack of knowledge on the
matter. Thus, assuming that U is true, V is neither true nor false. Are
we here perhaps, as Reichenbach maintains, facing a violation of the
excluded-third principle? To Weizsaecker what is certainly valid is the

restricted form: "If a sentence is decided, then it is either true or false," but it could also be "non-decided."[22]

The whole question is clarified by von Weizsaecker with the use of the Tarski distinction between object language and metalanguage: In quantum physics the experimental outcomes are part of physics, hence of the object language, but they are comunicable in classical terms, by the same natural language (metalanguage) through which the formalism of the theory is interpreted. If this part of the natural language is interpreted in turn, becomes impossible to describe the whole natural language. It is however possible to interpret the terms of classical physics over one part of the mathematical object language of quantum theory. This is what happens when classical physics is viewed as a limiting case of quantum theory. Hence the fusion of one part of the object language with the metalanguage and the necessity to impose restrictive conditions on the object language.

Von Weizsaecker suggests that one regard the relation between classical logic and complementarity logic as similar to the relation between classical mechanics and quantum mechanics. In this way,

> "classical logic would rather be the *a priori* of the method, which we are bound to use in the formulation of complementarity logic. However up to the point where our present knowledge reaches, complementarity logic should be considered as the true logic which includes classical logic as a limiting case sufficient in many instances."[23]

Complementarity logic is presented as a modification of propositional logic, where a proposition in agreement in the classical linguistic usage is to be interpreted with a contingent proposition. The physical concept of experiment finds correspondence in the logical concept of "question," and the one under study is called "fundamental question." For example, in the case of the double-slit experiment the fundamental question is, "Through which slit did the particle pass?". Assigning the truth values u and v to the two possible anwers, one rises by one logical level and the "metaquestion" results: "Which truth value have the possible anwers to the fundamental question?". In complementarity logic all the possible anwers to the metaquestion are normalized vectors (u,v). In the sense of classical logic, a metaquestion is an infinite alternative. In this way, complementarity logic is introduced in the object language with the help of one metalanguage which requires the two-valued logic.

The two sentences U and "U is true" belong two different levels and therefore have certainly a different meaning. In classical logic, however, they are equivalent in the sense that they are simultaneously true or false, while in complementarity logic the truth or falsity of U entails the truth or falsity for "U is true" but not vice versa. If "U is true" is false, then U can be indeterminate. It is possible then to establish a *quasi*-equivalence between fundamental sentence and metasentence which holds for the truth but not for the falsity of the two sentences. In this sense one may say that complementarity logic does not affect the classical concept of truth but only that of falsity.

Turning now to the fundamental question posed above, one finds here a simple alternative, that is a situation in which a decision must be taken between two equally probable possible events. Schroedinger's ψ waves are amplitudes of probability for electron arrival. Weizsaecker wants to demonstrate that wave representation is a necessary consequence of particle representation and of complementarity logic.

Following this type of logic, since complementarity is assumed as a property of logic, whenever particles are present, so must be probability waves. Quantum theory, seen as a change of logic, requires that the logical laws are deduced from quantum laws. Therefore the wave function behind the screen is uniquely determined by two complex numbers u and v which give, respectively, the amplitude and phase of the wave in correspondence with the slit. If ϕ_1 and ϕ_2 are normalized to 1, one has $\psi =$ $= u \phi_1 + v \phi_2$. The relation $uu' + vv' = 1$ holds, and the probabilities that the particle has gone through either slit are $w_1 = uu'$ and $w_2 = vv'$. If $u = 1$, then $v = 0$, and vice versa. Finally, if truth and falsity are represented by means of truth values 1 and 0 and if a sentence that describes a case as "pure" in the sense of quantum theory is called "elementary," then complementarity logic can be summarized in the following theorem: *Every elementary sentence can have, beside the truth values 1 and 0, a truth value given by a complex number.*

With reference to the general approach of von Weizsaecker's work, two remarks, closely connected with each other, are to be emphasized. The first concerns the particular characterization of quantum-mechanical propositions which should comform to quantum logic. It must be pointed out that the solution suggested by von Weizsaecker, in which Heisenberg concurs, consists in viewing the subject-object interaction between human observer and the world of phenomena as the subject of quantum physics. In this way the empirical content of physics can be regarded as the reflection of the initiative that an observer-scientist takes spontaneously and unpredictably in his relationship with the world. This is an interpretation of complementarity which cannot be reconciled with an interpretation in terms of hidden variables and on the basis of which one views complementarity logic as a nonclassical logic, while at the metatheoretical level the validity of classical logic is retained.

The second remark is that, with this procedure, von Weizsaecker

completly upturns Reichenbach's approach. The latter, in fact, wanting
to exclude from quantum language the propositions that are meaningless
from an observational point of view, distinguished between object language or observational language, and the quantum metalanguage. He does
see the logic of experimental propositions as two-valued and introduces
an intermediate truth value for the incompatible assertions of the theory. By contrast, von Weizsaecker considers as nonclassical only the
logic of events or of object language and seeks a consistent interpretation of "quantum logic" in a metalanguage in line with classical logic.
This allows him to refute the preliminary objection raised by Bohr against
a program of quantum logic and supported by the following argument:

> "It is decisive to recognize that, however far the phenomena
> transcend the scope of classical physical explanation, the ac-
> count of all evidence must be expressed in classical terms." [24]

Now, if it is right to interpret logical problems of quantum theory
as semantic problems, it seems appropriate to reconsider whether the
paradoxical situation detected at the basis of quantum theory is a consequence of the "classical" interpretation of new quantum concepts. In
other words, it seems advisable to take into account the hypothesis that
the most serious problems in the interpretation of quantum mechanics
originate in the attribution of "old" properties to "new" concepts.

In conclusion, it seems that by following the second of the two
distinct routes initially identified in relation to the problem of language in quantum theory, we have somehow been brought back to the first.
A possible alternative analysis could then be to reconsider "logical
problems" of quantum theory by starting this time from the definition
of an adequate requirement for meaning.

REFERENCES

1. B. C. van Fraassen, "The labyrinth of quantum logics," in C. A.
   Hooker, ed., *The Logico-Algebraic Approach to Quantum Mechanics*
   (Reidel, Dordrecht, 1975).
2. G. Birkhoff and J. von Neumann, "The logic of quantum mechanics,"
   ibid., p.12.
3. M. L. Dalla Chiara and G. Toraldo di Francia, *Le teorie fisiche*
   (Boringhieri, Torino, 1981), p.95.
4. G. Birkhoff and J. von Neumann, *op. cit.*, p.5.
5. *Ibid.*, p.16.
6. J. S. Bell, "On the problem of hidden variables in quantum mechanics,"
   Rev. Mod. Phys. 38, (1966).

7. J. Bub, *The Interpretation of Quantum Mechanics* (Reidel, Dordrecht, 1974), pp.62-63.

8. S. Kochen and E. P. Specker, "The problem of hidden variables in quantum mechanics," in *J. Math. Mech. 17* (1967).

9. J. Bub, *op. cit.*, p.66.

10. *Ibid.*, pp.65-71.

11. P. Février, "Les relations d'incertitude de Heisenberg et la logique," *Acad. Sc. (Paris) 204* (1937).

12. H. Reichenbach, *Philosophic Foundations of Quantum Mechanics* (University of California Press, Berkeley, 1965), p.145.

13. P. Suppes, "The probabilistic argument for a nonclassical logic of quantum mechanics," in C. A. Hooker, ed., *The Logico Algebraic Approach to Quantum Mechanics* (Reidel, Dordrecht, 1975), p.342.

14. P. Suppes, "Logics appropriate to empirical theories," *op. cit.*, p.334.

15. *Ibid.*, p.337.

16. See, for example: B. C. van Fraassen, "Analysis of quantum logic," in C. A. Hooker, ed., *Contemporary Research in the Foundations and Philosophy of Quantum Theory* (Reidel, Dordrecht, 1973); B. C. van Fraassen, "A modal interpretation of quantum mechanics," in E. Beltrametti, ed., *Current Issues in Quantum Logic* (Plenum, New York, 1981).

17. M. L. Dalla Chiara and G. Toraldo di Francia, *op. cit.*, p.89.

18. *Ibid.*, pp.99-105.

19. Cf. M. L. Dalla Chiara, "Some metalogical pathologies of nondistributive logics," *J. Phil. Logic 6*, 331 (1977).

20. M. L. Dalla Chiara and G. Toraldo di Francia, *op. cit.*, p.102.

21. C. F. von Weizsaecker, "Komplementaritaet und Logik," *Naturwissenschaften 42*, 20 (1955).

22. *Ibid.*, p.26.

23. *Ibid.*, p.28.

24. N. Bohr, "Discussion with Einstein," in P. Schilpp, ed., *Albert Einstein: Philosopher-Scientist* (Open Court, La Salle, Illinois, 1949), p.209.

REMARKS ON THE HYPOTHESIS OF HIDDEN VARIABLES

F. Pollini
Centro di Ricerca in Epistemologia e Storia della Scienza
Via Aldini 22 - 47023 - Cesena

ABSTRACT. On the grounds of a formal approach to physical theories, we propose a logical analysis of the hidden variables hypothesis, based on a closer examination of the theorems of Jauch and Piron, Kochen and Specker.
A very general definition of deterministic theory is discussed showing how any deterministic reinterpretation of quantum mechanics is incompatible with the mathematical structure of the present theory.

1. INTRODUCTION

Since the twenties and thirties the hypothesis of hidden variables has been constantly present within the debate going on about the paradox of physical reality, enlivening it from its very beginning up to the present day.

Due to the long duration and to the many-faceted character assumed through the times by this concept, it is not easy to give a clear-cut definition and evaluation of it: The various theories are hardly homogeneous, in the sense that they are set apart by the distinct approaches they utilized, by the different philosophical meanings attached to them, and by the types of argument they use to support their theses. However, on the whole they can all be referred to as attempts at completion (hypotheses of completion) of quantum theory, which, while preserving characteristic formal and mathematical properties of the latter, extend its explicative and predictive power.

Various present day authors have singled out Einstein as the most important — even if not historically, the first — supporter of a hypothesis of completion, this recognition béing based on a well-known interpretation of the Einstein-Podolsky-Rosen paradox[1].

For various reasons, however, the attribution to Einstein is misplaced: Indeed none of Einstein's papers contains an acceptance of the hypothesis of hidden variables. The very EPR paradox is a demonstration of the incompleteness of quantum theory, but is a far cry from a hypothesis effecting its completion[2].

As M. Jammer remarks, even though the EPR incompleteness argument

343

G. Tarozzi and A. van der Merwe (eds.), The Nature of Quantum Paradoxes, 343–352.
© 1988 by Kluwer Academic Publishers.

has been one of the most important factors favouring the modern development of the theory of hidden variables, it would be wrong to consider Einstein as a supporter if not altogether the most important advocate of hidden variables. However, it is remarkable that anyone should ascribe to Einstein an argument, which has long been used by the partisans of the orthodox interpretation of quantum theory to reject a classical completion hypotheses, that is, to demonstrate the completeness of the new theory.

The hypothesis of hidden variables appeared first in systematic exposition, within a context of lively debate surrounding contrasting theses and concepts of logic, physics and philosophy, related to quantum mechanics. In the fundamental 1932 book *Mathematische Grundlagen der Quantenmechanik*, which became know especially after the appearance of its 1955 english version, von Neumann, making reference to the logical properties of the statistical interpretation of thermodynamics, expressed an *ad hoc* hypothesis of hidden variables in order to demonstrate that such a hypothesis was inconsistent with the standard Hilbert space formulation of quantum theory. The result seemed conclusive: The statistical interpretation "is the sole consistent interpretation of quantum theory, i.e., of the whole of our experiments on elementary processes".

Within the debate on the paradox of physical reality, the hypothesis of hidden variables is thus first of all an argument which supports the validity of the "orthodox" interpretation of physical reality.

Actually, other papers have referred to the hidden-variable argument in order to support completion hypotheses (such as, for example, the contextual theory of hidden variables proposed, in 1952, by Bohm) which were consistent with the "orthodox" interpretation or in order to reject a "local theory" which could not be reconciled with the statistical predictions of quantum theory (Bell inequality, published in 1961) and which was based on the local principle discussed in the EPR paradox.

From a logical point of view, the most important contribution to the debate appearing after von Neumann's are those due to Jauch-Piron (similar to Gudder in their approach), and Kochen-Specker, whose original formulations appeared respectively in 1963 and 1968.

In both papers one has the following definition of hidden variables: A set of classical states which allow a reinterpretation of quantum states. Furthermore, a similar type of logical sequence appears in both sources: The demonstration of algebraic theorems, formulated in original terms, allows one to deduce the incompatibility between this hypothesis and the mathematical properties of quantum theory. What distinguishes the two approaches are the algebraic structures involved in the demonstrations (Kochen and Specker utilize the algebra of observables, Jauch and Piron the algebra of events) as well as the scope and relevance of the results. As I will hopefully show in what follows, the hypothesis of hidden variables according to Jauch-Piron does not require an extension of the events of quantum theory but only a different evaluation of them; Kochen and Specker, on the other hand, propose a deterministic reinterpretation of observables and states of quantum theory.

2. INDETERMINISTIC PHYSICAL THEORIES AND HIDDEN VARIABLES HYPOTHESIS

Every physical theory is liable to an idealization: This is identified with a structure

$$T = \langle L, A, D, K \rangle$$

comprising a formal language L, a set of specific axioms of T expressed in L, a logic which can be described be a set D of inference rules, and, finally, a class K of physical models of T^3.

The class A of specific axioms of T comprises the set A_T of specific physical axioms T and the set A_M of mathematical axioms of the mathematical subtheory T_M of T.

As far as the semantic part of the theory is concerned, we note that in the present case a physical model M of a theory T is viewed as a physical structure, associated with the langage L which causes the axioms of the theory to become true. A physical structure is made up of $M = \langle M_o, S, Q_1-Q_u, \rho \rangle$, with M∈k where:

(a) M_o is the mathematical part of M, that is, in general the standard model of the mathematical subtheory T_M of T (hence of the mathematical axioms of T).

(b) $\langle S, Q_1-Q_u \rangle$ is the operative part of M, where S is a set of physical situations: That is physical systems in given states for which it is possible to operatively define the physical quantity of model M characterized by Q_1-Q_u.

Insofar as the explicative function of the models is concerned, it is important to consier a few assumptions with relation to the measurement of a physical quantity:

(1) It is assumed that, in every physical situation S, it is possible to set up more than a measurement of the same quantity; the various results of the measurement of Q_1, in S are identified as Q_{11}^S, Q_{12}^S,...

(2) A given result Q_{11} can be represented either by an interval or real number I_{11} (the amplitude of which depends on the accuracy εQ_1 of the measuring device) or, in the general case, upon a probability distribution P_{11}^S or real numbers. The correcteness of the probability distribution depends both on the accuracy εQ_1 of the measuring device and on the accuracy εP_1 with which the probability is estimated.

(3) An ideal value of Q_1 is either a real number belonging to the interval I_{11} of a function which represents a probability distribution comprised in P_{11}. In the first case we talk about ideal results of ideal values of the first kind; in the latter, we talk about ideal values of the second kind. Let us denote with s(t) the state of the physical system s at the momento t, and with S' the set of all possible states which s can assume in M.

(c) ρ is a function that creates a correspondence between a mathematical interpretation in M and every concept in the operative part of M. In particular, for every physical system $s \epsilon S$, $\rho(S^s)$ represents the set of all ideal states that s can assume.

Let us further assume that, for every physical system s, there exists at least one ideal translation τ such that, in case s(t) is defined, $\tau(s(t)) \epsilon \rho(S^s)$; this means that, for every state of a physical systems s at moment t, there exists at least one ideal state, i.e., a mathematical entity in M_o, which we denote by associated with the state $s \epsilon S^s$.

Let us further denote by Q_i the mathematical entity $\rho(Q_i)$ in M_o, associated under ρ, with a given physical quantity in M; moreover, let us denote by H the set of all mathematical entities in M_o which describe (or can describe) physical measurable quantities of M; it is the set of observables of M.

On the basis of this formulation, one proceeds to define the statements of the language L of M and the concept of truth of the statements in a physical situation.

To this idealization of a physical theory, which will be made concrete for quantum theory, we must add a few specific definitions, so we can start talking about the properties, that are under investigation here.

Let us indicate by λ_M the set of all ideal states which characterize the model M and by θ the set of all observables of M, where an observable O is a function O: $\lambda_M \to M$, defined over the set λ_M of the states and having values comprised in the set M of the probability measurements over R.

Intuitively, if $P_{O\tau}$ is the probability amplitude which the observable O associates with state $\tau \epsilon \lambda_M$, and Q is a physical quantity corresponding to O, then, for every Borelian $B \epsilon R$, $P_{Q\tau}(B)$ identifies the probability that the result of a measurement of Q in the state τ is comprised in B .

It is possible to define over the set θ of observables of a model M an algebra called *the algebra of the observables*:

$$A_P^\theta = \langle \theta, {}^*\theta, + , \cdot \hat{\,} , 1, 0 \rangle$$

where θ indicates the relationship of commensurability[4], $+ \cdot$ indicate the taking of a sum and product respectively, limited by the commensurability relationships, $\hat{\,}$ represents multiplication by a real number, 1 and 0 denote the elements unity and zero of θ.
The algebra A_P^θ is a partial algebra over R, while $A_B = \langle \theta_I, {}^*\theta, + , \cdot \hat{\,} , 1, 0 \rangle$ of the idempotent observables of θ is a partial Boolean algebra.
The set λ_M of all the ideal states of M can be described in a different way according to the properties that intuitively concern the degree of accuracy with which the physical quantities can be measured.

Of considerable interest is the case where $\lambda_M = \Omega_M \cup \Delta_M$, that is, λ_M is composed of states without dispersion or of mixtures of states without dispersion, since, in every model M that satisfies this requirement, every probabilistic value can be interpreted as a

limitation of knowledge whose mathematical representation consists of
weight functions defined over the ensemble power of λ_M and having values
in the range $(0,1)$.

 *Similarly, all the functions of events are dichotomic functions of
probability or mixtures of dichotomic functions*[5].

Clearly the case described represents a deterministics situation:
On this basis we will enunciate the formal definitions of determinism
(and of indeterminism) of a model, and of a physical theory, hence the
definitions of evaluation and of deterministic reinterpretation of an
indeterministic theory.

A physical theory T is *called deterministic with respect to model M*
(belonging to K) if every state of M is a dispersionless state or a
mixture of dispersionless states.

A *physical theory T is called deterministic* if and only if T is
deterministic with respect to all its models. Otherwise it is
indeterministic (or *probabilistic*).

Let us now consider an indeterministic model $M_I = \langle M_o, S, Q_1, Q_u, \rho \rangle$.
M_I *allows a deterministic evaluation* if and only if all the functions of
events defined over V are dichotomic functions or mixtures of dichotomic
functions.

In this case, which corresponds to the Jauch and Gudder approach,
one remains within the limits of indeterministic theory, hypothesizing a
classical evaluation of the events afforded by an hypothetical ensemble
M of "hidden" events.

The concept of a *deterministic reinterpretation of an indeterministic model* is more far-reaching. M_I affords a deterministic
reinterpretation if there exists a deterministic model
$M' = \langle M'_o, S', Q_1 - Q_u, 1' \rangle$ such that:

1) $\tau_p \forall \lambda_M \cdot\cdot \tau'_D \exists \lambda_{M'} \ 0 \ \forall_\theta \ 0' \exists \theta'$

(where τ_p denotes a pure state, τ'_D a mixture of dispersionless states,
and $0'$ an observable of M'):

$B \forall R^\wedge \ (0(\tau_p)(B) = 0'(\tau'_D)(B))$,

In this way, state τ'_D is the deterministic reinterpretation of τ_p;
we consequently identify τ'_D with $R\text{-Det}(\tau_p)$:

2) $\omega \forall M \ \omega' \exists {}'M \ 0 \forall_\theta \ 0' \exists \theta'$,

(where ω indicates a *dispersionless* state):

$f_0(\omega) = f_{0'}(\omega')$

Thus the state ω is the deterministic reinterpretation of ω.

The condition imposed insures that the probabilistic values typical
of the original model M_I are preserved in model M': Indeed, the values
associated by observable 0 with the pure state τ_p are equal to the
values associated by observable 0', univocally correspondent to 0, to
the mixture of dispersionless states τ'_D corresponding to τ_p. The

mathematical formalism of the deterministic model M' is not necessarily the one typical of the theory to which the reinterpreted model belongs; this means that the definition of deterministic reinterpretation does not require that the reinterpreting model be a model of the theory but rather only that it be compatible from a mathematical point of view with its formalism[6].

The condition under discussion states that the value (real number r) of the function f_0, in the dispersionaless state ω of M_I is equal to the value of the corresponding function f_0,, in the state ω of M', which it reinterprets M_I.

Model M' consequently does not exclude probabilistic values, but it inteprets them as "ignorance", that is, as a limitation to knowledge; therefore model M' recovers all the probabilistic characteristics of M_I, albeit attributing to them a different meaning; it furthermore preserves the deterministic core of M. In other words, M' is a suitable extension of the deterministic part, if any, of M_I, so that it is possible to interpret all probabilistic values of M_I as a limitation to knowledge and not, rather, as a peculiar character of the theory; this is the concept of "hypothesis of hidden variables'.

From the definitions introduced it is possible to express in purely mathematical terms the hypothesis of hidden variables here discussed.

Theorem of Deterministic Evaluation:

If the indeterministic model M_T of T allows a deterministic evaluation, then the set W of all proability functions defined over the complete and atomic orthomodular lattice V of M_I is semideterministic.

Theorem of Deterministic Reinterpretation:

If the indeterministic model M_I of T admits a deterministic reinterpretation, then there exists a homomorfism $h : A^\Theta \to A_C$ of the partial algebra over R consisting of the set Θ of all observables of M_I in a commutative algebra A_C.

Corollary: If the indeterministic model M_I of T admits a deterministic reinterpretation, then there exists a homomorfism $h : A_B^\Theta I \to$ of the Booleen partial algebra $A_B I$, formed the set of all idempotent observables of M_I in a Boolean algebra A.

3. QUANTUM THEORY AND HIDDEN VARIABLES HYPOTHESIS

Let us last consider a formal presentation of quantum theory

$$QM' = \langle L, A, D, K \rangle$$

consisting of an abstract language L, a set of axioms specific of quantum theory, that is, mathematical axioms (axioms of functional analysis) and physical axioms ($A = A_M \cup A_F$), a logic D (in particular, classical logic), and a set K of models M of quantum theory.

Every model MϵK, composed of the standard model M_0 of functional analysis, a set S of physical situation, physical quantities Q_1-Q_n, and the mathematical interpretation function ρ satisfies the following requirements:

(R1) For every physical systems s in S $\rho(s)$ is a Hilbert space H in M_0 and $\rho(S^s)$ is the set of all statistical operators in H. Let P_ψ be the projection operators corresponding to the normalized vectors ψ of H: The statistical operator W, whenever formed by P_ψ, represents a pure state. If this is not the case W represents a mixture. On the basis of the bi-univocal correspondence between the ensemble of all projection operators P_ψ and that of all normalized vectors ψ in H, one can say that a pure state is represented by a normalized vector ψ of H; therefore, a necessary and sufficient condition for W to be a statistical operator corresponding to a pure state is $\underline{W} = P_\psi$.

Let us now consider every operator statistical $\underline{W}$ as a function of time. If $s(t_i)$ identifies the physical system s at time t and τ and ideal translation such that, for every moment t_i (for which s(t) is defined), $\tau(s(t_i)) \epsilon \rho(S^s)$ (in the case of quantum mechanics one has $\tau(s(t_i))=W_i$ with $t_i \epsilon [t_1..t_2]$), then $\tau(s)(\underline{t}_i)=W(\underline{t}_i)$, where $\underline{t}_i$ is a variable corresponding to the physical quantity time.

(R2) If s_1 and s_2 are two physical systems and $s_1 \boxtimes s_2$ represents the composite system, then $\rho(s_1 \boxtimes s_2) = \rho(s_1) \boxtimes (s_2)$, where $\boxtimes$ indicates the tensor product of Hilbert spaces.

If, for a given translation τ, $\tau(s_1(t))=W_1$ (of $\rho(s_1)$ and $\tau(s_2(t) = W_2$ (of $\rho(s_2)$), then there exists a statistical operator $W = W_1 \boxtimes W_2$ (of $W \triangleq \rho(s_1) \boxtimes \rho(s_2)$)
such that

$$\tau(s_1(t) \boxtimes s_2(t)) = W$$

(R3) For every physical quantity Q_i and every physical system s, $\rho(Q_i)$ is a self-adjoint operator in $\rho(s)$. The *eigenvalues* of $\rho(Q_i)$ represent all the ideal values of the first kind that the quantity Q_i can assume in s.

(R4) If $\tau(s(t)) = W$ and B is a Borelian set of real numbers, the real number $T_r WP\rho(Q_i)$ (where Tr is the trace operator and $P^{\rho(Q_i)}$ is the spectral measure associated with the self-adjoint operator $\rho(Q_i)$) represents the probability that, taking a measurement in s(t) of the physical quantities Q_i, one obtains an ideal value (of the first kind) which is contained in B.

Let this probability be called Prob $_W^{Qi}$ (R). Then one has Prob $_W^{Qi}$ (R) = Tr WP $^{\rho(Q_i)}$ (B)), which represents the general form of the statistical algorithm of quantum theory.

In the particular case for which $\tau(s(t))$ is an ideal pure state of the form $\Sigma r_i \psi_i$ (where ψ_i are normalized eigenvectors of $\rho(Q_i)$ with corresponding eigenvalues r_i) and B consists of a single number $\{r_k\}$, it follows that Prob $_W^{Qi}$ $\{r_k\} \triangleq |c_k|^2$.

(R5) Let $\tau(s(t)) = \Sigma c_j \Psi_j$, where $\{\Psi_j\}$ is a set of eigenvectors of $\rho(Q_1)$. Let us suppose that, after taking a measurement of the first kind (a measuring procedure, applied to the physical quantity Q_1, which preserves the first result Q_{11}^s obtained in the physical situation s), one gets the result $r_k + \varepsilon_{Q1}$ for the quantity Q_1 (where r_k denotes an eigenvalue of $/Q_1/$ corresponding to Ψ_k and ε_{Q1} the accuracy of the instrument used to measure Q_1). Right after, measurement (at time $t_1 > t$), the state of system designated by $s(t_1)$, will be translated in Ψ_k, that is, one will have $\tau(s)(\underline{t}_1) = \Psi_k$.

(R5) Asserts that after a measurement the physical system collapses into the state characterized by the eigenvalue obtained in the measurement; this is von Neumann's projection postulate.

The set of specific physical axioms of quantum mechanics can be expressed by means of a single metatheoretical scheme corresponding to the Schoedinger equation

$$W(t) = \exp(-iH\underline{t}) \; W(\underline{t}_o) \; (\exp iH\underline{t}) \quad ,$$

where H represents the Hamiltonian of the physical system described by W, $W(\underline{t})$ and $W(\underline{t}_o)$ denote the statistical operators that represent the states of s at time t and time t_o.

The set Q of observables, that is, of self-adjoint operators of a model M_s of quantum mechanics, forms a partial algebra over B, where the relationship of commensurability is defined in terms of commuting operators, while the set Q_I of self-adjoint and idempotent operators of M_s forms a partial Boolean algebra.

Since in quantum mechanics it is possible to demonstrate as a theorem *the indeterminacy* relation in, for instance, the form

$$\text{Exp } (\Delta Q)^2 \; \text{Exp } (\Delta P)^2 \geq \frac{1}{2} h,$$

where Exp (ΔQ) is the expectation value of the physical quantity corresponding to the operator ΔQ (c.c., the variation or dispersion of the quantity position Q) and exp (ΔP) analogously identifies the dispersion of the quantity momentum P, one may conclude that in every model of quantum mechanics there is no dispersionless state.

Quantum Theory, according to the Definitions formulated previously, is consequently indeterministic

Going back now th the definition of evaluation and deterministic reinterpretation and to the related theorems, let us express in the quantum language the concepts already referred to in the idealization of physical theories.

Theorem of Deterministic Evaluation of Quantum Theory

If the indeterministic model M_I of quantum mechanics admits a deterministic evaluation, then the set F of all functions $P_w : \theta_H \times R \rightarrow$

[0,1], defined over the Cartesian product of θ_H the set of all observables of the Hilbert space H of M_0, and the set of Borelians, is semideterministic.

Theorem of Deterministic Reinterpretation of Quantum Theory:

If M_I is an indeterministic model of quantum mechanics which admits a deterministic reinterpretation, then there exists, for every Hilbert space H associated via ρ with a physical system sϵS, a homomorfims h:$A_\rho \to A_C$ of the partial algebra over R formed by the set θ_H of self adjoint operators defined over H in a commutative algebra over R A_C.

Alternatively, on the basis of suitable mathematical theorems, which will be not be quoted here, if M_I is an indeterministic model of quantum mechanics which admits a deterministic reinterpretation, then there exists, for every Hilbert space H associated via ρ with a physical system sϵS, a homomorfism h: $A_B^S \to B$ of the partial transitive Boolean algebra

$$A_B^S \ \langle S, *_\theta, \xi, \cup, \cap, 1, 0\rangle$$

formed by the set S of all the closed subspaces of H in a Boolean algebra B .

The single application of two mathematical results allows us now to demonstrate the contradictoriness of the conditions indicated with respect to quantum theory.

The Jauch-Piron Theorem

The complete orthomodular atomic lattice of quantum mechanics does not admit a semideterministic set of probability functions.

The Kochen-Specker Theorem

The partial Boolean algebra formed by the closed subspaces of a Hilbert space with more then three dimensions does not admit homomorfism in Boolean algebra.

It follows that quantum theory does not admit deterministic evaluations, nor deterministic reinterpretation of the local probabilistic models fulfilling Bell's requirements.

NOTES

1. This is the thesis maintained in the article by J. F. Clauser, M. A. Horne, A. Shimony, and R. A. Holt, "Proposed experiments to test local hidden-variable theories", *Phys. Rev. Lett.* 23, (1969).
2. M. Jammer, *The Philosophy of Quantum Mechanics* (Wiley, New York, 1974), p. 254.
3. For this part of the presentation, I refer to the papers by M. L. Dalla Chiara and G. Toraldo di Francia: "A logical analysis of physical theories", *Riv. Nuovo Cimento Serie II*, 3, (1973); "The logical dividing line between Deterministic and Indeterministic Theories", *Studia Logica* XXXV, (1965); *Le teorie fisiche* (Torino,

Boringhieri, 1981).

4. Two observables O_1, O_2 of θ are commensurable if there exists an observable O and a pair of Borel functions s_1 and s_2 such that $O_1 = g_1(O)$ and $O_2 = g_2(O)$.

5. Indicating by γ the Cartesian product $\theta x R^{\wedge}$, an element $(O,B) \in \gamma$, with $O \in \theta$ and $B \in R^{\wedge}$, is called event of the model M of T. An event, insofar as constituted by an observable O and a Borelian R, is an abstract entity which can be empirically intepreted as follows: The value of the measurement of the quantity Q corresponding to O is found in R.

Every state $\tau \in \lambda_M$ identifies a function which assigns to every event a probability value and a function of events P_τ, defined over γ. If in γ one defines a relation of equivalence among events... (O_1,B), (O_2,C), $(O_1,B) \sim (O_2,C)$ iff $\tau \in \lambda_M$ $(P_\tau(O_1,B) = P_\tau(O_2,C)))$, and denotes with V the set quotient of γ module: then the algebraic structure $\underline{V} = \langle V, \leq, \perp, 1, 0 \rangle$ is called algebra of events of the model M.

Every function of events defined over V of $\underline{V}$ is a probability function.

6. A theory admits deterministic evaluation or deterministic reinterpretation if it satisfies such property with respect to every indeterministic model M $\in$ K.

7. The demonstrations of these theorems are formulated to great length in the articles by Jauch-Piron and Kochen-Specker quoted in the references.

PART 6

HISTORICAL DEVELOPMENTS OF THE EINSTEIN-
BOHR CONTROVERSY

HISTORICAL CONSIDERATIONS ON THE CONCEPTUAL EXPERIMENT BY EINSTEIN, PODOLSKY AND ROSEN

B. Carazza
Department of Physics, University of Parma
Parma, Italy

ABSTRACT. The contents of the original paper by Einstein, Podolsky and Rosen is reproposed, together with the conceptual experiment conceived by them and with the almost immediate reply by Bohr. It is pointed out that the central issue of the discussion lays in the philosophic standpoint which divides realism and positivism. The subject brought forward by Einstein is disconcerting. Difficulties disappear, according to Bohr, provided we give up the idea of describing the properties of microscopic objects regardless of the experimental equipment used for taking measurements on them. Looking at the paper of EPR and to the reply by Bohr, it is inferred that the nonrelativistic quantum mechanics in the Copenhagen interpretation is a phenomenological theory and that for the physicists of the time it was possible to think of a broader fundamental theory capable of reaching the same predictive results of traditional quantum mechanics. Perhaps this statement can be made also nowday. However, the hoped broader theory cannot be reproposed in terms of classical mechanics or of classical categories and concepts; somehow, it will be necessary to take into account a microscopic reality definitely unrepresentable in intuitive space time terms.

Some time ago I was told a story about Fermi who happened to walk by a conference hall where a lecture was being held. He stood for a while looking at the blackboard and listening to presentations. Then he asked one of the participants what the subject was. He was told that they where talking about "Fermi interaction." I think that the same would happen to Einstein, should he enter a meeting where the issue raised by his paper written in cooperation with Podolsky and Rosen[1] was being discussed.

It is not useless therefore to resume the subject from the very beginning, going in detail through the original work by Einstein, Podolsky, and Rosen (EPR), as well as through the reply offered by Bohr immediately afterwards. I hope this will enable us to focus on the essential question that was then raised and discussed and is still essential to date. I am interested in it as theoretical physicist, but also from historical point of view, since it seems to me that the history of

G. Tarozzi and A. van der Merwe (eds.), The Nature of Quantum Paradoxes, 355–369.
© 1988 by Kluwer Academic Publishers.

physics offers a way for reflecting on the foundations of our discipline.

I wish to underline that I deem it more opportune to speak of EPR's *conceptual experiment* rather than talking of a paradox. The authors themselves did not consider it as such, at least with reference to their paper. Let me digress from our main subject and speak about the so called conceptual or thought experiments. However one might choose to view them from a philosophic or methodologic point of view, they played an important role in the development of physics. Their main feature is that the conclusions that may be based on them can be sustained regardless of whether the experiment being considered was actually carried out or not. On the other hand, in many instances a conceptual experiment is so characterized precisely because, as a matter of fact, it cannot possibly be carried out.

In general, thought experiments have a logical feature[1], giving rise to a statement which at first sight may seem surprising, to be the consequence of some assumptions and principles already taken for granted and shared, and therefore acceptable. This is the case of Stevin who, when considering his well known "endless chain," connects the inclined plane laws with the impossibility of perpetual motion. Or they can lead to conclusions from a theory unforeseen as yet, and at times unthought of, or even evidence possible internal contradictions and limitations of the theory itself.

All this with no determinant help from mathematical tools, and without utilizing a formal-logic apparatus. The subject proposed originally by Einstein, Podolsky, and Rosen in the last part of their work seems to belong to this latter class of thought experiments. Now, let us examine in detail the contents of said paper.

Einstein, Podolsky, and Rosen start by assessing the distinction between the objective reality, which must be independent of any theory, and the physical concepts by which any theory operates. Such concepts are viewed as valid, and with them the theory, only when the latter can be deemed correct and when the description of the world it gives can be regarded as complete. A theory can be said to be correct when the predictions ensuing from it are in accordance with experiments and measurements. With regard to the second feature, the following requirement will apply: "Every element of the physical reality must have a counterpart in the physical theory."this being viewed as the "condition of completeness" of the latter. The scope delimited by the authors, as can be inferred from the very title of the work, is to examine the completeness of quantum mechanics.

The above statement is not clear without further specification of what are the elements of physical reality. To this end the following criterion is adopted: "If, without in any way disturbing a system, we can predict with certainty (i.e., with probability equal to unity) the value of a physical quantity, then there exists an element of physical reality corresponding to this physical quantity." Regarded as a sufficient "condition of reality," this criterion is considered to be in agreement with classical as well as quantum mechanical ideas of reality.

The concepts expressed up to this point are then illustrated in the case of quantum mechanics considering a particle with one degree of freedom. Taking into account the conditions met in quantum mechanics, in the general case of two physical quantities corresponding to noncommuting operators, the authors conclude that:

"either (1) the quantum mechanical description of reality given by the wave function is not complete, or (2) when the operators corresponding to two physical quantities do not commute, the two quantities cannot have simultaneous reality."

In other words, the description offered by quantum mechanics cannot at the same time be realistic and complete.

After the above preliminaries, the authors consider two systems, I and II. It is assumed that they have interacted for a certain interval of time $0 \leq t \leq T$, and that after said interval no more interaction takes place between them. The initial conditions being known, Schroedinger's equation will allow us to determine the wave function Ψ of the whole I+II system for any time $t > T$, but generally -- due to the interaction -- this function will not be factorizable, so that a well defined state cannot be assigned to either I or II.

Consider two operators·A and B relative to I and their respective eigenvectors and eigenvalues:

$$A u_n (x_1) = a_n u_n (x_1)$$

$$B v_n (x_1) = b_n v_n (x_1)$$

The authors express Ψ alternatively as:

$$\Psi = \Sigma \psi_n (x_2) u_n (x_1)$$

and

$$\Psi = \Sigma \varphi_s (x_2) v_s (x_1)$$

where x_1 and x_2 stand for the set of variables used to describe the two systems. The total state vector is therefore expanded with respect two different systems of orthogonal axes relative to system I, the development coefficients --indicated by ψ_n and φ_s respectively -- being functions of variables relative to system II. If we now consider the so called wave-packet reduction process, and assume that we have executed a measurement of the observable A, finding, for instance, a resulting value equal to a_k, then the whole system, immediately after the measurement, will be left in the state $\psi_k (x_2) u_k (x_1)$. If, on the other hand, the measurement is performed on the observable B, resulting in a value b_r, say, then, after this operation, the state of the system I+II will be $\varphi_r (x_2) v_r (x_1)$.

Both states into which the two above measurements projected the wave function of the total system are factorized, so that it is

possible to assign to II a well defined state vector. However, in case
the foregoing measurements are made when I and II no longer interact,
no real change can take place in II in consequence of operations
carried out on I. It is then concluded that "it is possible to assign
two different wave functions (in our example, ψ_k and φ_r) to the same
reality (the second system after the interaction with the first)."

After this disconcerting conclusion, the argument that leads to
it is resumed and further refined. Systems I and II are now
specifically defined as two particles in one dimension, and it is
assumed that the wave function of the whole system, after the
interaction has ceased, is as follows:

$$\Psi\,(x_1,x_2) \;=\; \int_{-\infty}^{+\infty} e^{(2\pi i/h)\,(x_1-x_2+x_0)p}\,dp$$

where x_0 is a constant. $\Psi\,(x_1,x_2)$ is a simultaneous eigenvector of the
relative coordinate $x_1 - x_2$, i.e. of the distance between the
particles, and of the total momentum. So in the state considered these
two variables have well defined values, namely x_0 and zero. The
observables A and B of the previous example are here identified with
the momentum and position operators, respectively, of particle I. As we
can see, on writing down the above expression using a different
notation:

$$\Psi\,(x_1,x_2) \;=\; \int_{-\infty}^{+\infty} \psi_p\,(x_2)\,u_p\,(x_1)\,dp$$

the state vector in question is already given as an expansion on a continuous basis of momentum eigenvectors of the first particle; and the
coefficients of this expansion, which depend on x_2, are nothing but the
eigenvectors of the momentum operator for the second particle.

On using as a basis the eigenfunctions of the position operator,
i.e. of $\delta\,(x_1-x)$, indicated as $v_x\,(x_1)$, the state vector pertaining to
the total system can also be expanded as follows:

$$\Psi\,(x_1,x_2) \;=\; \int_{-\infty}^{+\infty} \varphi_x\,(x_2)\,v_x\,(x_1)\,dx$$

This time, the coefficients of the expansion, $\varphi_x(x_2)$, are the
eigenvectors of the position operator for the second particle. Let us
look at the previous expression of the state vector. By measuring the
momentum of the first particle, we can effect a reduction of the wave
packet causing the state to be projected into one of the products
$\psi_k\,(x_2)\,u_k\,(x_1)$. The other particle is thus left in a state
corresponding to a well defined value of its momentum. It is supposed
moreover that, just before the measurement, the two systems no longer
interact. This can evidently be taken for granted, from the point of
view of the authors, if the distance x_0 is sufficiently large.
Therefore, it will be possible to predict with certainty the value of

the momentum belonging to the second particle without disturbing it at all. On the other hand, we can think of carrying out a measurement of x_1, and in this case, looking at the last expression of Ψ (x_1, x_2), it is clear that we can predict with certainty the value of the position coordinate for the second particle, still without interacting with it.

The three authors then reach the conclusion, following the reality criterion stated before, that two physical quantities related to noncommuting operators, as in the case here considered, can have simultaneous reality. Returning to the necessary choice between the completeness of the description given by the wave function and the simultaneous reality of two quantities as such, we are forced to admit that the quantum mechanical description of reality given by the wave function is not complete.

Let us briefly summarize the structure of the paper we examined. Following initial considerations regarding the validity of a theory, a necessary condition is formulated for the theory to be considered complete. In order to clarify the content of the said condition, a criterion of reality is further stated. Taking into account quantum mechanics, the authors then infer that the description given by the wave function cannot be at the same time realistic and complete when two observables corresponding to noncommuting operators are considered. Finally, the conceptual experiment follows relative to the inferences that can be drawn in regard to the physical properties of a system -- without disturbing it -- by performing measurements on a second system that previously interacted with it.

When specializing the state vector of the whole system, the conclusion is reached that observables such as position and momentum of a particle in one dimension can have simultaneous reality. However, since they correspond to noncommuting operators, in view of the alternative between reality and completeness inferred before for the description given by the wave function, it is concluded that this description is not complete.

It is evident that in the latter part of the paper we are dealing with a conceptual experiment, for these reasons: it is not necessary to actually make the measurements that are mentioned in order to reach the conclusion above.

It frequently happens, as we said, that conceptual experiments cannot be really carried out. We would actually be in such a situation, if two or more physical quantities were considered, quoting EPR, "as simultaneous elements of reality only when they can be simultaneously measured or predicted," should we intend to proceed accordingly. In line with the usual[2] rules of quantum mechanics, it is indeed not possible to predict simultaneous well defined values for both the position Q and momentum P of the second particle.

But Einstein and his collaborators did not agree with this conclusion. The reality of P and Q would depend, assuming the above point of view,

"upon the process of measurement carried out on the first system, which does not disturb the second system in any way.

No reasonable definition of reality could be expected to permit this."

Apart from the conclusion with regard to the incompleteness of quantum mechanics, the outstanding part of Einstein, Podolsky, and Rosen's work is the consideration of the conceptual experiment we described and the attention of physicists focused on this. On the one hand, it was resumed and presented in the form of a paradox, while on the other hand, following subsequent developments, it inspired the idea of experiments intended to be actually carried out, i.e., the present so called EPR-type experiments, such as those conceived to verify Bell's inequality.[3]

Let us now examine Bohr's reply. After a Letter to the Editor of *Nature*[3], he published a more detailed paper, which will be considered here, in the same magazine where Einstein had made the first move, i.e. the *Physical Review*[4]. We shall not examine this paper in detail, but we'll try to point out the general lines and underline the peculiarity of ideas being offered.

Bohr summarized here some points of view, which already ripened and found expression on other occasions, on which the so called "Copenhagen interpretation" of quantum mechanics is substantially based --precisely the interpretation that had been attacked by Einstein and his collaborators. It is difficult to characterize Bohr's philosophic standpoint; however I can do no better than to consider him in this instance as a positivist. Under these circumstances, one of the ilk of Duhem, suspicious as he was of any conceptual experiments, might have replied to EPR's thesis in a rather intolerant way, rudely alleging the very objection that the above authors had tried to stave off in advance, that is, the impossibility of stating empirically well-defined simultaneous values for both the quantities Q and P. In the end, Bohr concludes by sayng just this, but his argument appears to be richer and more detailed (it being proposed as an exposition that aimed at being clear as to how the formalism of the new mechanics was to be interpreted rationally and consistently).

The position taken by Bohr consists after all in the refusal of what B. d'Espagnat[5] calls "hypothesis H," which is expressed as follows:

"There is a sense in speaking of the existence of any microscopic system, whether or not there are measuring instruments capable to interact with it. Such a system can equally have certain physical properties, regardless of the existence or presence of said instruments."

The acceptance of the hypothesis H marks the philosophic position of realists. For Bohr instead, to the extent of determining in a non-ambiguous manner the physical properties of microscopic systems and the properties related to them, only the reference to measuring processes makes sense.

For him it is not possible to talk of the position and momentum of a particle, independent of the presence of empirical procedures capable of determining their values. On the other hand, due to the finite interaction between object and measuring agencies, conditioned by the very existence of the quantum of action, it is impossible to find an experimental apparatus capable of assigning simultaneous and well-defined values for two noncommuting quantities. This impossibility concerns the Heisenberg uncertainty relations, which directly follow from the quantum chanical formalism.

In the first part of the paper we are considering, Bohr shows the said impossibility discussing the procedure of measurements, by means of diaphragms with one or several slits, in the case of a system consisting of a single particle. By means of the same experimental equipment he further discuss the example treated by Einstein, Podolsky, and Rosen. The conclusion is that also in this case the specific procedure suited to measure, although indirectly, one of the variables for the second particle cut ourselves off any possibility of predictions regarding the value of the other variable.If there is no meaning in speaking of the physical properties of any objects without specifying also the experimental circumstances by which we determine them, it will follow that the reality criterion we saw is ambiguous in the expression "without in any way disturbing the system."

In the problem considered, there is "no question of mechanical disturbance of the system under investigation during the last stage of the measuring procedure." But, even at this stage, the conditions under which we operate will affect the second particle and must be specified:

"Since these conditions constitute an inherent element of
the description of any phenomena to which the term 'physical
reality' can be properly attached, we see that the
argumentation of the mentioned authors does not justify
their conclusion that quantum-mechanical description is
essentially incomplete."

We are free to choose measuring either the position or the momentum of the first particle, and therefore to predict the corresponding variable for the second one without mechanically interacting with it. But either choice corresponding to specific experimental circumstances will exclude the other one, exclude therefore the possibility to speak simultaneously of realities corresponding to the two physical quantities in question.

In the debate opposing Bohr and Einstein, Podolsky, and Rosen, we see two clearly different positions confronting realists, on one side, and positivists, on the other, regarding the consideration of "physical reality" (and, as a consequence, of the meaning to be attributed to the term "objectivity").

Bohr's paper ends with this observation: "... this new feature of natural philosophy means a radical revision of our attitude as regards physical reality..." Indeed, one of the most interesting aspects of the paper in question is the surfacing --in this connection-- of the relational conception of quantum states, as it was defined by M.Jammer[6]. The latter refers to scholastic theories of space, noting that, for Thomas Aquinas and Bonaventura, the "ubicatio" of the extremity of a

solid rod undergoes a radical change as soon as a part of the material
rod is removed.

Thus, for Bohr, who insists that the conditions defining any possible type of predictions and of measurements are also to be specified,
the combination of a micro-object with one particular measuring apparatus differs essentially from the combination of the same object with
another experimental arrangement for observation. Staying with the
analogy, realists like Einstein should be compared to Suarez who instead speaks of "ubi" in connection with the rod extremity, assuming
for it a reality independent of the circumstances under which it was
considered.

Even more interesting, and strictly connected with the previous
idea, is the notion according to which a microscopic system is conceptually and actually inseparable from the instrument utilized for determining its relevant properties and with which it interacts: The object
being investigated and the observing apparatus form a single indivisible system.

The concept of a closed or isolated system recurs regularly and
is of capital importance in physics. It presupposes the possibility to
consider the system in question as non interacting with the rest of the
physical world, and hence separable from it. In the famous experiment
of Newton's bucket, for instance, the separability postulate is en-
forced when one maintains that the relative motion of stars, in that
they are so far, has nothing to do with the behavior of water in the
rotating bucket.

And in the case we are examining here, the two particles in the
example being discussed are supposed to be separated and noninteracting
once they have sufficiently receded from one another. The possibility
tacitly taken for granted, of viewing the macro-edifice of phenomena as
constructed by starting from microbricks that are distinct entities,
played an essential role in the development of physical sciences. And
the research of primary elements, of elementary components conceivable
only in that they are separable as individuals distinct from the rest,
appears to be a constant of scientific thought, starting with the ancient atomists. Bohr questions just these prejudices.

He indeed, in the paper under discussion, narrowly sustains the
non-separability of what we consider --obviously improperly following
this scheme of things-- the objects of the microscopic world from the
macroscopic instruments with which they interact. However, it would
seem natural to extrapolate this statement, maintaining that in the
atomic and subatomic world it is no longer possible to mark reality by
well-defined individual entities, thus setting a limit to the possibility of further anatomizing nature. Moreover, Bohr himself on other occasions considered as a decisive element of the new mechanics what he
called the individual character of quantal processes, such as the emission of radiation from an atom, recognizing a "feature of wholeness"
for these processes at the microscopic level.

Let us also touch upon the general viewpoint, termed
"complementarity," with which the name of Bohr is especially associated
and that was set forth for the first time during a congress held in
Como in honor of Alessandro Volta. In his reply to EPR, Bohr takes the

opportunity to express this viewpoint in more detail than on previous occasions. But it is not easy to draw --from this or from other works of the author-- an explicit and sharply defined formulation of the "principle of complementarity." In connection with this, we are reminded of Plato's Fhilebus, where Socrates observes that if you cannot grasp absolute Good by means of one idea, you may try to grasp it with three: beauty, proportion, and truth --that obviously concur to the same end. I understand here that the three ideas which can be termed complementary in the common language, can be used together without limitation[4].

With regard to the meaning to be attributed to Bohr's complementarity, we must state beforehand that at the Como congress, he underlined[7] what in modern textbooks is called the wave-corpuscle duality and which was later emphasized by some authors, i.e., the fact that in certain instances, as in the case of diffraction, the behavior of light can be interpreted as being due to the propagation of electromagnetic waves, while in other cases, as in the photoelectric and Compton effects, light reveals a corpuscular feature which finds adequate expression only in the concept of the photon put forward by Einstein. The same can moreover be said with regard to material particles such as the electron, recalling the then-recent discovery of the selective reflection of electrons from crystals, which enhanced the original idea by de Broglie to associate a wave with electrons, while other phenomena would characterize them as corpuscles.

In this discussion, the radiation in empty space as well as isolated material particles are considered as idealizations, and to the extent that we adhere to the classical concepts of wave and corpuscle, we are faced with an inevitable dilemma, which is regarded as the very expression of experimental evidence.

"We are not dealing [he states] with contradictory, but with complementary pictures of the phenomena, which only together [but not simultaneously, let us add] offer a natural generalization of the classical mode of description."

In the work we are examining here, the Danish scientist takes into account, rather than the couple wave-corpuscle, the couple position-momentum, and their combination is said to characterize the image of the physical world as it is depicted by classical physics, and "in this sense [he says] they can be considered as complementary." These two notions, we reiterate, concur together with no mutual preclusion to define the state of a particle from the classical point of view. But, in the area of quantum mechanics, such variables along the same axis correspond to noncommuting operators, and any experimental procedures used to define in a non-ambiguous way the one or the other can not be compatible. The same can be said regarding any other pair of classical variables corresponding to two noncommuting operators.

Let us suppose at this point, that in order to describe phenomena, we must use concepts of classical physics, either because they are indispensable in practice, for instance when communicating empirical results, or because we acknowledge their epistemologic supremacy. The resulting language, when applied to quantum phenomena, taking into account Bohr's complementarity conception, will originate a new logic characterized by restricted sentential connectibility. As we see, the conceptual experiment of EPR brings to the attention of philosophers and of epistemologists old and new points of view in connection with the reality concept and offers new perspectives in the field of semantics and formal logic. I wish to stress this fact.

If it is true that some physical concepts originated in meta-physics, it is noteworthy that modern physics recently offered new clues for philosophic meditation, proposing to reconsider --for in-stance-- the old question of "principium individuationis" and the no-tion of substance or substratum, from scholasticism through Descartes, Arnauld, and others to the "Grundsatz der Beharrlichkeit" of Kant; and, with regard to identical particle systems or symmetry considerations, the Leibnitz's principle of sufficient reason. All this restores the ancient alliance of physicists and philosophers and revives the term "natural philosophy," confirming the hauling function of scientific culture in the modern intellectual tradition. The situation I refer to allows us to foresee a profitable relationship between scientists and philosophers. But, in this connection, I must note that some time philosophers, imitated by physicists, claimed --especially with regard to quantum mechanics-- that the results of the new physics, even unduly extrapolated, would confirm this or that particular philosophic preju-dice. Such an attitude is certainly not constructive.

After an initial, rather lively, discussion, joined in by the founding fathers, regarding the interpretation of the mathematical for-malism on which the new mechanics was based, the so-called Copenhagen interpretation in the end prevailed. The latter was also called "orthodox," and it was Bohr, as we said, who mainly contributed to its sanctioning.

The young physicists of the time in general adopted the Copen-hagen interpretation (CI) with little hesitation, no doubt because --we must admit-- it offers a rational and consistent way of thinking which is apt to eliminate several difficulties. Besides, no valid alterna-tives were available at that time. But the CI was followed probably also because it was more profitable simply to adopt the new formal recipebook, which in time was revealed to be very fruitful in dealing with new phenomena, rather than to meditate on foundations of the new theory. The subsequent development of quantum field theories and the exploration of new subatomic phenomena, I think somewhat, set this problem aside[5].

It would be interesting therefore to consider with detachment the situation as it was at the time, trying to express an impartial judgment regarding the debate between people like Einstein, Schroedinger, and Ehrenfest, on one side, and the supporters of the now prevailing orthodox point of view, on the other.

To this extent I would like to offer my small contribution, lim-ited to the examination of the debate as it would appear by only read-ing the two papers we have been dealing with, holding strictly to their contents with no reference to any previous or subsequent comments.

This objective in undoubtedly very limited, however it will bring into focus the central issue, since --as I believe-- the essence of the question lies in the confrontation between realists and positivists. This is well illustrated by the witty exchange of retorts between EPR and Bohr.

The debate between the two attitudes involved can be traced to the opposition between two different ways of conceiving the scope of a physical theory, and the features it must show, confronting the contemplation of the phenomenologic, versus the so called fundamental, theories.

The two categories actually meet, even though their opposing supporters may be differently motivated, the two basic requirements of science: the cognitive requirement, on one side, and, the requirement inspired by the slogan "scientia propter potentiam," on the other.

Einstein papers are very clear and usually contain a linear speech. But this one, written in cooperation, is somewhat different. A physicist accustomed to the orthodox point of view will feel uneasy upon reading it. Probably this is due to the impression that the authors did not entirely divest themselves of a number of prejudices of classical mechanics. In any case I have this feeling.

It would, moreover, have been more opportune to talk explicitly in that paper about attributes and features of objects, defined by theoretical terms corresponding to empirical values, instead of using a vague term such as "elements of physical reality" correlated to "physical quantities." The vagueness of definitions, apart from not sounding well to the logician and the philosopher, is difficult to accept, even for those who are used to the brevity and clearness of mathematical statements.

But everything considered, we must admit that, once we accept hypothesis H and assume that the two particles at a sufficient distance no longer interact, as the criterion of reality is satisfied, the EPR reasoning is conclusive.

Furthermore, abandoning the mental attitude I am used to, I can't help but being shocked by the fact that, once the state of an individual system is unequivocally defined by the measurement of a complete set of observables, it is not possible to predict the results of a subsequent measurement, except in probabilistic terms. We could naively think that not all conditions were given to define the initial state of things; or that, during the free evolution of the system, this was perturbed in a manner unknown to us; or even that, when we speak for instance of an electron, we refer to it as we would refer to "man," in that electrons are all different from one another for secondary qualities, in the same way as a man is different from any other man, even though the class is represented by one term. Since neither the lack of determinism nor the EPR argument does in the least trouble any adherent of the orthodox point of view, I cannot but think that "something is rotten in the state of Denmark."[8]

On the other hand, the Copenhagen interpretation has the unquestionable merit of being consistent and of interpreting rationally the theoretical scheme, which in turn developed rationally, starting from the methods of the old quantum theory. This scheme has been, and still is, a striking success in framing all atomic and sub-atomic phenomena, and the empirical data demonstrates that quantum mechanics provides a deeper understanding of nature.

But if we strip Bohr's discourse of the philosophic background, I would think that what is left is equivalent to taking a purely phenomenological and operational position. As is known, operationalism requires that any terms utilized to describe the physical properties of a system are not defined unless and until a precise set of empirical operations capable of measuring them are identified.

I say that Bohr upholds the operational viewpoint when he insists on the necessity to give, in the way described above, an unambiguous meaning to the notions of momentum and position. My thesis is strengthened by the fact that he himself, rather mischievously, drew attention to the analogy between his statements and the operational definitions of length and time intervals invoked by Einstein in his paper on special relativity.

And if we neglect all emphatic statements like that concerning the "radical revision of our attitude as regards physical reality," then the essential use of operations in his reply, referring to diaphragms and slits, will further confirm my opinion.

After all, it would seem reasonable, if we strictly adhere to the text we are considering, that quantum mechanics in the current interpretation is to be considered nothing but a phenomenological theory, which captures the regularity of the connection that exists between the readings of a set of macroscopic measuring instruments at any one time and those of an analogous set at a subsequent time.

Now, let us examine another point. Of course, our experience demonstrates that the electron will behave at times like a wave and at times like a corpuscle, depending on experimental conditions. But, if we put aside what Bohr seems to consider a necessity, i.e., using classical categories such as position and velocity when speaking of the electron (and this would appear as an a priori epistemologic position), then the above empirical proof will lose impact against realism. Bohr actually refers to the attributes of an object (even though for him -- probably-- the object in itself has no meaning, but it is simply defined by the relevant attributes), and if we cannot speak at the same time of the position and the momentum, because their values are not defined simultaneously with certainty, then nobody can prevent us from thinking that this happens because the two terms do not exactly reflect the attributes of the object. From the considerations above, we would be allowed to imagine a fundamental theory that would overcome the present set of problems. Of course, this new theory should be broader than, and offer the same empirical predictions as the one it seeks to replace.

There was a time when people like Mach opposed the use of the "atom" idea, branding it as a metaphysical concept. Yet, phenomenological thermodynamics was fruitfully interpreted in fundamental terms precisely through use of this suspect idea in conjunction with statistical mechanics. The realistic attitude and the aspiration for a fundamental theory that might overcome the Copenhagen version of quantum mechanics can be mistaken for a yearning --that seems impossible of fulfillment--

for the return to a space-time description, typical of classic physics. All the more so because, in their times, opponents of the orthodox point of view, such as Einstein, Schroedinger, and Ehrenfest revealed in different shades a similar wish. And since the lesson that can be drawn from the success of quantum mechanics is that, at a certain level, the space time description proper of classical physics is no longer effective, I would like to say here that the identification of the classic conceptual scheme with realism and the search for a fundamental theory in connection with atomic phenomena are not necessary.

To this end, I wish to go back to the seventeenth and eighteenth centuries, noting how long the road was, starting from the notion of "body", discussed, amongst others, by Newton and vaguely defined to the extent possible in physics until --with Euler-- we arrive at the notion of "masspoint" which pervades all of classical mechanics. The notions, suitable for the description of the outside world as suggested by everyday experience and discussed from a metaphysical standpoint in Galileo and Newton's times, were elaborated and clarified; they became theoretical terms representing the conceptual basis of pre-quantum mechanics.

The description of an object by means of its position coordinates and momentum was successful, starting from Newton's cosmology, and was subsequently applied to other cases, notably the kinetic theory of gases. But it is actually surprising that such a description, suitable for dealing with huge objects --when compared to our scale of dimensions-- like planets, should still be valid in reference to the microscopic components of a gas. This is so unless we support the metaphysical position according to which conceptual categories that one requires to understand the outside world are given once and for all, and we claim to adjust things to our mind. If instead we recognize the "adaequatio intellectus ad rem," and agree that concepts in physics are worked out starting from a dialectic relationship between nature and ourselves, we shall expect that, by the time we deepen our knowledge of it, conceptual categories will have to be modified, without renouncing therefore a realistic position.

From our point of view, there is nothing disgraceful in this. If --via our senses-- we acquire concepts that are able to describe the macroscopic world, obviously we shall not be able to utilize them in order to identify new descriptive categories in the microcosm, of which we have no direct experience. In this context, however, there remains mathematics. Just as some time ago the telescope revealed a new world, the new telescope provided by the mathematical tool will allow us to invent and refine new categories for the description of the sub-atomic world, and in this sense mathematics already helped in evidencing the regularity of quantum mechanical phenomena.

Therefore, we must rely on mathematics, trying to discern amongst the various abstract languages the one that is best suitable for describing in a realistic way the quantum world. Of course, the new language will be anything but intuitive, if we consider as intuitive the notions that are common to all who live in our daily world. But the mathematical language is in turn intuitive when we use it as the theoretical physicist does.

     In conclusion, I think I can maintain, with reference to the EPR
paper and to Bohr's reply, that, at that time, the idea of a realistic
and fundamental theory for atomic phenomena was supportable. This theory would overcome the traditional formulation of quantum mechanics and
would be equivalent to it, from an empirical point of view. In all
probability the relevant characteristics would have involved giving up
some conceptual categories of classical mechanics. But what can we say
about this subject fifty years later? Even taking into account new developments that took place in the meantime in this regard, I feel I can
advocate the same conclusion, as long as I ignore high energy phenomena
or the quantum theory of fields. However, it is clear that nonrelativistic quantum mechanics is a limit of some more accurate description
of phenomena, and therefore we cannot ignore the lesson of the quantum
field theory and the phenomena it refers to. At this point, new difficulties arise.

     Even though we shall avoid thinking emphatically --as Bohr did--
that natural philosophy is faced with a radical change in the conception of reality, what could we say of the reality of electron itself,
when we think of electron-positron annihilation or of the fact that the
same two particles can be generated pairwise by the reverse process? If
we also ponder Newton's admonition not ascribe to an object anything
other than what does neither increase nor decrease, i.e., to endow instead physical elements of reality with a character of permanence, then
the notion of the real existence of the electron takes on an aspect
different from the usual one.

     Another fact that should lead us to reflect comes from the symme-
try properties of a system of identical particles, which denies individuality to a component of such a whole. This fact leads to consequences that can be verified already in classical statistics, where it
is necessary to resort to the so called correct Boltzmann counting, and
it is significant that symmetrization postulates should become a theorem in field theory. Finally, we cannot ignore the image of the physical electron as it is given to us by the renormalization process in
quantum electrodynamics.

     In conclusion, I think that perhaps the best way to devise a more
satisfactory nonrelativistic theory of phenomena at the atomic level is
not the straightforward one, but one arising from a reflection on the
foundations of the field theory. And I suspect that the resulting fundamental objects will be abstract mathematical entities.

NOTES

1. The logical status of thought experiments is discussed by W.Yourgrau
   in Ref.2.
2. That is, the quantum mechanics in the Copenhagen interpretation
   which Einstein and coworkers indeed call "usual".
3. See Ref.6.
4. I just quoted Socrates' statement since it implies a metaphysical
   conception, identifying in particular beauty with truth, of great

importance, which, as a heuristic principle, has stimulated research and offers an acceptance criterion for theories.

5. In the meantime, however, the question was not entirely forgotten, and we had interesting developments with regard to the foundations of quantum mechanics, and the problem raised by EPR, from the epistemologic standpoint as from the side of theoretical physics.

REFERENCES

1. A.Einstein, B.Podolsky, and N.Rosen, *Phys. Rev. 47*, 777 (1935).
2. *Proceedings of the Tenth International Congress of History of Science*, Vol.I (Hermann, Paris, 1964), p.359.
3. N.Bohr, *Nature 136*, 65 (1935).
4. N.Bohr, *Phys. Rev. 48*, 696 (1935).
5. B. d'Espagnat, *Conceptions de la physique contemporaine* (Hermann, Paris, 1965).
6. M.Jammer, *The Philosophy of Quantum Mechanics* (John Wiley, New York, 1974).
7. *Atti del Congresso Internazionale dei Fisici*, Vol.II (Nicola Zanichelli, Bologna, 1928), p. 565.
8. Hamlet, I. IV.

CONTINUITY AND DISCONTINUITY: THE EINSTEIN–BOHR CONFLICT OF IDEAS
AND THE BOHR–FOCK DISCUSSION

Valerio Tonini
La Nuova Critica
Via F. Denza,48
00197 Roma, Italy

ABSTRACT

The debate between Einstein, Bohr, Fock and others on the issues of continuity vs. discontinuity and determinism vs. indeterminism finds an epistemological solution in the principle of conjugation based on the concepts of structural relativity (1946) and of irreversibility of temporal action (1948). The acts of absorption and emission of a quantum, making up the structure of the Universe are objective in nature. Nevertheless, the structure of the global reality can never be represented in its innumerable process varieties as a unique explicative system and can be adequately tested through the conjugate and alternative use of four distinct paradigmatical classes of systems: deterministic, probabilistic, indeterministic and informational.

1. THE ORIGIN OF THE EINSTEIN–BOHR CONFLICT OF IDEAS.

Continuity versus discontinuity: This is the issue that tormented Einstein all his life, and he bitterly disagreed with the physicists of the Copenhagen school, headed by Niels Bohr, on this question. This conflict emerged from a sharp discussion that Einstein had with Bohr in 1927 during the Fifth Physical Conference of the Solvay Institute on "Electrons and Photons."

The philosophical crux of the difference of opinion between Einstein and Bohr lay in this: Einstein believed in the possibility of scientific thought formulating a more and more complete and consistent universal model of truth, while Bohr insisted on the continual change in physical conceptions and thus on the continuing need to revise our mental picture of the world by continually adjusting it to the empirical knowledge gradually attained. Yet again in 1961 L. Rosenfeld, of the Copenhagen school, in his essay entitled "Le conflit épistemologique entre Einstein et Bohr" (1) wrote:

"Einstein, like Pygmalion, is fascinated by the formal
beauty of his own mental creation and he raises it to the

371

level of a universal model, provoking a "mystification" which, on the other hand, Einstein himself found it impossible to justify when faced with the radical in-deterministic criticism of quantum physics."

Thus a whole generation of scientists found itself at a perplexing epistemological crossroads, having to decide between a still-classical conception of <u>necessary causes</u>, from which stemmed the very positiveness of scientific argument, end the discovery of something aleatory which time and again put everything in doubt. In effect, those were the years of "the crisis in European sciences", so strongly expressed in the works of Husserl and Jaspers, which have become emblematic of the troubled consciousness of the men working during that period. Einstein also experienced this drama, he was continually torn between a basic realism, which nourished his ingenious, creative imagination, and a certain rather superficial and philosophically not very clear attachment to Ernst Mach's ideas. And it is precisely because of this inner conflict that Albert Einstein's intellectual life so interests us. Making an in-depth study of it means living through all the anxieties of a century of understanding the bitterness of the words with which Einstein ended his life:

"Die letzeren, fluechtigen Bemerkungen sollen nur dartun wie weit wir nach meiner Meinung davon entfernt sind, eine irgendwie verlaessliche begriffliche Basis fuer die Physik zu besitzen." (2)

Between 1905 and 1916 Einstein had already found great difficulty in resolving some inconsistencies in his first, special relativity theory (which he himself described as "eine Kinderei") and in learning the mathematical tools of absolute differential or tensorial calculus, already formulated by the Italian mathematicians Ricci, Levi-Civita and Palatini, whose contribution to the theory of general relativity was fundamental (3). However, Einstein's philosophical position with respect to that of the group of "theoretical quantum physicists" was to give rise to dramatic moments of total reciprocal incomprehension.

2. THE E. CARTAN - A. EINSTEIN LETTERS, 1929-32

On January 7, 1930, Einstein wrote to the great Belgian mathematician Elie Cartan in order to point out the inherent difficulties in the "singularity-free" solutions of field equations:

"The worst is that our theoretical physicists do not wish to collaborate, but rather abuse me because they have no feeling for the naturalness of this approach (except for Langevin!)." (4)

Einstein had a long discussion with Cartan on what was termed the "degree of generality" of mathematical formulas, in order to be able to judge their "degree of arbitrariness" and thus to justify one particular choice out of all possible solutions. This is the radical problem in all scientific choices, which has only found a satisfactory, epistemological solution in recent years via the precise, realist formulation of the principle of conjugation, which has definitively quashed diatribes, such as Rosenfeld's, which only reflect a rigid reductive and exclusive position that no longer makes sense.

In effect, the antinomy between realist hypotheses and Machian suggestions revealed a psychological rather than an analytical problem, so much that, even in the last letters he exchanged with Max Born, Einstein spoke of the reciprocal incomprehension of the problem.

Born remarked:

> "...these last letters show how two intelligent people are unable to understand each other when discussing a concrete problem. Each of them was convinced that the other was wrong, because each of them set out from a different standpoint which seemed to him so incontestable that he could not accept the other's point of view." (5)

Paul Valery, who listened to Eistein in Paris on November 12, 1929, noted how he was guided, above all, by his intuition, "the standpoint of architecture".Paul Valery was the author of that splendid dialogue "Eupalinos ou l'architecture", where Phèdre says to Socrates:

> "Les plus sages et les mieux inspirés des hommes
> veulent donner a leurs pensées une harmonie et
> une cadence qui les défendent des altérations
> comme de l'oublie."
> "Folie!" replies Socrates.

We shall shortly see, in the Bohr-Fock discussion, how the problem assumed a concrete, and I should say vital, aspect; but perhaps, meanwhile, it is better to underline the difficulty in mutual comprehension that is already seen to emerge from the Einstein-Cartan letters, in spite of the common search for a unity of formulation of the laws of nature, aimed at resolving the antinomies which emerged all the more, the more one sought to get to the root of the problems, which, at a certain point, became "paradoxical."

Cartan had formulated his theory of "generalised spaces" entailing a "theory of systems in involution" and he wrote of this in a letter to Einstein on December 3, 1929, in which he considered his theory:

> "tout a fait appropriée au probleme que vous avez posé"
> whereas, "le degré de généralité du schema géometrique
> correspondant a votre système de 22 équations est un
> peu faible, les anciennes théories classiques de la
> gravitation et de l'électromagnétisme donnant a la Phy-
> sique un degré de généralité plus considerable. Je serai

tres heureux d'avoir votre opinion là-dessus. Si je fai-
sais fausse route, je désirerais que vous me disiez." (6)

There appears throughout the Einstein-Cartan correspondence a
certain reticence with regard to what was called the "pure theory of
gravitation," formulated by Einstein in 1916, which did not tie in well
with the new "types of physics" or "modes of being of science," which
were proposed via new semantic paths (A. Tarski) or purely mathematical
paths (the spaces of affine connection of H. Weyl, A.S.Eddington and
E.Cartan; the topology; the geometry of baths by L.P. Eisenhart and
O. Vehlen, etc.). None of these formulations succeeded in resolving the
question of why the Einstein theory of relativity was incompatible with
quantum physics.

3. THE COPENHAGEN SCHOOL AND DIALECTIC PHYSICS

What was Einstein searching for? The link between the general
invariability of the laws of physics and the space-time continuum. This
is the central theme of his work "Die Grundlage der allgemeine
Relativitaetstheorie" (1916). The fundamental idea of the general theory
of covariant quantities had to be reached via the definition of certain
"tensors," characterised by the fact that the transformation equations
of their components were such that they expressed general covariances
without having to introduce any further hypothesis to define
physical reality and to explain events, or, in other words, the
facts that occur in it. But as E.T. Bell was to point out, no
satisfactory justification had been given for connecting the
implications of mathematical logic with the physical world. Neither the
logistic programme developed by B. Russell and A.N. Whitehead in
Principia Mathematica (1910-13), nor the theory of ramified types,
nor the intuitionism of Brouwer, had revealed the capacity to solve a
problem in which the developements of Machism had produced a chaos that
the neopositivism of the Vienna Circle would never be able to remedy, in
spite of the valiant attempts made by O. Neurath and R. Carnap in the
Encyclopedia of Unified Science (Chicago, 1938). Nevertheless, we
must acknowledge the great merits of the Vienna Circle neopositivist
movement. Without this movement, science would not have been able to
continue that fundamental self-criticism which finally led to the
construction of a strong systemic epistemology designed to give a
real physical meaning to the formulations of that structural
relativity which, in my opinion, allowed the determinism-indeterminism
antinomy to be resolved in realistic terms (7). The arduous task of
solving this problem was carried out during the decade immediately after
the Second World War; in this connection, I maintain that the discussion
between Bohr and Fock is exemplar, and I should like to recount it
briefly. At that time, between 1952 and 1955, a lively debate had begun
between Soviet scientists and the proponents of "western physicalism."

This fact particularly interested me, because I discovered how certain
realist interpretations that the Soviet scientists gave, both to the
theory of relativity and to quantum theory, coincided surprisingly, and

I should say almost to the letter, with the ideas that I myself
developed during that period in opposition to the old school of Machian
empirio-criticism (8).

My first encounter with the Soviet scientists took place in Zurich
in 1924, when I happened to discuss with M.E. Omelyanovskij his pamphlet
"Against Subjectivism in Quantum Mechanics" (1953). This meeting
therefore occurred three years before the discussion between Bohr and
Fock in Copenhagen, which must be considered a fundamental milestone in
a fully comprehensive revival of what we would term today the
"objective" sense of the properties of the matter of which the universe
of observables is composed.

Omelyanovskij took as his starting point the principle that, by
"making use of the philosophy of dialectic materialism," one should
oppose the

> "idealism that Heisenberg and Bohr, Schroedinger and
> Dirac set out from, and which, without any doubt, has
> had a fatal influence on that quantum theory that they
> have created or elaborated." (9)

> Omelyanovskij also reproached:
> "V.A. Fock who, in his article The Fundamental Laws of
> Physics in the Light of Dialectic Materialism , examin-
> ing the discussion between Einstein and Bohr on the
> physical meaning of the wave function, only criticised
> Bohr's erroneous idealistic observation on "physical
> reality"; but Fock not only did not criticise Bohr's
> very principle of complementarity, from which his reason-
> ing on physical reality derives, he even tried to link
> it to the thesis of dialectic materialism."

From my conversation with Omelyanovskij, I realised that he was
principally concerned with keeping in line with the ideological and
political establishment and not with a strongly scientific position.
However, he seems to me to have shown some approval of the strong sense
of operative and systematic realism which inspired my interpretation
of Heisenberg's principle of non commutation of the operators.
Furthermore, from his remarks on Fock, I understood that the latter was
fully in command of his subject; thus, I tried to be immediately
informed of Fock's discussion with Bohr in Copenhagen in 1957.

4. THE BOHR-FOCK DISCUSSION

Omelyanovskij had observed that

> "Under the sign of the principle of complementarity,
> the Copenhagen school had not so much developed as ob-
> scured quantum physics" (10), and he lamented the fact that "The
> little knowledge that de Broglie, Schroedinger and
> the other physicists had of dialectic materialism did

not permit them to remain in a materialist position with
regard to the knowledge of microphenomena." (11)

"Unfortunately," Omelyanovskij said, the "idealist Copenhagen school
lays down the law in present-day bourgeois physics." However, I agreed
with Omelyanovskij's remark that

"Until the physical meaning of the quantities that
figure in a theory of physics is clarified, this theory
is not, substantially, a theory of physics, but only
its mathematical schema." (12)

since I myself was involved in an operative and structuralist research
aimed at proposing a renewed logica major seu applicata seu
materialis, which was intended to be a real processual semantics, at a
time when nearly all western thinkers thought that there only existed
the logica minor or logica formalis, which had become
mathematical logic. This is a subject that is now being dealt with in
the biological and neuronic field, but which was, at that time, totally
premature.

Fock understood -- and this is what Omelyanovskij reproached him
for -- that the characteristic trait of quantum physics was the fact
that

"We must already not simply talk of the object, but of
an object which has a reciprocal link with a determinate
type of apparatus." (13)

In the account that Fock gave of his visit to Copenhagen, there
emerged a precise sense of the value of experimentation and of apparatus
(the operative definition of the observables) and, furthermore, of the
difficulty in translating the principle of complementarity into
something that could really constitute a bridge between the experimental
and the theoretical planes.

Here is Fock's account of his discussion with Bohr   :

"As I was able to ascertain from my personal conversations with Niels Bohr, in effect his position is far
nearer the materialist conception than may appear
from reading his works on the fundamental problems of
quantum physics. First of all, Bohr maintains that
one must take nature such as it is. He expresses his
definite disagreement with the positivist standpoint
and fully recognises the objectivity of the properties
of atomic objects. As far as terminology is concerned,
Bohr is ready to renounce the use of the term 'uncontrollable interaction', which he considers unfortunate. Bohr also agrees that the general principle
of causality must be distinguished from determinism of
the Laplace type, and that this determinism alone contradicts the laws of atomic physics."

Fock added:

"The real meaning of the wave function began to be explained in Max Born's works on the statistical interpretation of quantum physics. The fundamental meaning
of the concept of probability was clarified. Niels
Bohr's ideas played a very important role in the clarification of this problem, particularly his ideas on
the fact that the quantum-mechanical description of the
properties of an atomic object should be combined with
the classical description of the means of observation
(of the experimental apparatus). But the excessive
emphasis on the function of the apparatus gives a
reason for reproaching Bohr for underestimating the
need for an abstraction and for forgetting that the objects of study are the properties of the micro-object
and not the characteristics of the apparatus. In effect
the properties of atomic objects, such as charge, mass,
spin type of energy operator and law of particle interaction with the domain are completely objective and may
be abstracted by means of observation, whereas, on the
other hand, such properties require new quantum-theo-
retical concepts to be formulated. This refers particularly to the formulation of the problem of many bodies.
Bohr's terminology also gives rise to incomprehension.
Thus he speaks of an "uncontrollable interaction," although an interaction seen as a physical process, is
always controllable. Bohr speaks of "uncontrollability"
to cover the lack of coordination that results when
using classical concepts outside their field of application. Furthermore, we can mention the antithesis
posited by Bohr between "principle of complementarity"
and "principle of causality". The principle of causality must be understood at the assertion of the existence of the laws of nature and, in particular, those
connected with the general properties of space and time
(finite speed of the propagation of actions, impossibility of acting on the past). With this interpretation,
quantum physics not only does not contradict the
principle of causality, but also gives it a new expression and extends its application to the laws of probability.
    The novelty of Bohr's ideas and their expression,
which is difficult to comprehend because he uses a
terminology which is not always fortunate, have given
rise to many misunderstandings and misinterpretations,
in the spirit of positivism. The most extreme positivist position was taken by P. Jordan; more serious
physicists like M. Born, W. Heisenberg and others,
were very attracted for a while by positivist concept-

ions, but now they are gradually moving away from them.
Thus, in one of his most recent works, published in a
collection in honour of Niels Bohr's seventieth birthday, W. Heisenberg already admits the objectivity of
the concept of quantum state." (14)

It is interesting to remember that during the conference of the
Academy of Science in Russia on "Philosophical Questions in the
Natural Sciences" (Moscow, 1959), where Fock revealed the results of his
discussion with Bohr, another Soviet mathematician, A.D. Aleksandrov,
pointed out how the fundamental relations of the theory of relativity
coincided with the structural interpretation that I myself had
formulated in 1945-1946. Thus, in the issue of Nuova Critica (9,
1960) where I published the translation of the two papers by V.A. Fock
and A.D. Aleksandrov, I raised the question of a now-necessary
comparison between the operational and realist rationality of a
systematic epistemology, in which I felt culturally immersed and which I
would like to describe as being essentially Italian, and the Soviet
culture, which aimed at going beyond "the pure conventionalism of the
modern Machists," (15) by expressing the concept that even a
micro-object is not an isolated "classical corpuscle," but has a
"singularity" interconnected with the structure that incorporates
it.

5. THE MICROPHYSICAL REALITY

The emergence of new physical quantities, quantum numbers, a more
and more refined theory of measurement, investigations into the purity
of the materials employed, and, on the other hand, the formulation of
universal constants such as the speed of light c, the elementary charge
e, and the electron mass m, required a logical and formal conceptual
representation that was capable of overcoming the so-called
uncontrollability in Niels Bohr's principle. Already a particular
character of objectivity and individualisation was established in
Pauli's principle of exclusion (16), and in the diversity between the
statistics of Pauli-Fermi-Dirac and that of Bose-Einstein. I would,
therefore, judge the conclusions on the microphysical reality reached by
Fock, after his discussion with Bohr, to be truly definitive, in the
cogent sense of the word. They can be summarised in quite convincing
terms as follows:
- 1. "The properties of the objects are always manifested in interaction
with other objects, particularly with the means of observation." (17)
- 2. The exactness of the description is limited by Heisenberg's
interactive relations.
- 3. The double (wave or corpuscular) "manifestation" of the electron
indicates that an elementary particle like the electron has its own
valence in relation to the situation in which it finds itself and
the experiment to which it is subjected. This double valence
characterises the behavior of the particle in its reaction to a
determinate type of apparatus.

- 4. The physical observable therefore depends on the reactivity of properties intrinsic to the object that react to the external conditions of controllability.

These considerations are clearly and obviously realistic. The classical method considered objects and processes in themselves, independently of the method and apparatus used to study them. Today we observe what is provoked. Fock recognises that, in this sense, Bohr is not only an outstanding physicist, but also an "outstanding thinker" who has characterised, with the principle of complementarity, the limitations which must be put on the mechanistic determination of abstract physics. Quantum physics excludes "Laplace's determinism" with regard to the interaction of microphysical particles, but "the principle of causality remains in so far as it is a unitary principle enunciated by the human mind." (18)

> "Complementarity refers to the limitations to be placed
>
> on the classical description of phenomena, but since the quantum formalism operates through the use of probability concepts, it does not renounce, at all, the description of the properties of individual objects."
> (19)

Thus the wave function expresses the distribution of the probabilities of a certain variable quantity and foresees what is potentially possible. This probability is determined by the internal properties of the object itself in the given concrete situation, and in this sense de Broglie's vain attempt at a "double solution" (20) is nonsense.

In this autobiography, published in 1949, Einstein wrote:

> "One avoids giving a simple "yes" or "no" answer to the question whether a function ψ of the quantum theory represents an effectively real situation, in the valid sense for a system of material points or for an electro-magnetic field. Why is this?" (21)

Why is there so much uncertainty in the different ways of conceiving the evolution described by the ψ function? Well, in an article "Déterminisme et Indéterminisme" published in Scientia in 1948, I analysed the interpretation that L. de Broglie had given to the probability wave, and noted a weak point which, in my opinion, provided the key to the right significance that must be attributed to the so called associated waves and to the indetermination that they are presumed to indicate. In effect, the operator in his first measuring operation does not "localize" a corpuscle, it does not "individuate" the corpuscle in a certain position in space: it only provokes a physical event, an observable action, which is an elementary act of absorption or emission of energy. Where, on the contrary, this operation is referred to as the localisation of a corpuscle at a point in space, this way of speaking is only metaphorical because, bearing in mind the operational principle whereby one needs to determine the values of at least two conjugated variables to define the physical state of any quantity, a single measuring operation cannot in any way individuate something that can be called "the physical object." When L. de Broglie, after the first measuring operation, asked himself: "What has this corpuscle localized in the first measuring operation finally become?" he was drawing a deceptive conclusion because, in effect, a corpuscle, in that first operation, had not been individuated at all. That first measuring operation had simply produced an event (action) in time dt, in a domain ds, which, when connected with the whole theoretical and experimental structure used, we consider indicative of the possible presence of a particle in a space-time "cell."

This event (action) creates possibility of other successive events, a possibility that may be represented by imaginary waves which, propagating at an imaginary speed c, spread out from that primitive event into the whole surrounding space. It is obvious that asking what became of the first "event" makes no sense whatsoever. Nobody would ask himself such a question, whereas it is natural -and in a sense legitimate- to ask oneself about the evolution of an elementary particle, an electron, or a photon, only when other "conjugated" observations, connected with a whole conceptual and experimental system, permit one to deal with "physical entities" as "nomological objects," individuated within the limits of the unavoidable errors of measurement that each observation entails.

The action of observing, the "event" provoked by a measuring act, is really unique, unrepeatable in time and space, and irreversible. This completed action is the origin of an infinite possibility of further events. But, if it is considered in isolation, it is utterly absurd to expect it to determine the future. Neither should it be forgotten that no physical state can ever be even approximately defined with a single observable; it requires at least two parameters (the duality principle).

Once an action has occurred, it may provoke a whole series of possible future events that may be linked to it via a relation of wave dispersion (with an imaginary speed c) which transmit the possibility of observable events into the surrounding space, the infinitesimal cell δs, where, in the infinitesimal division of time δt, the event was produced.

Thus we find ourselves in the presence of a possibility, which may be represented in the form of a dispersive wave, of a becoming or of a specifiable trajectory, and not in the form of a rigid mechanical determinism, since a single parameter could never define an evolution, and the Heisenbergian uncertainty, inherent in every measuring operation, prevents the simultaneous definition of conjugated parameters. One is therefore obliged to introduce a distribution function which represents a probability density. The associated wave of de Broglie is only an imaginary wave, the expression of possibilities that spread out in time: A wave of dispersion which does not bear energy and is deprived of a real physical existence. It can never give us, or at least only probabilistically, the future situation of a "corpuscle," seeing that it was not established for this purpose; it has nothing to do with energy states and dissolves itself by propagating with an unreal phase speed, faster than the speed of light. One can only imagine that it accompanies, like a dispersive halo, the effective trajectory of the "corpuscle," a trajectory that can be made to coincide with the group speed. The wave function ψ expresses the probability, after a certain time, of finding the system in a particular physical condition, which is in turn determinable, with a certain approximation, via a new observation. Nothing more, but, precisely since wave equations always describe a process in relation to time, they can always describe a certain order of causality, or rather the possibility of future events stemming from a certain event which can never be punctualized in an instant, though it can be individualized in a certain infinitesimal space-time cell. The antecedent event is the <u>necessary</u> condition for the <u>possible</u> determination of future events.

The <u>microphysical reality</u>, therefore, remains rationally defined by the fact that every physical observable is the product of an action: an act of absorption or emission of energy.

6. THE IRREVERSIBILITY OF TEMPORAL ACTION

Every observable (variation of state), being identified with an action, is conditioned by all the operative modes of relativity and uncertainty that accompany every action, even the most elementary (even Planck's quantum of action). But the action which gives rise to observations is, in itself, what it is, objectively; even if isolating an act of observation from the series of operations that define it, still, a conceptual operation, since every action, whether it is observed or not, is still the product of many concomitant and convergent actions. Every action is never isolated and isolatable, it is not independent; but many thousands of years of experience of innumerable interconnected actions allow one to express, without any doubt whatsoever, the following principle: The acts (elementary) of action (acts of the absorption or emission of a quantum of energy) of which all physical facts (the structure) of the universe are composed have an <u>objective character</u>.

By "objective character" we mean that every action or event is <u>real</u>, in the sense that, after the event, one cannot behave as though it had not occurred. The action is independent of the reference and is what it is. Even for the smallest observable action, the pair of variables necessary and sufficient for describing any action, is conjugated in the Hamiltonian sense underlined by Heisenberg's indeterminacy relations.

The many thousands of years of experience that permit the preceding realist statements are not only a mental process, but they belong to the whole of present-day physiology, at least since the beginning of the developement of a neuronic system. This is the field of inquiry in which new microbiological and neurological knowledge is emerging.

Science is a science of real actions and, in rooting itself in the experience of the real world, it is possible to formulate an understanding of reality which is expressed by the enunciation, in the sphere of the epistemology of systems, of the <u>principle of conjugation</u>, which in synthesis is formulated as follows: The structure of reality is never representable in its innumerable processual varieties by a single explicative system, but it is adequately controllable via the conjugated and alternative use of four distinct paradigmatic classes of systems: (1) determinist (regarding linear processes for which the decomposition and recomposition of the effects is admitted); (2) probabilist (sets that follow the laws of large numbers, to which the principle of equal probability is applied a priori); (3) indeterminist (systems that take into account the non-commutation of the operators); (4) informational (systems self-controlled by feedback and internal memorized information processing).

Let us retrace our steps a little: Einstein, in his discussion with Bohr, raised the question of the possibility of foretelling an individual phenomenon in the atomic field. Einstein maintained that this would only be possible if one challenged the applicability of Heisenberg's inequalities. But since this possibility does not exist, he judged the quantum theory to be "incomplete," without, however, being able to say why.

In effect, the vital sense, dictated by the experience of life, of the potentially possible, that is, of the operative activity that realizes determinate potential virtualities, had been completely lost. At the root of this powerful expression of classical human rationality, there must be an unequivocal awareness of the irreversibility of temporal action.

But then why such difficulty in holding together Einstein's determinist continuum and indetermination? Perhaps the reason lies in this: The experience of daily life, which confirms the principle of causality by the realization that we are not able to influence the past, was rejected as unscientific.

Today we are aware that only on a macrocosmic scale (that is the human dimension) can a physical phenomenon be described as an isolate trajectory. A physical phenomenon is always a complex process within a structure that involves it. This process can be represented, on abstraction, by choosing the opportune variables, via the different theories and models that we have grouped together into four paradigmatic classes.

Einstein sought in vain to generalise the principle of causal continuity by increasingly perfecting the tensorial calculus; but today we have clearly understood how the "trajectory", or rather the evolution of any "quantity", is given by the confluence, instant by instant, of innumerable lines of the universe and that the processes run through reality reacting to the incalculable implications of every diverse existences. I think it is still necessary to refer to a solid epistemological concept which clearly distinguishes the "structure" of reality from the "formal systems" which adequately represent it. The developement of this solid concept is the task of the present post-Husserlian scientific phenomenology.

NOTES AND REFERENCES

1 Rosenfeld's paper was presented at the Colloque de l'Académie Internationale de Philosophie des Sciences (16-18 Oct. 1961, Paris) entitled "Philosophie de la Physique", Brussels 1962. The participants were the physicists L.Brillouin, L.Rosenfeld, I. Prigogine, J.Geheniau, J.L.Destouches, O.Costa de Beauregard, J.P.Vigier and A.Tonnelat; the mathematicians M.Frechet, H.Freudenthal, G.Bouligand, and P. Bernays; the philosophers of science F.Gonseth, V.Tonini, S.Dockx, D.Dubarle, G.Hirsch, A.Metz and R.Poirier.

2 A.Einstein's last writings are contained in the volume Cinquant'anni di relatività, Editrice Universitaria, Firenze, 1955; editor M.Pantaleo. See V.Tonini, "Il realismo in fisica" in La Nuova Critica 1, 1955; "Il testamento scientifico di Einstein e la filosofia della fisica oggi", La Nuova Critica, 50,(1979).

3 See V.Tonini, "Le interpretazioni della nuova fisica: la teoria della relatività", paper presented at the National Conference of the Società Filosofica Italiana on "La filosofia della scienza in Italia nel Novecento", Bergamo, 15-17 March 1985; proceedings edited by E. Agazzi, to be published by Angeli.

4 E.Cartan and A.Einstein, Lettres sur le parallélisme absolu, 1929-1932 (Académie Royale de Belgique and Princeton University Press, Brussels, 1979), p.110.

5 A.Einstein, H.Born and M.Born, Briefwechsel 1916-1955, kommentiert von Max Born (Munich, 1969; Italian translation: Einaudi, Torino, 1973), letter dated 20.1.54; p.253.

6 E.Cartan, op.cit., p.22-30.

7 V.Tonini "Déterminisme et Indéterminisme", Scientia, March-April, 1948; Fondamenti metodologici della relatività strutturale , Centro Internazionale di Comparazione e Sintesi, Rome, 1950

8 V.Tonini, "Relatività non einsteiniana", Rendiconti del seminario della Facoltà di Scienze , Università di Cagliari, Vol.XVI, 4 1946; "Relatività strutturale", id. Vol. XVII, 4, 1947.

9 M.E.Omelyanovskij, "Il materialismo dialettico e il principio di complementarità di Bohr", La Nuova Critica 1, 1955.

10 Ibid., p.7.

11 Ibid., p.9.

12 Ibid., p.10.

13 V.A.Fock, Vastnik Leningradskogo Universiteta, No.4, 1949, quoted in Ref.9, p.13.

14 V.A.Fock, "L'interpretazione della meccanica quantistica", La Nuova Critica 9, 1960.

15 See D.I.Blokhinzev, Foundations of Quantum Mechanics, Moscow, 1949, quoted in Ref.9, p.23.

16 See V.Tonini, "Sul significato logico-operativo del principio d'esclusione," Scientia, April 1953.

17 V.A.Fock, Ref.13, p.71.

18 Ibid., p.88

19 Ibid., p.86

20 L.de Broglie, <u>Une tentative d'interprétation causale et non linéaire de la mécanique ondulatoire (La théorie de la double solution)</u>, Gauthier-Villars, Paris, 1956.

21 A.Einstein, <u>Autobiografia scientifica</u>, (Boringhieri- Einaudi, Torino 1958.

HOW ITALIAN PHILOSOPHY REACTED TO THE ADVENT OF QUANTUM MECHANICS IN THE THIRTIES

Vincenzo Fano

Dipartimento di Filosofia
Università di Bologna, Italia

ABSTRACT

An overview of the reactions of Italian philosophers to the advent of quantum mechanics in Italy between the two world-wars is presented. The different attitudes of the idealistic, positivist and marxist thinkers are examined. It is shown that, apart from A.Gramsci, M. Losacco and A. Pastore, Italian philosophers adopted an idealistic Kantian disposition toward the new discipline, in line with von Weizsäcker's interpretation.

1. INTRODUCTION

The crisis in classical mechanics caused many philosophical concepts to be introduced in physics, in particular concepts connected with philosophies of prevailing subjectivistic primacy. One often speaks about observer and observables, consciousness of the experimenter, or about complementarity of conceptual systems applied to the same physical reality. This prominence of the subject in the new theories of quantum mechanics and of relativity has been pointed out by physicists and philosophers alike from the very beginning.

In a 1941 paper,[1] Weizsäcker, an eminent physics pupil of Heisenberg with profound philosophical interests, attempted to demonstrate the connection of quantum mechanics with Kant's philosophy.

According to Weizsäcker, the whole of classical physics is "de facto a priori" with respect to quantum mechanics.[2] In fact, physics does not know anything about the atom in itself and only acquires some knowledge of it by means of experiments described classically. Such classical a priori is analogous to the intuition of space, time, and causality, which makes physics a possibility in Kant's gnoseology.

Quantum physics has originated a contradiction between corpuscular description and wave-like description of particles, which is solved by stating the unknowability of the particle in itself,

385

G. Tarozzi and A. van der Merwe (eds.), The Nature of Quantum Paradoxes, 385–401.
© *1988 by Kluwer Academic Publishers.*

opposed to its possible subjective categorization as corpuscle or as wave.

The analogy between Kant's philosophy and quantum mechanics can also be further exemplified. The relationship between physics and chemistry is the same as the antinomy of atomism; if we want on the one hand to know the properties of matter through the interaction of its smallest parts, namely the atoms, then these parts lose their space quality, so that speaking about their internal structure loses all meaning. On the other hand, if we really want to know what happens inside one atom, we have to destroy it, and thus we can not reconstruct the matter of perception. [3]

On the other hand, as Weizsäcker points out, Kant's gnoseology lacks a principle similar the Heisenberg uncertainty principle. [4] This is where the quantum mechanics a priori and Kant's a priori differ most profoundly. Indeed, the former has no logical necessity, while the latter is absolute and irremovable; it is not as if "every possible experiment 'were to be' described classically", but rather "every real experiment we know about 'is' described classically, and we wouldn't know how to do it differently" [5]. In other words there is no transcendental demonstration of the a priori character of the classical mechanics, but only an a priori relative to quantum mechanics. Similarly, one could define classical physics as an a priori for perception. [6]

Kant was not in the position to predict such a conceptual complexity, consequently it might be advisable to substitute the concept of a priori in his gnoseology with another one suitably weakened. Weizsäcker does not elaborate on this issue, and concludes as follows:

"From a purely physical point of view, Kant's 'Copernican revolution' merely brings about a change of names, from which nothing follows. Only quantum mechanics achieved the same level of abstraction through a physical route, and made use of the new freedom afforded by this for the formation of physical concepts, 'to create a physics which can no more be interpreted in a realistic manner'." [7]

We thus see that quantum mechanics becomes not only a confirmation of Kant's system, but also a sufficient condition for its further move toward the subjectivism.

It is our opinion that the philosophical reaction of Weizsäcker to quantum mechanics is paradigmatic, and we set out to search for it in the Italian philosophical literature of the thirties. We will see that attitudes similar to that discussed above were indeed present, yet with differences; however, completely opposite views or widely differing opinions were expressed as well.

Before we proceed with the historical review, we want to point out a difference of philosophical style which distinguish the Italian authors of that period from our contemporaries. In those days what was missing was a view of the trends of philosophical and scientific thought which were developing not very far from one's specific field of endeavour; everybody went his own way without taking the trouble to quote authors and ideas which might be similar or opposed to one's own ideas. One might well say: "To each philosopher one philosophy". This has caused our research to be more difficult, and this survey incomplete. To the best of our knowledge, for instance, neither B. Croce nor G. Gentile ever talked about quantum mechanics, and consequently we will not deal with these authors.

2. IDEALISM

Let us begin by discussing the strictly idealistic reactions to quantum mechanics. We cannot talk about an "epistemology of idealism" proper, but only about a philosophical conception of the aims of scientific activity. Idealism does not have a say as to the foundation of the various sciences, but this does not mean that it does not have a philosophy of science of its own.[8] However, quantum mechanics is a very inviting lure for idealistic philophers; the devaluation of reality in itself, which follows implicitly from the wave-particle dualism and from the Heisenberg uncertainty principle, can supply a solid confirmation for a thoroughly subjectivistic gnoseology; indeed there have been authors who have taken advantage of this possibility.

G. De Giuli, an author interested in the problem of science and a collaborator of the journal "Scientia", one of the few Italian magazines of international scope, has expressed his ideas in these terms. As early as 1931, this author published a note[9] in the journal "Rivista di Filosofia", which begins by strongly advising philosophers to deal with the most recent discoveries of science, inasmuch as they are not inclined toward positivism, but rather toward spiritualism, idealism, indeterminism, even mysticism.[10] De Giuli goes on victoriously: Even in the thought of Einstein and Bohr -- two authors who are almost invariably quoted together, even though at this very moment their disagreement about the concept of physical reality was rampant -- matter does not exist, and all the contradictions connected with the dualism between matter and spirit are annulled in a purely idealistic monism.[11]

The author refers here to the book by Eddington, "The Nature of the Physical World" and underscores his idealistic view. The materials for the construction of the world are nothing but relationships from which space, time, energy and motion, that is exclusively spiritual elements are derived. Probably, what the author has in mind when talking about relationships are actually measurements; furthermore, space, time, and motion are considered subjective elements in line with Kant's thought; on the other hand, Einstein celebrated $E=mc^2$, the form of which was known to everyone, albeit its meaning escaped the comprehension of most, was seen as a reduction of the whole matter to energy. While matter is certainly a material element, energy is "unmistakably spiritual". De Giuli follows again Eddington and states that the spirit is what constructs the world; even though the dichotomy between existence and non existence of matter cannot be solved, yet:

"what is sure is that the physical world is a world seen
from inside, measured by devices which are spiritual, and
which are ruled by spiritual laws; the substrate of
everything is of mental nature."[13]

The review is concluded with one more devaluation of science;
physics has achieved today this idealistic conclusions, while
philosophy had anticipated them two centuries ago with Hume; idealism
is certainly not a recent discovery, so is it not as if science had
been once more preceded by philosophy?

We see that in De Giuli's position the point of view of quantum
mechanics from Kant's perspective is brought to its extreme idealistic
consequences, while the awareness of the foundation problems is much
less accurate than Weizsäcker.

Our author confirms his position when reviewing a translation of
a book by Jeans.[14] In this case he wants to stress the need for
science to turn to philosophy, rather than focussing on the idealism
implicit in modern physics; indeed one has come to realize that truth
cannot be merely scientific truth or philosophic truth, but must
involve the man as a whole. Therefore, science and philosophy are not
allowed to ignore one another, but rather philosophy must take into
account the results of science, while science must deal with problems
of universal character, i.e., philosophical problems.[15] As usually
happens, when an internal collaboration between science and philosophy
is attempted, also this proposal is to the disadvantage to either one.
Until a synthesis is sought which does not allow an exchange of
principles between the two partners, both science and philosophy are
able to preserve their autonomy. But the very moment one desires that
a direct deduction of scientific results from philosophical
principles, or vice versa, is feasible, one of the two partners is
reduced to the other, if not completely then at least for the part
which has interacted. According to De Giuli, science is reduced to
philosophy; philosophy indeed must only take into account the results
of science, while the latter necessarily is connected to philosophical
problems.

The latter issue is taken up again by De Giuli in another
review.[16] This time the book reviewed is N. Abbagnano's "La fisica
nuova. Fondamenti di una teoria della scienza". As is well known,
Abbagnano is a pupil of Aliotta, one of the thinkers most concerned
with the problems of science, and his first approach to philosophy, in
the 1930's, takes off from epistemological questions. Abbagnano
analyzes the relationship between science and philosophy and reaches
the conclusion that the two disciplines must progress separately;
science poses limits to the transcendental claims of philosophy, and
philosophy searches, according to Kant's prescriptions, the a priori
conditions which make science a possibility. De Giuli claims that this
distinction is deceptive. Indeed, to determine the conditions which
make science a possibility amounts to giving limitations to science,
hence to making a philosophical problem out of a scientific problem.
Similarly, to state that science imposes problems on philosophy

amounts to recognizing the dependence of the former on the latter. However, later on De Giuli softens his reductionist views; in order for the scientist to have faith in philosophy, the latter cannot possibly go beyond science to become a super science, but must exhibit the friendly aspect of a science which, like the others and with the others, albeit from a different point of view, endeavours to supply man with the possibility to solve his problems.[17] The author stresses again the idealism of Eddington, Heisenberg, and Jeans, and disbelieves the claim of independence from the dichotomy idealism-realism of Abbagnano's thinking.

Although in this note the exploitation of quantum mechanics in favour of idealism is not explicitly declared, it is very easy to read it between the lines.

A similar position was expressed by F.Pagano in a lecture held at the Biblioteca Filosofica of Palermo and titled "Novecento".[18] He does not explicitly want to bring grist to the mill of idealism, yet he has a very idealistic concept of quantum mechanics. According to him, the 1900's are characterized by a thorough rationalism which appears manifest in all the fields of culture. Thus

"physics is completely absorbed into mathematics ... neither electrons nor waves exist ....[19] The universe is the thought of the physicist himself.[20] ... The electron ... asks the physicist: why take all the troubles to know my true reality? I will be as you like me: discontinuous in phenomena of photoelectric type, continuous in interference phenomena."[21]

The ultimate aim of Pagano, however, is to belittle science in favor of philosophy.[22] This continuous devaluation of science by idealistic philosophers must nevertheless be adequately comprehended from a historical point of view. As Agazzi pointedly remarks:

"The neoidealism of Croce and Gentile expressed itself as a very legitimate claim with respect to science-oriented positivism which loaded on science and on its methods the whole value of knowledge, debasing philosophy to simple quarrel of ideas, to inane effort to penetrate the 'unknowable', under the pretence of 'knowing' there, where, at very most, only room for fideistic stands or sentimental consent was left."[23]

In view of all this, no wonder philosophy should feel the need to claim its role usurped, retrenching the value of science and appealing to spiritualistic and idealistic thesis.

A more complex stand is taken by De Ruggiero, a pupil of Gentile, knowing scholar of the problem of science and of the concept of European and American thought. He was responsible for a column in the journal "La Critica",[24] were he wrote about various issues of contemporary philosophy, among them also relativity and quantum mechanics.

De Ruggiero's paper starts again with the advise that the philosopher should lift his eyes from his papers and take some notice of the "extraordinary bustle" going on in contemporary physics.[25] It appears that physicists have taken leave from good old prudent positivism and have opened up to philosophical concepts. This new interest for speculation is due to the sharpen ing conflict between continuous and discontinuous description of phenomena. The two possible approaches have always stood facing each other inside physics, either claiming to provide the ultimate explanation both of matter and of radiation. However, the unification of the concept of matter and radiation has caused a superposition of the two incompatible descriptions. At this point there are two possible solutions. The first is of Kantian style and recognizes the antinomy between continuous and discrete, as related to a mathematical antinomy of the "Critique of Pure Reason". No matter how long the analysis is carried out, the thought invariably finds itself facing a repetition of the aporia, which means that we attribute a necessary way of thinking to things themselves. This point of view is followed by phenomenalists, notably by Heisenberg, who is, however, completely oblivious of Kant.[26] The author does not seem satisfied with this alternative, even though he never explicitly say so.

The second alternative consists in stating the problem in metaphysical terms. To what extent is it possible to imagine a physics devoid of substance, of relationships, in other words, a physics either exclusively continuous or exclusively discrete? De Ruggiero appears to give his preference to the former possibility, on condition that substances are inserted at a later stage. But in this case, how can "rapports sans supports" exist? The answer is easy: Contemporary physics appears to leave more and more space to a sort of internal teleology, hence to the prominence of the whole over its constituent parts.[27] A new, more versatile picture emerges, where novelty, originality, creative energy have a space.[28] In other words, this physics

> "meets our philosophical requirements in a different, indirect way ... That is by this dismantling the structure of the elementary ingredients of the physical world, which the old science had made too rigid and compact, it allows to attribute even to the very lowest layers of reality some of those characters which, in their most developed stages, we express with the names of sensibility, memory, representation, consciousness, thought."[29]

But, we could ask ourselves, what keeps together this internal teleology of physical phenomena? De Ruggiero answer is to be found in the conclusion of his paper:

> "It is not so much the uncertainty of the atom as such, but rather its characteristics of being variously

determined according to the nature of the energy field, which yelds material for considerations of idealistic type."[30]

In other words, behind all physical phenomena we see the regulating mind, which is no more an external miracle, as in the atomistic mechanicism, but is endowed with an inner action on the various levels of cosmic life.

De Ruggiero appears concerned with the foundational problems of quantum mechanics, and in particular he gives a very modern digression on the history of the wave-particle dualism and certainly shows he has caught the essénce of the crisis of physics in the ontological incompatibility between continuous and discrete. However, he also attributes a spiritual "quid" to the concept of energy and, instead of focussing, the way De Giuli had done, on the role of the observer in the construction of the physical world, he points out that it is possible to interpret idealistically a purely, i.e., substanceless, mechanics, in which the field effects ruled by the sole possible substance, the thinking self, would dominate.[31]

We can thus conclude by endorsing the judgement Tarozzi[32] has given of the reaction of the Italian neoidealism to quantum mechanics: The most knowing philosophers soon detected the subjectivistic latency of the new theory.

3. POSITIVISM

As Garulli has clearly demonstrated[33], the epistemological debate which took place between the wars was not limited to idealistic philosophy. For instance, G. Tarozzi, A. Pastore, and A. Aliotta significantly contributed to enrich the scope of the Italian philosophy of science. Although most of these authors were related to the positivist tradition or to F. De Sarlo's realistic school, their analysis of the single foundational problem do not appear particularly precise. It is true that they generally show a wider scientific experience than their idealistic contemporaries and a first-hand knowledge on the latest discoveries of science, but they are not immune from philosophical far-fetched inferences, nor from improper exploitation of scientific results.

Annibale Pastore was originally a literary scholar; he studied logic as a pupil of G. Peano and was a teacher of A. Gramsci and L. Geymonat. In his case it is appropriate to talk about a real modern epistemology, since he attacks the problems of science armed with a good knowledge of mathematical logic and physics.

The work we find most closely related to our topic is the collection of articles "La logica del potenziamento",[34] where the title must be read in the sense of a search for a logic of the acquisition of knowledge. Pastore tries to formalize this problem and blames Peano for conceiving logic only as discursive thinking and not intuitive thinking. The logic of strengthening, in a certain sense,

might correspond to what is now called the logic of discovery. The
acquisition of the single elements of thought poses at least two
problems: (a) What ontological form do these elements have; (b) in
what relationship do they stand to the knowing subject. Pastore's
answer to the first question is a universal relativism, in which all
entities are conceived "sub specie relationis".[36] As to the second,
he introduces the distinction between universe (U) and discourse (D).
U represents the intuition of primitive systems, the real logical
structure of thinking, whereas D represents deduction, the syntactic
part or the mathematical part. When analyzing knowledge, no choice
must be made for either D or U, but a delicate balance between the two
components must be maintained. We now have the instruments for the
formulation of the fundamental problem of logic. Given a system D of
entities and propositions, how is it possible to constitute a system
D' which contains a further element q ? The logic of strengthening
demonstrates that it is possible to go from D to D' when U of D and U'
of D' exist. The relationship among the D's and the U's are the
logical equations.[37] It is therefore the latter which improve the
necessary condition for a new intuitive element to be added to the
monodimensionality of the logical discourse D. We could perhaps speak
about a formalization of the a priori synthetic judgement. However,
the introduction of strengthening in logic consists just in this
increase of intuitive elements necessary for the construction of the
new discourse.

The general form of the logical equations can be thus expressed
$D/U = D'/U'$. [38]

Pastore applies the concept of logical equation to the Lorenz
transformation, to Planck's quantum and to the Schrödinger equation.
We will not delve into the technical details of the demonstrations,
which are not always too clear and are interspersed with giant leaps
from logic to physics and, vice versa, which do not, at first sight
appear all consistent. Suffice it to say that for U and U' Pastore
substitutes in turn Galileian, relativistic, and quantal space-time,
in an attempt to match the different space-time intuitions with the
mathematical equations of the respective theories (D and D').

The author thus concludes his analysis:

"In every case, according to the D, U principle, the true
sense of the impossibility to choose an interpretation
which separate the corpuscle from the wave is that which
prevents us from separating space from time, position from
motion, matter from form, discontinuous from continuous,
and, in logic, discourse from universe, finally entities
from relation ships."[39]

Heisenberg's uncertainty principle is conceptually incorporated
in the logic of the universe of discourse as well. According to the
fundamental principle, entities are relationships, but on the other
hand, all relationships determine only a single class of entities;
every time a relationship is added, the extension of the class is
narrowed. Now, if we view as physical the relationship between

entities and relationships in the space-time, we are again facing Heisenberg's uncertainty theory. The determination of a single space-time relationship (velocity) implies uncertainty for a large class of entities (space-time position) and the other way around.

In such a state of affairs, on the basis of the logic theories of strengthening, one immediately understands that Heisenberg's law is a physical interpretation according to the logical model D, U of a fundamental exigence of thinking.[40]

It must be pointed out that Pastore's interpretation contains an asymmetry: While in the orthodox interpretation of quantum mechanics the concepts of wave and corpuscle are on a par, in his ontology the supremacy is given to relationships, hence to waves. Therefore he deduces an uncertainty which is not the same as Heisenberg's, but is more similar to the purely classical uncertainty of wave theory. It is an uncertainty which presupposes the primacy of the field interpretation and is expressed in the Heisenberg formulas, but without Planck's constant. He himself recognizes:

"Therefore by way of relationships the being remain invariably undetermined, from the point of view of the individuation."[41]

The same is not true for quantum mechanics, where it is possible to determine exactly the position of particle at the expense of the knowledge of its velocity.

On the positive side we must record, however, that Pastore recognized that Heisenberg's uncertainty does not imply indeterminism. If determinism amounts to dependence on initial conditions, quantum mechanics is deterministic; it does not matter whether the description of the physical state through the wave function is only statistical. If the laws of the corpuscle are statistical, the laws of wave mechanics are deterministic.[42] The logical strengthening requires a statistical calculation for the being and a causal calculation for the relationship.[43]

At this point the author examines the thesis expressed by Fermi,[44] according to which it is necessary to construct a physics which is corpuscular and wavelike at the same time, and explains the phenomena both of matter and of radiation. In the language of the logic of strengthening, this means that:

"The construction of the new D' in which the reconciliation advocated by Fermi, requires the invention of a new physical concept to substitute the pair corpuscle-wave. This is what the continuous fertile development of physics and of every scientific endeavour is made of."[45]

This position appears undoubtedly very modern.[46] It is not easy to assess conclusively Pastore's philosophy on quantum mechanics. It certainly constitutes the most complete analysis to be found in the philosophical writings of these years. We believe there is a strong neo-Kantian component in the authors thinking, which finds a heartening confirmation in the new physics. On the other hand, if we compare his interpretation with the one of Weizsäcker, we must say that the former does not make any concession to subjectivism: The way the object is an entity with respect to the subject, so is the subject with respect to the object. Hence the polarization thing in itself versus phenomenon holds no more, and not only is the object unknowable as objective, but the subject is unknowable as subjective as well.

Aliotta, as is well known, is a pupil of F. De Sarlo, and his thinking is consequently originally developed from problems of psychology. However, the spiritualistic component of his philosophy becomes more and more noticeable with passing years, until it prompts the writing of "La reazione idealistica contro la scienza"[47], a work influenced by Croce and by romantic idealism but which contains one of the most comprehensive and multifaceted visions of science to be found in the Italian philosophy of that time.

Actually, the part of Aliotta thinking in which we are interested is the one dating from the thirties, which culminate in a radical experimentalism expressed in a collection of essays "L'esperimento nella scienza, nella filosofia, nella religione".[48] In this work the experimental approach, viewed as the action of the experimenter on the matter to be observed, is brought to its most extreme consequences. On the one hand the experiment is conceived in the broadest sense; further on the author is even found saying that Christianity is experimentally confirmed since historically it is the religion which has lead to most successes. On the other hand, the importance he attributes to the action and the choice of the observer gives rise to an absolute relativism, which will find a confirmation in Einstein's relativity even before it does in quantum mechanics.[49]

The author formulates Heisenberg's uncertainty principle along the lines according to which the impossibility of simultaneously determining, with precision, the velocity and position of the particle depends on the observer's intervention, which presupposes the reality of the two entities but the impossibility of measuring them.[50] This opinion had been already largely criticized in these years and Heisenberg's and Bohr's formulation had became much more subtle than that, due to the compelling criticism of Einstein and others. In fact, it is more rigorous to say that it is the measuring apparatus which acts on the particle; and it makes sense to talk about the physical state of the latter only in connection with the instrument.[51] Aliotta, however, insists on the subjective aspect of the observer action:

"This inescapable action of the observer, which Heisenberg had brought into focus, confirms the theory of experiment we have outlined."[52]

The experiment is not a passive reproduction of reality but an active doing. We find in it the imprint of our physical and mental activity.

> "It is not possible to separate subjective from objective. ... A physical phenomenon is not the sensible datum of which the old coarse empiricism spoke, but is the significance we attribute to it interpreting it in the universal system of our logic. There is no determination independent from our acts, both mental and physical, which construct it."[53]

Aliotta is not an idealistic thinker, had indeed realist background, and his development moved him toward spiritualism. However, it would not be possible to give of quantum mechanics an interpretation more clearly subjectivistic than the one he has given. His point of view is not far from that of Weizsäcker, but he goes one further step-not quite so long-by eliminating the thing in itself.

A fully Kantian position is expressed, by contrast, in Abbagnano's already quoted work "La fisica nuova".[54] The young philosopher attacks the problem of "giving a gnoseology essay which makes the validity of new physics possible"[55]. This attitude forestalls an attention devoted to physics by philosophy that is quite a different in nature from that given by the authors hitherto discussed. The question is not to look for a confirmation of one's own philosophy, selecting the results of science which appear most suitable for this purpose, but to penetrate the structure of the new theory intimately so as to grasp these conceptual connections which can, and must, influence one's thinking. It is no coincidence that Abbagnano, as Santucci correctly remarks,[56] has not been influenced by Gentile's philosophy, and his point of view actually belongs to the next generation.

The author goes on to discuss the Copenhagen interpretation of quantum mechanics which, according to him, replaces the concept of physical entity with the concept of measurement, so that the structure of physical phenomena is finally constituted by the observer's action.[57] In order to overcome the aporias in which idealism and realism alike are caught when interpreting this state of affairs, it is necessary to have recourse to the transcendental; in other words,

> "If there is knowledge, there is a self, and if there is a thing or an object (common experience), either that or there is an observing system and a phenomenon controlled and measured (scientific observation); but in the very separation of self from the thing, of the observer from the fact, a separation occurring along an ever moving boundary, one reaches the unity of a total system, and such unity, which is the principle which organizes and rules the whole".[58]

A more clear-cut statement of Kantism, in line with Weizsäcker's position, could hardly be expected.

We conclude the survey of the opinions expressed by positivists or spiritualists-realists by discussing the point of view of Losacco, an author who was profoundly influenced by De Sarlo and Croce. He reaches at the end of his philosophical career a position of critical realism, defined thus by himself in his most important theoretical work, "Preludi al nuovo realismo critico".[59] Idealism is criticized through an invitation to turn back to the concrete datum of knowledge. Reality limits the freedom of the individual and shows his contingency, which implies the absolute necessity of one God. The appeal to the concreteness of reality involves the attempt to construct a philosophy of nature à la Schelling. According to Losacco this possibility has been also introduced by Driesch and Ostwald.[60]

The author is able to find a confirmation of his point of view in quantum mechanics. On one hand the return of the discontinuous in nature betrays a novel attention paid to the individual, the single concrete datum, at the expense of the abstract and the general.[61] On the other hand, the demonstration that causal explanation and statistical explanation only have an heuristic value and must be thought of as complementary does not allow him to focus on the subjective element. He accept the concept that the observation of atomic phenomena can alter them, in analogy with the case of introspection in psychology:

> "But once the due part is attributed to the observer-dependent subjective element, it is reasonable to assume that certain characteristics of the observed phenomena are independent and objective 'strictu sensu.' Indeed (as Jörgensen observed) the mutual dependence between the investigating means and the atomic phenomena is a mutual dependence between two classes of objects, rather than that between subject and object."[62]

In other words, it is not the action of the observer which modifies the physical state but the action of the measuring device:

> "Therefore one has, now as well as before, a justified reason to assume that objective existence of a physical Universe independent of any eventual knowledge we might have of it."[63]

It is our opinion that Losacco, well conversant with the thinking of Husserl, Brentano, Meinong, and Külpe, who was among the firsts to import the ideas of phenomenology, has captured here the gist of the matter, namely that the new physics does not imply a subjectivistic turn of thinking, but only a new and more complex role of the object.

4. MARXISM

We want finally to discuss a few pages devoted by Gramsci to our topic. Even though he never was deeply involved with the philosophy of science, still he made some very significant remarks on this subject.[64]

Undoubtedly his idealistic and antipositivist background caused Gramsci to reduce science to ideology and superstructure. In his view, science has taken the place of the old religion presenting itself as dogmatism, as the misticism of the exterior reality. On the other hand, he always made a clear distinction between the positivist theology and the experimental method. Marxism cannot accept the latter as the one real form of knowledge, but still recognizes it has part of the conceptual framework of a full form of communism. Objective reality in itself and for itself, independent of man and of history, is a useless metaphysical creed. However, it is possible to talk about human as well as historical objectivity, which is based on the mutual subjective agreement among all men, and this kind of objectivity can be grasped by mean of the experimental method. In other words, among all possible subjectivities, science is the most objective, and its importance for the progress of humanity is not to be denied.[65]

Gramsci takes the lead from Eddington's "Nature of the Physical World"[66] to pose for himself the problem of the new character of the experimental method in physics. However, he seems to get mixed up over the use of the microscope in biology and the observation of particles in physics. But his quoting Camis and his reference to "the funambolist thinking of certain scientists, especially British, with regard to the 'new physics'"[67], leave little doubt that what the author has in mind is quantum mechanics.

The problem is presented as follows:

"Perhaps the matter seen under the microscope is no longer really objective matter, but a creation of the human spirit which does not exist objectively or empirically?"[68]

The answer is much too clear:

"The position of man remains inaltered, none of the fundamental concepts of life is shaken, let alone upturned."[69]

Furthermore, if it was true that "they (the infinitely small phenomena) cannot be considered independent of the subject who observes them"[70], it would follow "that science can no longer exist the way it has been conceived until now, but must be transformed in a series of acts of faith in the statements of every experimenter"[71]. Also "they (the infinitely small phenomena) would not be 'observed', but 'created' and they would fall under the same domain of pure phantastic intuition of the individual"[72]. It would also be difficult to talk about reproducibility of the experiment and "it would not be a case of 'solipsism' but rather of 'demiurgy', or whichcraft."[73]

The reasons for the dependence of the observer are not inescapable: they lie indeed in the technical problems posed by this experiments and in those "having to do with the description and the objective representation of the phenomena observed"[74]. The reasons for this being: (1) Scientists are didactically trained to only talk about macroscopic phenomena; (2) the common language is insufficient, since it was coined exclusively for macroscopic phenomena; (3) the sciences of the infinitely small are still young; (4) experiments are extremely complex. The author concludes, therefore, saying that "we are dealing, all things considered, with a passing, initial stage, of a new scientific epoch."[75]

Gramsci has not caught how radical the problem of observation in quantum mechanics was and does not distinguish it clearly from a question of technical ability and precision, considerations which are adequate only for classical physics. On the other hand, for all his idealistic background, and consequent leaning toward subjectivism, he does not absolutely accept the concept of the observer's action upon reality, and he throws a light, with exceptional scientific common sense, on the limitations of the language, of techniques and theories with regard to new discoveries. This point of view definitely does not make any concession to the interpretation à la Weizsäcker.

We conclude by stressing the prominence of Kantian interpretations of quantum mechanics in Italian philosophy, in spite of a few significant exceptions: Gramsci, Losacco and partly Pastore. On the other hand, it must be noticed that, in our country, quantum mechanics did not have a cultural propagation as large as relativity did. This is so because of the scant participation of Italian scientists in the construction of the mathematical framework of quantum mechanics, as compared to that of relativity, and to the great logical complexity of the latter. These are probably the two main reasons why among our philosophers the interpretations of Einstein's theories are much more common and better documented.[76]

REFERENCES AND NOTES

1. C. F. von Weizsäcker, "Das Verhältnis der Quantenmechanick zur Philosophie Kants", Die Tatwelt,66-98, (1941).
2. Ibidem.
3. Ibidem.
4. Ibidem.
5. Ibidem.
6. Ibidem.
7. Ibidem, author's italics.
8. See also E.Agazzi, ed., "La filosofia della scienza in Italia nel '900" (Franco Angeli, Milano, 1986). Henceforth referred to as Agazzi (1986).
9. G. De Giuli, "Scienza e idealismo", Riv. di Fil., 53-56 (1931), a review of A.S.Eddington, "La nature du monde physique" (Payot, Paris, 1929) and of J.Parodi, "Du positivisme à l'idéalisme". La philosophie d'hier (Vrin, Paris, 1930).
10. Ibidem, p. 53.

11. Ibidem, p. 54.
12. See Ref. 9.
13. Ibidem, p. 55.
14. G. De Giuli, Riv. di Fil., 87-89 (1935), a review of J.Jeans, "I nuovi orizzonti della scienza", foreword by G. Gentile Junior. (Sansoni, Firenze, 1934). Both Jeans and Gentile express similar ideas.
15. Ibidem, p. 88.
16. G. De Giuli, Giornale critico della filosofia italiana 107-8 (1936), a review of N.Abbagnano, "La fisica nuova. Fondamenti di una teoria della scienza" (Guida, Napoli, 1934).
17. Ibidem, p. 108.
18. F.Pagano, "Novecento", Logos 255-67 (1936). We have also utilized the excellent short presentation of the same by M.Frasca Spada, in Agazzi (1986), pp. 474-75.
19. Ibidem, p. 257.
20. Ibidem, p. 258.
21. Ibidem, p. 259.
22. Ibidem, p. 267.
23. Agazzi (1986), pp. 24-5.
24. The column written by De Ruggiero from 1927 on was entitled "Note sulla più recente filosofia europea e americana", and was later collected in the volume, "Filosofi del '900" (Laterza, Bari, 1934).Our quotations are from the latter edition.
25. Ibidem, p. 197.
26. Ibidem, p. 204.
27. Ibidem, p. 206.
28. Ibidem, p. 212.
29. Ibidem, p. 212.
30. Ibidem, p. 212.
31. See also G.Gentile Junior. in the Italian translation of Jeans's book quoted in Ref. 14.
32. G.Tarozzi, see introduction to this volume.
33. E.Garulli, "L'epistemologia filosofica fra le due guerre", in Agazzi(1986), pp. 213-37.
34. A.Pastore, "La logica del potenziamento" (Rondinella,Napoli, 1936). We will talk primarily about the article, "Introduzione alla teoria delle equazioni logiche", a communication presented to "Classe di scienze morali della R.Accademia delle Scienze dell'Istituto di Bologna" on December 27, 1934.
35. Ibidem, p. 7.
36. Ibidem, p. 6.
37. Ibidem, p. 72.
38. Ibidem, p. 74.
39. Ibidem, p. 82.
40. Ibidem, p. 83.
41. Ibidem, p. 83.
42. See also E.Nagel, "The Structure of Science"(Harcourt, New York, 1961).
43. See Ref. 34, p. 85. The determinism of the causal explanation of waves has been demonstrated by Ehrenfest.

44. E. Fermi, "Introduzione alla fisica atomica"(Zanichelli, Bologna, 1928).

45. See Ref. 34, p. 87.

46. See also F.Selleri, "Gespensterfelder", in "Wave-Particle Dualism", S.Diner et al., eds. (Reidel, Dordrecht, 1984); G.Tarozzi, "From ghost to real waves: A proposed solution to the wave-particle dilemma", ibidem. For an excellent review of today's problems of quantum mechanics see F.Selleri and G.Tarozzi, "Quantum mechanics reality and separability", Riv. Nuovo Cimento 4, 1-53 (1981).

47. A.Aliotta, "La reazione idealistica contro la scienza", 1912, reprinted by C.Carbonara (Libr. Scient. Ed., Napoli, 1970).

48. A. Aliotta, "L'esperimento nella scienza, nella filosofia, nella religione" (Perella, Napoli, 1936). Reprinted in "Opere complete", VII (Cremonese, Roma, 1954).

49. A. Aliotta, "La teoria di Einstein e le mutevoli prospettive del mondo" (Sandron, Palermo, 1922).

50. "Opere complete", VII, p. 112.

51. See also M.Jammer, "Philosophy of Quantum Mechanics" (Wiley, New York, 1974).

52. "Opere complete", VII, p. 114.

53. Ibidem, p. 115.

54. See Ref. 16.

55. See Ref. 16, p. VII.

56. A.Santucci, "Esistenzialismo e filosofia italiana" (Bologna, Il Mulino, 1959), p. 79.

57. See Ref. 16, p. 45.

58. Ibidem, p. 102. We have utilized the excellent summary by E.Garulli, Ref. 33, pp. 224-26.

59. M. Losacco, "Preludi al nuovo realismo critico" (Modena, 1938).

60. Ibidem, p. 339.

61. Ibidem, p. 341.

62. Ibidem, pp. 345-6.

63. Ibidem, p. 346.

64. See also P.Rossi, "Immagini della scienza" (Editori Riuniti, Roma, 1977), p. 246.

65. Ibidem, pp. 227-47.

66. A.Gramsci, "Quaderni dal carcere", V.Gerratana, ed. (Torino, Einaudi, 1975), p. 1451. Eddington's work is the same as quoted by De Giuli in Ref. 9.

67. Ibidem, p. 1452.

68. Ibidem, p. 1451.

69. Ibidem, pp. 14512-52.

70. Ibidem, p. 1452.

71. Ibidem, p. 1452.

72. Ibidem, p. 1454.

73. Ibidem.

74. Ibidem.

75. Ibidem.

76. E.Agazzi, in Agazzi (1986), p. 30, expresses this judgement with respect to the epistemological debate of scientists, but it is

possible to extend it to philosophers, as well, with some
cautions. However, there have also been physicists who gave a
philosophical analysis of quantum mechanics. See, for instance:
G.Furlani, "La concezione del mondo fisico nella scienza
moderna", Riv. di Fil. 365-85 (1932); E.Fermi, "La fisica
moderna", Nuova antologia 65, 137-45 (1930); E.Persico, "Aspetti
logici di questioni fisiche", in Atti dell'VIII Congresso
Nazionale di filosofia, 1933 (Società filosofica romana, Roma,
1936). All these opinions are sympathetic to the new theory. An
interesting exception is represented by the criticism of the
great mathematician G. Castelnuovo, in "Determinismo e
probabilità", Scientia 53, 1-12 (1933); "Il principio di
causalità", Scientia 60, 61-8 (1936); "Les vues philosophiques
d'un grand physicien", Scientia 59, 40-3 (1936).

ITALIAN STUDIES IN THE FOUNDATIONS OF QUANTUM PHYSICS.
A BIBLIOGRAPHY (1965-1985)

Margherita Benzi
Facoltà di Lettere e Filosofia, Università di Bologna
Bologna, Italy

The following bibliography collects contributions to the foundations of quantum mechanics made by Italian authors from approximately 1965 to 1985, where under this heading we include both new physical, mathematical, and logical results, as well as pieces of philosophical reflection about quantum mechanics.

As our starting point we have chosen the mid-sixties, which ushered in the growth of interest in the foundations of physics that has taken place in Italy, subsequent to the debate on the theory of measurement, the proof of the irrelevance of von Neumann's theorem regarding the impossibility of a deterministic completion of quantum mechanics, and the discovery of Bell's type inequalities, which moved the problem concerning the compatibility between quantum mechanics and local hidden-variable theories into the experimental arena.

A further development of the interest in field under discussion was generated by the 1970 Varenna summer school of the Italian Physical Society, dealing with the "Foundations of Quantum Mechanics" followed by the organization, in Erice (1974), Varenna (1978), Udine (1979), Perugia (1982), Bari (1983), and Urbino (1985) of international conferences whose proceedings have been printed by various publishers, like Reidel, Academic Press, North Holland, and Springer.

For much older works, the reader may consult V. Somenzi's bibliography in *Philosophy in the Mid-Century: A Survey*, R., Klibansky, ed. (La Nuova Italia, Florence, 1958), vol. I, pp. 281, 284.

The list in no way pretends to be exhaustive. Due to the fact that different kinds of sources -- current periodicals, on-line systems DIALOG, private communications -- have been used, the references are not perfectly uniform: Some of them specify both the first and the last page of the quoted article, whereas others mention just the first page.

The bibliography comprises also articles and books which have been co-authored and co-edited by Italian and non-Italian scholars.

G. Tarozzi and A. van der Merwe (eds.), The Nature of Quantum Paradoxes, 403-425.
© *1988 by Kluwer Academic Publishers.*

ARTICLES

1.   Accardi, L., "Non relativistic quantum mechanics as a non-
      commutative Markov process", *Adv. Math. 20*, 329-366 (1976).
2.   Accardi, L., "Noncommutative Markov chains associated to a
      preassigned evolution: an application to the quantum theory of
      measurement", *Adv. Math. 29*, 226-243 (1978).
3.   Accardi, L., "Quantum Markov processes", in Blaquiere, A., *et
      al.*, eds., *Dynamical Systems and Microphysics*, (Springer,
      Berlin, 1980).
4.   Accardi, L., "On the quantum Feynman-Kac formula", *Rendic. Sem.
      Mat. e Fis. di Milano 48*, 135-179 (1980).
5.   Accardi, L., "Topics in quantum probability", *Phys. Rep. 77*,
      169-192 (1981).
6.   Accardi, L., "Probabilità e teoria quantistica", *Physis 23*, 485-
      524 (1981).
7.   Accardi, L., Fedullo, A., "On the statistical meaning of complex
      numbers in quantum mechanics", *Lett. Nuovo Cimento 34*, 161-172
      (1982).
8.   Accardi, L., "Foundations of quantum probability", *Rendic. Sem.
      Mat. dell'Univ. di Torino*, n. speciale, 249-270 (1982).
9.   Accardi, L., "On the connection between the probabilistic and
      the Hilbert-space description of dynamical systems", *Lett. Nuovo
      Cimento 8*, 585-589 (1983).
10.   Accardi, L., "The probabilistic roots of quantum mechanical
      paradoxes", in Diner, S., *et al.*, eds., *The Wave-Particle
      Dualism*, (Reidel, Dordrecht, 1984), pp. 297-330.
11.   Agazzi, E., "Fisica galileiana e fisica contemporanea", in *Nel
      quarto centenario della nascita di Galileo Galilei*, (Vita e
      pensiero, Milano, 1966), pp. 1-51.
12.   Agazzi, E., "Struttura ed evoluzione nelle scienze
      dell'infinitamente piccolo: fisica dell'atomo e biologia
      molecolare", *Riv. Filos. Neoscol. 61*, 657-665 (1969).
13.   Agazzi, E. "The concept of empirical data. Proposal for an
      intentional semantics of empirical theories", in Przelecki, M.
      et al., eds., *Formal Methods in the Methodology of Empirical
      Sciences*, (Reidel, Dordrecht 1976).
14.   Agazzi, E., "Physics as philosophy and as the paradigm of
      science", *Epistemologia*, special issue for the Centenary
      Celebration of Albert Einstein, *3*, 135- (1980).
15.   Agazzi, E., "Logic and the methodology of empirical sciences",
      in E. Agazzi, ed., *Modern Logic. A survey*, (Reidel, Dordrecht,
      1981), pp. 255-282.
16.   Agazzi, E., "Time and causality" in Dalla Chiara, M.L., ed.,
      Italian Studies in the Philosophy of Science, (Reidel,
      Dordrecht, 1981), pp. 299-321.
17.   Agazzi, E., "Determinismo, indeterminismo e causalità", *Synesis*,
      1/2-3, 13-36 (1984).
18.   Agazzi, E., "La questione del realismo scientifico", in
      Mangione, C., ed., *Scienza e filosofia. Saggi in onore di*

Ludovico Geymonat (Garzanti, Milano, 1985), pp. 171-192.

19.   Agazzi, E., "Commensurability, incommensurability and Cumulativity in scientific knowledge", *Erkenntnis 22*, 51-77 (1985).

20.   Ali, S.T., Ghirardi, G.C., "Unstable systems and measurement processes", *Nuovo Cimento A 24*, 220- (1974).

21.   Ali, S.T., Fonda, G.C., Ghirardi, G.C., "Pertinence of the semigroup law in the history of unstable elementary particle", *Nuovo Cimento A 25*, 135 (1975).

22.   Amaldi, E., "Radioactivity, a pragmatic pillar of probabilistic conceptions", in Toraldo di Francia, G., ed., *Problems in the Foundations of Physics*. Proceedings of the International School of Physics "Enrico Fermi", Course LXXXII, held at Varenna, 25th July - 6th August 1977. (North Holland, Amsterdam, 1979).

23.   Andrade e Silva, J., Selleri, F., Vigier, J.P., "Some possible experiments on quantum waves", *Lett. Nuovo Cimento 36*, 503, (1983).

24   Aquilanti, V., "Atomic collision experiments at the borderline between classical and quantum mechanics", in Diner, S. et al., eds, *The Wave-Particle Dualism* (Reidel, Dordrecht, 1984).

25.   Augelli, V., Garuccio, A., Selleri, F., "La mécanique quantique et la realité", *Ann. Fond. Louis de Broglie 1*, 154 (1976).

26.   Baldo, M., Recami, E., "Comments about recent letters on spacelike states", *Lett. Nuovo Cimento 2*, 643-646 (1969).

27.   Baracca, A., Bergia, S., Bigoni, R., Cecchini, A., "Statistics of observations for "proper" and "improper" mixtures in quantum mechanics", *Riv. Nuovo Cimento 14*, 169-188 (1974).

28.   Baracca, A., Bohm, D.J., Hiley, B.J., Stuart, A.E.G., "On some new notions concerning locality and nonlocality in the quantum theory", *Nuovo Cimento 28 B*, 453-466 (1975).

29.   Baracca, A., Bergia, S., Livi, R., Restignoli, M., "Reinterpretation and extension of Bell's inequality for multivalued observables", *Int. J. Theor. Phys. 15*, 473 (1976).

30.   Baracca, A., Bergia, S., Cannata, F., Ruffo, S., Savoia, M., "Is a Bell-type inequality a good test of quantum mechanics?", *Int. J. Theor. Phys. 16*, 491 (1977).

31.   Baracca, A., Cornia, A., Livi, R., "Quantum mechanics, "first kind" states and local hidden variables: three experimentally distinguishable situations", *Nuovo Cimento B 43*, 65-72 (1978).

32.   Baracca, A., Lunardini, A., Ruffo, S., "On the tests of quantum theory: how many different theoretical frameworks?", *Lett. Nuovo Cimento 22*, 281-288, (1978).

33.   Barchielli, A., Lanz, L., Lupieri, G., "A new treatment of macroscopic observables in quantum mechanics", *Phys. Rev. A 99*, 77-102 (1979).

34.   Barchielli, A., Lanz, L., Prosperi, G.M., "A model for the macroscopic description and continual observations in quantum mechanics", *Nuovo Cimento B 72*, 79-121 (1982).

35.   Barchielli, A., "Continual measurements for quantum open systems", *Nuovo Cimento B 174*, 113-138 (1983).

36.   Barchielli, A., Lanz, L., Prosperi, G.M., "Statistics of

continuous trajectories in quantum mechanics: operation-valued stochastic processes", *Found. Phys. 13*, 779-812 (1983).

37.   Barchielli, A., "Continuous observations in quantum mechanics. an application to gravitational-wave detectors", *Phys. Rev. D 32*, 347-67 (1985).

38.   Barchielli, A., Lupieri, G., "Quantum stochastic calculus operation valued stochastic processes, and continual measurements in quantum mechanics", *J. Math. Phys. 26*, 2222-30, (1985).

39.   Barone, F., Galdi, G.P., "On the questions of atomicity and determinism in Boolean systems", *Lett. Nuovo Cimento 24*, 179-182 (1979).

40.   Barone, F., "On implemented state automorphisms with the logico-algebraic approach to deterministic mechanics", *Lett. Nuovo Cimento 30*, 155-159 (1981).

41.   Barone, F., Grassini, R., "Logicoalgebraic approach to Lagrangian systems", *Int. J. Theor. Phys. 22*, 829-836 (1983).

42.   Bedford, D., Selleri, F., "On Popper's new EPR-experiment", *Lett. Nuovo Cimento 42*, 325 (1985).

43.   Beltrametti, E.G., Cassinelli, G., "Quantum mechanics and p-adic numbers", *Found Phys. 2*, 1-7 (1972).

44.   Beltrametti, E.G., Cassinelli, G., "On the logic of quantum mechanics", *Z. Naturwiss. A 8*, 1516-1530 (1973).

45.   Beltrametti, E.G., Cassinelli, G., "Logical and mathematical structures of quantum mechanics", *Riv. Nuovo Cimento 6*, 321-404 (1976).

46.   Beltrametti, E.G., Cassinelli, G., "Properties of states in quantum logic", in Toraldo di Francia, G., ed., *Problems in the Foundations of Physics* (North Holland, Amsterdam, 1979).

47.   Beltrametti, E.G., Cassinelli, G., "Problems of the proposition state-structure of quantum mechanics", in Dalla Chiara, M.L., ed., *Italian Studies in the Philosophy of Science* (Reidel, Dordrecht, 1981).

48.   Beltrametti, E.G., Cassinelli, G., "The logic of quantum mechanics", in *Encyclopedia of Mathematics* (Addison-Wesley, New York, 1982).

49.   Benedetti, A., Teppati, G., "The decision problem for mathematical structures of quantum theory", *Lett. Nuovo Cimento 2*, 695-696 (1971).

50.   Benedettin, G., Galgani, L., "Transition to stochasticity in a one-dimensional model of a radiant cavity", *J. Stat. Phys. 27*, 153-169 (1982).

51.   Benedettin, G., Galgani, L., Giorgilli, A., "Boltzmann's ultraviolet cutoff and Nekhoroshev's theorem on Arnold's diffusion", *Nature 311*, 444-446 (1984).

52.   Bergia, S. Cannata, F., Ruffo, S., Savoia, M., "Group theoretical interpretation of von Neumann's theorem on composite systems", *Am. J. Phys. 47*, 548 (1979).

53.   Bergia, S., Cannata, F., Cornia, A., Livi, R., "On the actual measurability the density matrix of a decaying system by means of measurements of the decay products", *Found Phys. 10*, 723

(1980).

54. Bergia, S., Cannata, F., "Higher-order tensor and tests of quantum mechanics", *Found. Phys. 12*, 843 (1982).

55. Bergia, S., "On the possibility of extending the tests of quantum mechanical correlations, in Diner, S. et al. (eds.), *The Wave-Particle Dualism*, (Reidel, Dordrecht, 1984).

56. Bergia, S., "Il fotone come predizione della termodinamica statistica. Relazione su invito al congresso nazionale dell'A.I.F.", *La Fisica nella Scuola 16*, 178 (1984).

57. Bergia, S., "Il problema della modellizzazione nella fisica quantistica", in Massafra, S., Minazzi, F., (eds.) *Il problema delle scienze nella società contemporanea*, (Angeli, Milano 1985.

58. Bergia, S., Cannata, F., Monzoni, V., "Explicit examples of theories satisfying Bell's inequalities: do they miss their goal prior to contradicting experiments?", *Found. Phys. 15*, 145 (1985).

59. Bergia, S., Cannata, F., Giorgini, B., Zamboni, V., "A solvable model for irreversible quantum phenomena", *Lett. Nuovo Cimento 43* (1985), 113.

60. Bertolini, G., Diana, E., Scotti, A., "Correlation of annihilation g -Ray polarization. *Nuovo Cimento B 63*, 651-665 (1981).

61. Bocchieri, P., Prosperi, G.M., "Recent developments in quantum ergodic theory", in Bak, T.A. (ed.), *Statistical Mechanics: Foundations and applications, Proceedings of Meeting*, 11-15 Jul. 1966, Copenhagen. W.A. Benjamin (New York) 1967, pp. 17-31.

62. Bocchieri, P., Scotti, A., Bearzi, B., Loinger, A., "Anharmonic chain with Lennard-Jones interaction, *Phys. Rev. A 2*, 2013-18 (1970).

63. Bocchieri, P., Loinger, A., "The Rayleigh-Jeans law is incompatible with classical electrodynamics", *Lett. Nuovo Cimento 1*, n. 17 (1971), 709-710.

64. Bocchieri, P., Loinger, A., Valz-Gris, F., "Classical electrodynamics of a one-dimensional Hohlraum. Invalidity of the equipartition law. *Nuovo Cimento B 19*, 1-14 (1974),.

65. Bocchieri, P., Loinger, A., "Nonexistence of the Aharonov-Bohm effect", *Nuovo Cimento A 47*, 475-482 (1978).

66. Bocchieri, P., Loinger, A., "Locality in quantum electromagnetism", *Nuovo Cimento A 59*, 121-133 (1980).

67. Bocchieri, P., Loinger, A., "Charges in multiply connected spaces", *Nuovo Cimento A 66*, 164-172 (1981).

68. Bocchieri, P., Loinger, A., "Comments on the Letter "On the Aharonov-Bohm effect" of Boersch et al., *Lett. Nuovo Cimento 30*, 449-450 (1981),.

69. Bocchieri, P., Loinger, A., "Incompatibility of the Aharonov-Bohm effect with the quantum laws". *Lett. Nuovo Cimento 35*, 469-472 (1982).

70. Bocchieri, P., Loinger, A., "Quantum laws and Aharonov-Bohm effect", *Lett. Nuovo Cimento 39*, (1984).

71. Borchi, E., Pelosi, G., "A quantum mechanical derivation of the

general uncertainty relation for real signals in communication theory". *Signal Process 2*, 289-292 (1980),.

72. Bressan, A., "On wave functions in quantum mechanics". *Rend. Sem. Mat. Univ. Padova 60*, 77-98 (1978).

73. Bressan, A., "On wave functions in quantum mechanics II. On fundamental observables and quantistic states". *Rend. Sem. Mat. Univ. Padova 61*, 221-228 (1979).

74. Bressan, A., "On wave functions in quantum mechanics. III: A theory of quantum mechanics where wave functions are defined by means of surely fundamental observables," *Rend. Sem. Mat. Univ. Padova 71*, 365-392 (1979).

75. Bressan, A., "On physical possibility", in Dalla Chiara, M.L., ed., *Italian Studies in the Philosophy of Science*. (Reidel, Dordrecht 1981), pp. 197-214.

76. Caianiello, E.R., "Hermitian metrics and the Weil-London approach to the 'quantum theory'". *Lettere al Nuovo Cimento 25*, n. 8 (1979), 225-229.

77. Caianiello, E.R., "Some remarks on quantum mechanics and relativity", *Lettere al Nuovo Cimento 27*, n. 3 (1980), 89-96.

78. Caianiello, E.R., "Geometrical 'identification' of quantum and information theories". *Lett. Nuovo Cimento 38*, 539-543 (1983).

79. Caianiello, E.R., "Maximal acceleration as a consequence of Heisenberg's uncertainty relations. *Lett. Nuovo Cimento 41*, 370-72 (1984).

80. Caianiello, E.R., "Geometry from quantum mechanics". *Nuovo Cimento B, 59*, 350-366 (1980),.

81. Caianiello, E.R., "Remarks on the maximal acceleration hypothesis", *Lett. Nuovo Cimento 34*, 112-114 (1982).

82   Caldirola, P., "Teoria della misurazione e sistemi ergodici nella meccanica quantistica", *Scientia 99*, 297-319 (1964).

83. Caldirola, P.,"Statistical mechanics of noncommutative systems", *Nuovo Cimento*, 46 B,172-181 (1966).

84. Caldirola, P., "Physics and Philosophy in Italy", in Klibansky, R., ed., *Contemporary Philosophy. A Survey*, vol. II (La Nuova Italia, Firenze, 1969), pp. 223-231.

85. Caldirola, P., "On the introduction of a fundamental interval of time in quantum mechanics", *Lett. Nuovo Cimento 16*, 151-155, (1976).

86. Caldirola, P., "On the finite-difference Schroedinger equation", *Lett. Nuovo Cimento 17*, 4611-464 (1976).

87. Caldirola, P., Recami, E., "Causality and tachyons in relativity", in Dalla Chiara, M.L., ed., *Italian Studies in the Phylosophy of Science*, (Reidel, Dordrecht 1981).

88. Caldirola, P., "Introduction of the chronon in the theory of electron and the wave-particle duality", in Diner, S., et al., eds., *The Wave-particle Dualism* (Reidel, Dordrecht, 1984).

89. Cantoni, V., "Generalized 'transition probability'", *Comm. Math. Phys.*, 44, 125-128 (1975).

90. Cantoni, V., "The Riemanian structure on the states of quantum-like systems", *Comm. Math. Phys. 56*, 189-193 (1977).

91. Cantoni, V., "Geometric Aspects of quantum systems", *Rend. Sem.*

Mat. Fis. Univ. Milano 48, 35-42 (1978).

92. Cantoni, V., "Generalized transition probability, mobility and symmetries", *Comm. Math. Phys. 87*, 153-158, (1982/83).

93. Capasso, V., Fortunato, D., Selleri, F., "Von Neumann's theorem and hidden-variable models", *Riv. Nuovo Cimento 2*, 149-199 (1970).

94. Capasso, V., Fortunato, D., Selleri, F., "Sensitive observables of quantum mechanics", *Int. J. Theor. Phys. 7*, 319 (1973).

95. Carazza, B., "Il problema del corpo nero e le origini della meccanica quantistica", in *Contributi alla storia della meccanica quantistica. Quaderni di storia e critica della Scienza*, N.S., 7, 17-33 (1976), (Domus Galileiana, Pisa, 1976).

96. Carazza, B., Guidetti, G.P., "Ehrenfest e il principio adiabatico", in *Contributi alla storia della meccanica quantistica. Quaderni di storia e critica della scienza. N.S. 7*, 79-102 (1976).

97. Carazza, B., Casartelli, M., D'Elia, A., "Segal entropy and the principle of least interference", *Phys. Lett. A 62*, 205-206 (1977).

98. Carazza, B., Guidetti, G.P., "La nascita dell'equazione di Klein-Gordon", *Arch. Hist. Exact Sc. 22*, 373-383 (1980).

99. Casartelli, M., Galgani, L., "Compatibility of the Schroedinger and the Born interpretations of the wave function," *Phys. Lett. A 50*, 217-218 (1974).

100. Casati, G., Guarnieri, I, "Aharonov-Bohm effect from the 'hydrodynamical' viewpoint," *Phys. Rev. Lett. 42*, 1579-81 (1979).

101. Casati, "Chaos in quantum mechanics", in Haken H., ed., *Evolution of Order and Chaos in Physics, Chemistry and Biology*, (Proceedings of the International Symposium on Synergetics held at Schloss Elman, April 26- May 1, 1982) Springer, New York, 1982).

102. Casati, G., Guarnieri, "Chaos and special features of quantum systems under external perturbations," *Phys. Rev. Lett. 50*, 640-643 (1983).

103. Cassinelli, G., Truini, P., "Toward a generalized probability theory: conditional probabilities,", in Toraldo di Francia, G., ed., *Problems in the Foundations of Physics* (North Holland, Amsterdam, 1979).

104. Cassinelli, G., Olivieri, G., "The statistics of unbounded observables in Hilbert-space quantum mechanics," *Nuovo Cimento B 84*, 43-52 (1984).

105. Cattaneo, G., Nisticò, G., "Orthogonality and orthocomplementations in the axiomatic approach to quantum mechanics. Remarks about some critiques," *Math. Phys. 25*, 513-31 (1984).

106. Cavalleri, G., "Propagator of stochastic electrodynamics," *Phys. Rev. D, 23*, 363-372 (1981).

107. Cercignani, C., Galgani, L., Montaldi, E., "Zero-point energy in classical nonlinear mechanics", *Phys. Lett. A 38*, 303-304, (1972).

108. Cercignani, C., Galgani, L., Montaldi, E., Sirtori, M., "Quantization à la Nerst for the plane rotator", *Lett. Nuovo Cimento 40*, 235-239 (1984).

109. Cerofolini, G.F., "Quantum mechanics and gravitation," *Lett. Nuovo Cimento 23*, 509 (1978).

110. Cerofolini, G.F., "Speculation on gravitation-matter interaction," *Lett. Nuovo Cimento 26*, 125 (1979).

111. Cerofolini, G.F., "Quantum and subquantum mechanics," *Nuovo Cimento B 58*, 286 (1980).

112. Cerofolini, G.F., "On the nature of the subquantum medium," *Lett. Nuovo Cimento 29*, 305 (1980).

113. Cerofolini, G.F., "Questions of method in a large class of improperly-posed problems," *Bull. Acad. Roy. Belg.66*, 499, (1980).

114. Cerofolini, G.F., "Derivation of the generalized Fokker-Planck equation for particles with Zitterbewegung," *Lett. Nuovo Cimento 34*, 424, (1982).

115. Cerofolini, G.F., "On the formal equivalence between a reformulation of Bohm and Bub's hidden-variable theory and subquantum mechanics," *Lett. Nuovo Cimento 35*, 457 (1982).

116. Cerofolini, G.F., "Gravitational length, strength of interactions and the superheavy particle," *Lett. Nuovo Cimento 38*, 240 (1983).

117. Cerofolini, G.F., "Derivation of the generalized Fokker-Planck equation from the Boltzmann transport equation for particles with Zitterbewegung," *Nuovo Cimento B 79*, 59 (1984).

118. Cini, M., De Maria, M., Mattioli, G., Nicolò, F., "Wave packet reduction in quantum mechanics: a model of a measuring apparatus," *Found. Phys. 9*, 479 (1979).

119. Cini, M., "Quantum theory of measurement without wave packet collapse," *Nuovo Cimento B*, 73, 27 (1983); in Tarozzi, G., and Van der Merwe, A. eds., *Open Questions in Quantum Physics* (Reidel, Dordrecht 1985), p. 185.

120. Cini, M., "Cultural tradition and environmental factors in the development of quantum electrodynamics", *Fundamenta Scientiae 3*, 229 (1982).

121. Cini, M., Serva, M., "Stochastic theory of emission and absorption of quanta," Nota interna n. 847 (1985), Dipartimento di Fisica, Università La Sapienza, Roma.

122. Cirelli, R., Cotta-Ramusino, P., "On the isomorphism of a quantum logic with the logic of the projections in a Hilbert space," *Int. J. Theor. Phys. B 9*, 11-29 (1973).

123. Cirelli, R., Gallone, F., "Algebra of observables and quantum logic," *Ann. Inst. Henri Poincaré A 19*, 297-331 (1973).

124. Cirelli, R., Cotta-Ramusino, P., Novati, E., "On the isomorphism of a quantum logic with the logic of the projections in a Hilbert space II," *Int. J. Theor. Phys. 11*, 135-144 (1974).

125. Cirelli, R, Gallone, F., Gubbay, B., "An algebraic representation of continuous superselection rules", *Math. Phys. 16*, 201-213 (1975).

126. Comi, M., "On the photon position observable", *Nuovo Cimento 59*

A, 1211-1233 (1980).

127. Conte, E., "On retrocollapse in quantum mechanics," *Lett. Nuovo Cimento 31*, 380-382 (1981).

128. Conte, E., "A predictive model of collapse-retrocollapse of quantum mechanics," *Lett. Nuovo Cimento 32*, 286-288 (1981).

129. Corleo, G., Gutkowski, D., Masotto, G., "Are Bell's inequalities sufficient conditions for local-hidden variable theories?", *Nuovo Cimento B 25*, 413-424 (1975).

130. Cornia, A., Lunardini, A., Ruffo, S., "Hidden variables and proper mixtures for multivalued observables: some results for the correlation function", *Lett. Nuovo Cimento 22*, 161-166 (1978).

131. Crisciani, F., Ghirardi, G.C., Rimini, A., Weber, T., "Quantum limitations for spin measurements on systems of arbitrary spin," *Nuovo Cimento B 64*, 338 (1981).

132. Crovini, L., Galgani, L., "On the accuracy of the experimental proof of Planck's radiation law", *Lett. Nuovo Cimento 39*, 210-214 (1984).

133. Cufaro-Petroni, N., "On the observable difference between proper and improper mixtures I," *Nuovo Cimento B 40*, 235 (1977).

134. Cufaro-Petroni, N., "On the observable difference between proper and improper mixtures II," *Nuovo Cimento B 40*, 381 (1977).

135. Cufaro-Petroni, N., Vigier, J.P., "On two conflicting physical interpretations of the breaking of restricted relativistic Einstenian causality by quantum mechanics," *Lett. Nuovo Cimento 25*, 151 (1979).

136. Cufaro-Petroni, N., Vigier, J.P., "Stochastic derivation of Proca's equation in terms of a fluid of Weyssenhoff tops endowed with random fluctuation at the velocity of light," *Phys. Lett. A 73*, 289 (1979).

137. Cufaro-Petroni, N., Vigier, J.P., "Markov process at the velocity of light: the Klein-Gordon statistics," *Int. J. Theor. Phys. 18*, 807 (1979).

138. Cufaro-Petroni, N., Vigier, J.P., "Causal superluminal interpretation of Einstein-Podolsky-Rosen paradox," *Lett. Nuovo Cimento 26*, 149 (1979).

139 Cufaro-Petroni, N., Garuccio, A., Selleri, F., Vigier, J.P., "Sur la contradiction entre la théorie quantique classique (idéalisée) de la mesure et la conservation du carré du moment angulaire total dans le paradoxe d'Einstein-Podolsky-Rosen," *C.R. Acad. Sci. Paris, B 290*, 111 (1980).

140. Cufaro-Petroni, N., Maric, Z., Zivanovic, Dj., Vigier, J.P., "Baryon octet magnetic moments in an integer-charged-quark oscillator model," *Lett. Nuovo Cimento 29*, 565 (1980).

141. Cufaro-Petroni, N., Maric, Z., Zivanovic, Dj., Vigier, J.P., "Stable states of a relativistic bilocal stochastic oscillator: a new quark-lepton model," *J. Phys. A 14*, 501 (1981).

142. Cufaro-Petroni, N., Vigier J.P., "Stochastic derivation of the Dirac equation in terms of a fluid of spinning tops endowed with random fluctuations at the velocity of light", *Phys. Lett. 81 A*, 12 (1981).

143. Cufaro-Petroni, N., Droz-Vincent, Ph., Vigier, J.P., "Action at distance and causality in the stochastic interpretation of quantum mechanics," *Lett. Nuovo Cimento 31*, 415 (1981).

144. Cufaro-Petroni, N., Vigier, J.P., "Stochastic model for the motion of correlated photon pairs," *Phys. Lett. A 88*, 272 (1982).

145. Cufaro-Petroni, N., Vigier J.P., "Dirac's aether in relativistic quantum mechanics," *Found. Phys. 13*, 253 (1983); Barut, A.O., Van der Merwe, A., Vigier, J.P., eds., *"Quantum, Space and Time: the Quest continues,"* (Cambridge University Press, Cambridge, 1982).

146. Cufaro-Petroni, N., Vigier, J.P., "Stochastic interpretation of relativistic quantum equations", in van der Merwe, A., ed, *Old and New Questions in Physics, Cosmology and Theoretical Biology, Essays in honour of W. Yourgrau* (Plenum, New York, 1983).

147. Cufaro-Petroni, N., Vigier, J.P., "Causal action at distance interpretation of the Aspect-Rapisarda experiments," *J. Phys. Lett. A 93*, 383 (1983).

148. Cufaro-Petroni, N., Vigier, J.P., "Random motions at the velocity of light and relativistic quantum mechanics," *J. Phys. A 17*, 599 (1984).

149. Cufaro-Petroni, N., Kypranidis, A., Maric, Z., Sardelis, D., Vigier, J.P., "Causal stochastic interpretation of Fermi-Dirac statistics in terms of distinguishable non-locally correlated particles," *Phys. Lett. A 101*, 4 (1984).

150. Cufaro-Petroni, N., Gueret, Ph., Vigier, J.P., "A causal stochastic theory of spin-1/2 fields," *Nuovo Cimento B 81*, 243 (1984).

151. Cufaro-Petroni, N., Gueret, Ph., Vigier, J.P., "Form of a spin dependent quantum potential," *Phys. Rev. D 30*, 495 (1984).

152. Cufaro-Petroni, N., Dewdney, C., Holland, P., Kyprianidis, A., Vigier, J.P., "Elimination of negative probabilities within the causal stochastic interpretation of quantum mechanics," *Phys. Lett. A 106*, 368 (1984).

153. Cufaro-Petroni, N., "A causal fluidodynamical model for the relativistic quantum mechanics," in Tarozzi, G., and Van der Merwe, A., *Open questions in quantum physics*, (Reidel, Dordrecht, 1985).

154. Cufaro-Petroni, N., Gueret, Ph., Kyprianidis, A., Vigier, J.P., "An alternative derivation of the spin dependent quantum potential", *Lett. N. Cim. 42*, 362 (1985)`.

155. Cufaro-Petroni, N., Dewdney, C., Holland, P., Kyprianidis, A., Vigier, J.P., "Causal space-time paths of individual distinguishable particle motions in N-body quantum systems: elimination of negative probabilities," *Lett. Nuovo Cimento 42*, 285 (1985).

156. Cufaro-Petroni, N., Gueret, Ph., Kyprianidis, A., Vigier, J.P., "Second-order wave equation for spin-1/2 fields," *Phys. Rev. D 31*, 3157 (1985).

157. Cufaro-Petroni, N., Dewdney, C., Holland, P., Kyprianidis, A., Vigier, J.P., "Realistic physical origin of the quantum

observable operator algebra in the frame of the causal stochastic interpretation of quantum mechanics: the relativistic spin-0 case", *Phys. Rev D 32*, 1375 (1985).

158. Dalla Chiara, M.L., "A general approach to non-distributive logics," *Studia Logica 25*, 139-162 (1976).

159. Dalla Chiara, M.L., "Quantum logic and physical modalities," *J. Phys. Logic 6*, 391-404 (1977).

160. Dalla Chiara, M.L., Toraldo di Francia, G., "A formal analysis of physical theories," in Toraldo di Francia, G., ed., *Problems in the Foundations of Physics* (North Holland, Amsterdam, 1979).

161. Dalla Chiara, M.L., "Logical foundations of quantum mechanics," in Agazzi, E., ed., *Modern Logic: A Survey* (Reidel, Dordrecht, 1981).

162. Dalla Chiara, M.L., "Some metalogical pathologies of quantum logic," in Beltrametti, E., van Fraassen, B., eds., *Current Issues in Quantum Logic* (Plenum, New York, 1981)

163. Dalla Chiara, M.L., "Is there a logic of empirical sciences?" in Dalla Chiara, M.L., ed., *Italian Studies in the Philosophy of Science* (Reidel, Dordrecht, 1981).

164. Dalla Chiara, M.L., Metelli, P.A., "Philosophy of quantum mechanics," in *Contemporary Philosophy: A New Survey*, vol. 2 (Martin Nijhoff, Dordrect, 1982).

165. Dalla Chiara, M.L., "Physical implications in a kripkian semantical approach to physical theories," in *Logic in the 20th century*, special issue of *Scientia* (1983).

166. Dalla Chiara, M.L., "Some logical problems suggested by empirical theories," in Cohen, R.S., and Wartofsky, M.W., eds., *Language, Logic and Method* (Reidel, Dordrecht, 1983).

167. Dalla Chiara, M.L., "Some fundamental problems in mathematics suggested by physics," *Synthese 62*, 303-315 (1985).

168. Dalla Chiara, M.L., "Names and descriptions in quantum logic," in Mittelstaedt, P., and Stachow, E.W., eds., *Recent Developments in Quantum Logic*, (B.I. Wissenschaftsverlag, Mannheim, 1985).

169. Dalla Chiara, M.L., "Quantum Logic", in Guenthner, F., and Gabbay, D. eds., *Handbook of philosophical Logic*, vol. III (Reidel, Dordrecht, to appear).

170. Dalla Chiara, M.L., "The relevance of quantum logic in the domain of non-classical logics," in *Proceedings of the 7th Congress "Logic, Methodology and Philosophy of Science"* (North Holland, Amsterdam, 1986).

171. Dalla Chiara, M.L., Toraldo di Francia, G., "Individuals, kinds and names in physics," VS. 40, 29-50 (1985).

172. Dalla Chiara, M.L., Toraldo di Francia, G., "Individuals, Properties, and Truths in the EPR-paradox," in Lahti, P., Mittelstaedt, P., eds., *Foundations of Modern Physics* (World Scientific, Singapore, 1985).

173. Daneri, A., Loinger, A., Prosperi, G.M., "Quantum theory of measurement and ergodicity conditions," *Nucl. Phys. 33*, 297 (1962).

174. Daneri, A., Loinger, A., Prosperi, G.M., "Further remarks on the

relations between statistical mechanics and quantum theory of measurements," *Nuovo Cimento B 44*, 119-128 (1966).

175. De Natale, F., "Il paradosso della realtà fisica. Sul Convegno di Cesena (11-13 Aprile 1985)", *Paradigmi 3*, 455-465 (1985).

176. De Falco, D., De Martino, S., Siena, S., "Position-momentum uncertainty relations in stochastic mechanics," *Phys. Rev. Lett. 49*, 181-183 (1982).

177. Degasperis, A., Fonda, L., Ghirardi, G.C., "Does the life-time of an unstable system depend on the measuring apparatus?", *Nuovo Cimento A 21*, 471 (1974).

178. Di Salvo, E., Viano, G.A., "Uniqueness and stability in the inverse problem of scattering theory," *Nuovo Cimento B 33*, 547-565 (1976).

179. Falciglia, F., Iaci, G., Rapisarda, V.A., "Do information propagate inside a quantum-mechanical system?", *Lett. Nuovo Cimento 26*, 327-331 (1979).

180. Falciglia, F., Garuccio, A., Pappalardo, L., "Rapisarda's experiment: on the four-coincidence experimental 'FOCA 2', a test for nonlocality propagation," *Lett. Nuovo Cimento 34*, 1-4 (1982).

181. Falciglia, F., Garuccio, A., Iaci, G., Pappalardo, L., "Optical-transmittance measurement method in the FOCA 2 experiment," *Lett. Nuovo Cimento 37*, 65 (1983).

182. Falciglia, F., Fornari, L., Garuccio, A., Iaci, G., Pappalardo, L., "A new approach to testing the separability in microphysics: Rapisarda's experiment," in Diner, S., *et al.*, eds., *The wave particle dualism*, (Reidel, Dordrecht, 1984), pp. 397-412.

183. Fano, V., "Il limite fra complementarità e dialettica: la filosofia della fisica di Bohr", *Ann. Disc. Fil. Un. Bologna*, 41, 177-207 (1983).

184. Faraci, G., Gutkowsky, D., Notarrigo, S., Pennisi, A.R., "Angular correlation of scattered annihilations photons, to test the possibility of hidden variables in quantum theory," *Appl. Phys. 5*, 63 (1974).

185. Faraci, G., Gutkowsky, D., Notarrigo, S., Pennisi, A.R., "An experimental test of the EPR paradox," *Lett. Nuovo Cimento 29*, 607 (1974).

186. Faraci, G., Pennisi, A.R., "Polarization of the annihilation photon of triplet positronium," *Nuovo Cimento B 31*, 289-296 (1976).

187. Favella, L.F., "Brownian motion and quantum mechanics," *Ann. Inst. Henri Poincaré 7*, 77-94 (1967).

188. Ferrari, E., "On the scattering from an experimental beam of wave-packets," *Rivista Nuovo Cimento 7*, 1-26 (1983).

189. Ferretti, B. "On the possibility of a macroscopically causal quantum relativistic theory," *Atti Acc. Sci. Ist. Bologna. Cl. Sci. Fisiche. Rendiconti 10*, 41-81 (1962/63).

190. Fonda, L., Ghirardi, G.C., "Contribution to the decay theory of unstable quantum systems," *Nuovo Cimento A 67*, 257 (1971).

191. Fonda, L., Ghirardi, G.C., "Contribution to the decay theory of unstable quantum systems II," *Nuovo Cimento A 6*, 553 (1971)

192. Fonda, L., Ghirardi, G.C., "Some remarks on the origin of the deviation from the experimental decay law of an unstable particle," Nuovo Cimento A 7, 180 (1972).

193. Fonda, L., Ghirardi, G.C., Rimini, A., Weber, T., "On the quantum foundations of the exponential decay law," Nuovo Cimento A 15, 689 (1973).

194. Fonda, L., Ghirardi, G.C., Rimini, A., "Evolution of quantum systems subject to random measurements," Nuovo Cimento B 18, 1 (1973).

195. Fonda, L., Ghirardi, G.C., Rimini, A., "Decay theory of unstable quantum system", Rep. Prog. Phys., 41, 587 (1978).

196. Fortunato, D., Selleri, F., "Sensitive observables on infinite-dimensional Hilbert spaces," Int. J. Theor. Phys. 15, 333 (1976).

197. Fortunato, D., Garuccio, A., Selleri, F., "Observable consequences from second-type state vectors of quantum mechanics," Int. J. Theor. Phys. 16, 1 (1977).

198. Galgani, L., Scotti, A., "Planck-like distributions in classical nonlinear mechanics," Phys. Rev. Lett. 28, 1173-1176 (1972).

199. Galgani, L., Scotti, A., "Recent proress in classical nonlinear mechanics," Nuovo Cimento ", 189-209 (1972).

200. Galgani, L., "Meccanica classica e meccanica quantistica," Scientia 110, 469-482 (1975); "Classical mechanics and quantum mechanics," Ibid., pp. 483-494.

201. Galgani, L., "On the problem of the dynamical foundations of classical statistical mechanics and its possible relevance for the foundation of quantum mechanics," in Carciovei, A., ed., Lectures on Statistical Mechanics, (Poiana Brasov School, 1979) (Bucarest, 1980), 527-544.

202. Galgani, L., "On Nerst's deduction of Planck's law and the problem of the Arnold diffusion," Nuovo Cimento B 62, 306-314 (1981).

203. Galgani, L., "Statistical mechanics of weakly coupled oscillators presenting stochasticity thresholds," Lett. Nuovo Cimento 31, 65-72 (1981).

204. Galgani, L., "Ordered and chaotic motions in Hamiltonian systems and the problem of energy partition," in Buchler, J.R., ed., Chaos in Astrophysics, (NATO Advanced Research Workshop, Palm Coast, April 1980) (Reidel, Dordrecht, 1984).

205. Gallonea, F., Sparzani, A., "Segal quantization of dynamical systems," J. Math. Phys. 20, 1375-84 (1979).

206. Garola, C., "Propositions and orthocomplementation in quantum logic", Int. J. Theor. Phys. 19, 639-378 (1980).

207. Garuccio, A., Selleri, F., "Nonlocal interactions and Bell's inequality", Nuovo Cimento B 36, 176 (1976).

208. Garuccio, A. Scalera, G., Selleri, F., "On local causality and the quantum-mechanical state vector", Lett. Nuovo Cimento 18, 26 (1977).

209. Garuccio, A.,"Generalized inequalities following from Einstein locality", Lett. Nuovo Cimento 23, 559-565 (1978).

210. Garuccio, A., Selleri, F., "On the equivalence of deterministic

and probabilistic local theories", *Lett. Nuovo Cimento 23*, 555 (1978).

211. Garuccio, A., Selleri, F., "Action at a distance in quantum mechanics", *Epist. Lett.* 24, 1 (1979)

212. Garuccio, A., Macarrone, G.D., Recami, E., Vigier, J.P., "On the physical nonexistence of signals going backwards in time, and quantum mechanics", *Lett. Nuovo Cimento 27*, 60-64 (1980).

213. Garuccio, A., Selleri, F., "Systematic derivation of all the inequalities of Einstein locality", *Found. Phys. 10*, 209 (1980).

214. Garuccio, A., Vigier, J.P., "Possible experimental tests of the causal stochastic interpretation of quantum mechanics", *Found. Phys. 10*, 799 (1980).

215. Garuccio, A., Vigier, J.P., "Description of spin in the causal stochastic interpretation of Proca-Maxwell waves: theory of Einstein's 'ghost waves'", *Lett. Nuovo Cimento 30*, 57-63 (1981).

216. Garuccio, A., Rapisarda, V.A., "Bell's inequalities and the four coincidence experiment", *Nuovo Cimento A 65*, 269-297 (1981).

217. Garuccio, A., Popper, K.R., Vigier, J.P., "Possible direct physical detection of de Broglie's waves", *Phys. Lett. 86 A*, 397 (1981).

218. Garuccio, A., Rapisarda, V.A., "A comparison between deterministic and probabilistic local hidden-variable theories", *Lett. Nuovo Cimento 30*, 443-448 (1981).

219. Garuccio, A., Rapisarda, V.A., Vigier, J.P., "Superluminal velocity and causality in EPR correlations", *Lett. Nuovo Cimento 32*, 451-456 (1981).

220. Garuccio, A., Rapisarda, V.A., Vigier, J.P., "New experimental set-up for the detection of de Broglie's waves", *Phys. Lett. 90 A*, 17 (1982).

221. Garuccio, A., Falciglia, F., Iaci, G., Pappalardo, L., "Magnetic-Field effects on the polarization of photon couples emitted in an atomic cascade", *Lett. Nuovo Cimento 38*, 52 (1983).

222. Garuccio, A., Selleri, F., "Enhanced photon detection in EPR type experiments", *Phys. Lett. 103 A* 99 (1984).

223. Garuccio, A., Kyprianidis, A., Sardelis, D., Vigier, J.P., "Possible experimental test of the wave packet collapse", *Lett. Nuovo Cimento 39*, 225 (1984).

224. Garuccio, A., Dewdney, C., Kyprianidis, A., Vigier, J.P., Gueret, Ph., "Time dependent neutron interferometry: Evidence in favour of de Broglie's waves", *Lett. Nuovo Cimento 40*, 481 (1984).

225. Garuccio, A., Kyprianidis, A., Vigier, J.P., "Relativistic quantum potential: the N-body case", *Nuovo Cimento 83 b*, 135 (1984).

226. Garuccio, A., Dewdney, C., Kyprianidis, A., Vigier, J.P., "Energy conservation and complementarity in neutron single cristal interferometry", *Phys. Lett. 104 A*, 325 (1984).

227. Garuccio, A., Dewdney, C., Kyprianidis, A., Vigier, J.P., "Anomalous photoelectric effect: Quantum potential theory vs. effective photon hypothesis", *Phys. Lett. 105 A*, 15 (1984).

228. Garuccio, A., "Third kind measurements and wave-particle
      dualism", in Tarozzi, G., van der Merwe, A., eds., *Open
      Questions in Quantum Physics*, (Reidel, Dordrecht, 1985).

229. Garuccio, A., Dewdney, C., Gueret, Ph., Kyprianidis, A., Vigier,
      J.P., "Time dependent neutron-interferometry: Evidence against
      wave-packet-collapse?", *Found. Phys. 15*, 1031 (1985).

230. Ghirardi, G.C., Rimini, A., Weber, T., "Is the density matrix
      description of statistical ensemble exhaustive?", *Nuovo Cimento
      29 B*, 135 (1975).

231. Ghirardi, G.C., "Quantum dynamical semi-groups and the reduction
      process", *Nuovo Cimento 30 B*, 133 (1975).

232. Ghirardi, G.C., Rimini, A., Weber, T., "Implications of the
      Bohm-Aharonov hypothesis", *Nuovo Cimento 31 B*, 177 (1976).

233. Ghirardi, G.C., "Answer to the comment on 'Is the density matrix
      description of statistical ensemble exhaustive?'", *Nuovo Cimento
      33 B*, 457 (1976).

234. Ghirardi, G.C., Rimini, A., Weber, T., "Some simple remarks
      about quantum nonseparability for systems composed of identical
      constituents", *Nuovo Cimento 39 B*, 130 (1977).

235. Ghirardi, G.C., Rimini, A., Weber, T., "A reformulation and a
      possible modification of quantum mechanics and the EPR paradox",
      Nuovo Cimento 39 B, 130 (1977).

236. Ghirardi, G.C., Omero, C., Rimini, A., Weber, T., "The
      stochastic interpretation of quantum mechanics", *Riv. Nuovo
      Cimento 1*, 1 (1978).

237. Ghirardi, G.C., Omero, C., Weber, T., Rimini, A., "Small-time
      behaviour of quantum nondecay probability and Zeno's paradox in
      quantum mechanics", *Nuovo Cimento 52 A*, 421 (1979).

238. Ghirardi, G.C., Omero, C., Weber, T., "Quantum vs. classical
      laws for sequential-decay process", *Nuovo Cimento 52 A*, 443
      (1979).

239. Ghirardi, G.C., Omero, C., Rimini, A., Weber, T., "Quantum
      dynamical semigroups description of decay processes and Zeno's
      paradox in quantum mechanics", in *Proceedings of the VII
      International Colloquium on Group Theoretical Methods in
      Physics, Kiry at Anavim, 1979; Ann. Israel Phys. Soc.2 (1979)*.

240. Ghirardi, G.C., Weber, T., "On some recent suggestions of
      superluminal communication through the collapse of the wave
      function", *Lett. Nuovo Cimento 26*, 599 (1979).

241. Ghirardi, G.C., Rimini, A., Weber, T., "A general argument
      against superluminal transmission through the quantum mechanical
      measurement process", *Lett. Nuovo Cimento 27*, 293 (1980).

242. Ghirardi, G.C., "Nonseparability in quantum mechanics", in
      Blaquerie, A., Fer, F., Marzollo, A., eds. *Dynamical Systems and
      microphysics*, (Springer, New York, 1980).

243. Ghirardi, G.C., Miglietta, F., Rimini, A., Weber, T.,
      "Limitations of quantum measuring processes", in *Proceedings of
      the IX International Colloquium on group theoretical methods in
      physics, Cocoyoc, Mexico, 1980*, (Academic Presse, New York,
      1981).

244. Ghirardi, G.C., Rimini, A., "Some topics in the quantum theory

of measurement", in Blaquerie, A. et al., *Dynamical Systems in Microphysics,* (Springer, New York, 1980).

245. Ghirardi, G.C., Rimini, A., Weber, T., "Limitations on quantum measurements I: Determination of the minimal amount of ideality and identification of the optimal measuring apparatuses", *Phys. Rev. D 24,* 347 (1981).

246. Ghirardi, G.C., Miglietta, F., Rimini, A., Weber, T., "Limitations on quantum measurements II: Analysis of a model example", *Phys. Rev. D 24,* 353, (1981).

247. Ghirardi, G.C., Rimini, A., Weber, T., "Rotational invariance and spin measurements", *Physica A 114,* 241, (1982).

248. Ghirardi, G.C., Rimini, A., Weber, T., "Quantum evolution in the presence of additive conservation laws and the quantum theory of measurement", *J. Math. Phys. 23,* 1792 (1982).

249. Ghirardi, G.C., Rimini, A., Weber, T., "Value preserving quantum measurements: impossibility theorems and lower bounds for the distortion", *J. Math. Phys. 24* 2454 (1983).

250. Ghirardi, G.C., Weber, T., "Quantum mechanics and faster-than-light communication: methodological considerations", *Nuovo Cimento B 78,* 9 (1983).

251. Ghirardi, G.C., Rimini, A., Weber, T., "A model for unified quantum description of macroscopic and microscopic systems". Preprint IC/84/240, invited talk to the II Workshop on Quantum Probability and Applications, Heidelberg 1984, to appear in *Lecture Notes in Mathematics,* (Springer, New York).

252. Gozzini, A., "On the possibility of realising a low intensity interference experiment with a determination of the particle trajectory", in Diner, S., et al., eds., *The Wave-particle Dualism* (Reidel, Dordrecht, 1984).

253. Guccione, S., "Quantum logic and the two-slit experiment", in Dalla Chiara, M.L., ed., *Italian Studies in the Philosophy of Science* (Reidel, Dordrecht, 1981).

254. Guerra, F., "On the connection between Euclidean-Markov field theory and stochastic quantization, in *C*-Algebras and their Applications to Statistical Mechanics and Quantum Field Theory (Proceedings of the International School of Physics "Enrico Fermi". Course LX, Varenna 1973,* pp. 190-202 (North Holland, Amsterdam, 1976).

255. Guerra, F., Ruggiero, P., "New interpretation of the Euclidean-Markov field in the framework of physical Minkowski space-time", *Phys. Rev. Lett. 31,* 1022-5 (1973).

256. Guidone, M.,"De Broglie and wave-particle dualism", *Scientia 117,* 629-633 (1982).

257. Gutkowski, D., Masotto, G., Valdes, M.V., "On the sufficiency of Bell's conditions", *Nuovo Cimento B 50,* 323-343 (1979).

258. Lanz, L., Prosperi, G.M., Sabbadini, A., "Time scales and the problem of measurement in quantum mechanics", *Nuovo Cimento 2 B,* 184-192 (1971).

259. Livi, R., "An experimental test of quantum mechanics in molecular predissociation", *Nuovo Cimento B 48,* 272-286 (1978).

260. Loinger, A., "Comments on a recent paper concerning the quantum

theory of measurement", *Nucl. Phys. 108*, 245-249 (1968).

261. Maiocchi, R., "Le considerazioni epistemologiche di Paul Langevin sulla meccanica quantistica ed i loro riflessi nella cultura francese dell'anteguerra", *Scientia 110*, 493-538 (1975).

262. Marshall, T.W., Santos, E., Selleri, F., "Collective effects in the atomic cascade experimental tests of Bell inequalities", *Lett. Nuovo Cimento 38*, 417 (1983).

263. Marshall, T.W., Santos, E., Selleri, F., "Local realism has not been refuted by atomic cascade experiments", *Phys Lett. A 98*, 5 (1983).

264. Marshall, T.W., Santos, E., Selleri, F., "On the compatibility of local realism with atomic cascade experiments", in Tarozzi, G., van der Merwe, A., eds. *Open Questions in Quantum Physics*, (Reidel, Dordrecht, 1985) pp. 87-101.

265. Matteucci, G., Pozzi, G., "New diffraction experiment on the electrostatic Aharonov-Bohm effect", *Phys. Rev. Lett. 54*, 2469-72 (1985).

266. Merli, P.G., Missiroli, G.F., Pozzi, G., "On the statistical aspect of electron interference phenomena", *Am. Phys. 44*, 7 (1976).

267. Mignani, R., Recami, E., "How to interpret advanced solutions", *Lett. Nuovo Cimento 18*, 5-9 (1977).

268. Notarrigo, S., "Polarization correlation of annihilation radiation", *Progress in Scientific Culture. The Thinkshops on Physics. Interdisc. J. Ettore Maiorana C. 4*, 452 (1976).

269. Notarrigo, S., "A Newtonian separable model which violates Bell's inequality", *Nuovo Cimento 83 B*, 73 (1984).

270. Olkhovsky, V.S., Recami, E., "About collision-processes lifetimes and causality", *Nuovo Cimento 63 A*, 814-826 (1969).

271. Olkhovsky, V.S., Recami, E., Gerasimchuck, A.J., "Time operator in quantum mechanics I: Nonrelativistic case", *Nuovo Cimento 22 A*, 263-278 (1974).

272. Pascazio, S:, "A careful estimation of photon rescattering in atomic-cascade experimental tests of Bell's inequality", *Nuovo Cimento 5 D*, 23-39 (1985).

273. Pascazio, S., "Experimental tests of Bell's inequalities. Are all local models really excluded?", *Phys. Lett. 111*, 339-342 (1985).

274. Petruccioli, S., "Mechanical model and correspondance rules in the construction of the atomic theory", *Physis 29*, 555-579 (1981).

275. Piccioni, O., Bowles, P., Enscoe, C., Garland, R., Mehlop, W., "Is the Einstein-Podolsky-Rosen paradox demanded by quantum mechanics?", in Tarozzi, G., van der Merwe, A., eds., *Open Questions in Quantum Physics*, (Reidel, Dordrecht, 1985).

276. Pignedoli, A., "Sui fondamenti della meccanica quantica", *Atti Accad. Naz. Sci. Lett. Arti Modena* (1972).

277 Pignedoli, A., "Meccanica classica e meccanica quantica di fronte alla filosofia del linguaggio", *Atti Accad. Naz. Sci. Lett. Arti*, Modena (1977).

278. Popper, K.R., Garuccio, A., Vigier, J.P., "An experiment to

interpret E.P.R. action at a distance: The possible detection of real de Broglie waves", *Epist. Lett.* *30*, 21 (1981).

279. Prosperi, G.M., Scotti, A., "Ergodic theorem in quantum mechanics", *Nuovo Cimento 13*, 1007-1012 (1959).

280. Prosperi, G.M., "Quantum theory of measurement", in *Encyclopedic Dictionary of Physics*, Vol. II. Suppl. II (Pergamon Press, Oxford, 1967).

281. Prosperi, G.M., "Macroscopic physics and the problem of measurement in quantum mechanics", in *Foundations of Quantum Mechanics*, IL corso SIF (Academic Press, New York, 1971), pp. 97-126.

282. Rapisarda, V.A., "On the measurement by dichotomic analysers of the polarization correlations of optical photons emitted in atomic cascade", *Lett. Nuovo Cimento 33*, 437-444 (1982).

283. Recami, E., "Black body radiation and generalized theory of physical dimensions", *Lett. Nuovo Cimento 2*, 297-304 (1971).

284. Recami, E., "About new space-time symmetries in relativity and quantum mechanics", *Nuovo Cimento A 23*, 205-215 (1976).

285. Recami, E., "An introductory view about superluminal frames and tachyons", in Recami, E., ed., *Tachyons, Monopoles, and Related Topics* (Proceedings of the first section of the interdisciplinary seminars held at Erice, 1-15 September 1976), (North Holland, Amsterdam 1978)

286. Recami, E., Macarrone, G.D., "Are classical tachyons slower than light quantum particles?" *Lett. Nuovo Cimento 37*, 345-352 (1983).

287. Rietdijk, C.W., Selleri, F. "Proof of a quantum mechanical non local influence", *Found. Phys. 15*, 303 (1985)

288. Rosa, R., "L'interpretazione della probabilità in termini di propensità in K.R. Popper", *Statistica 1*, 99-115 (1979).

289. Ruffo, S., "A comparison between nonlocal models and quantum mechanics", *Lett. Nuovo Cimento 20*, 221-226 (1977).

290. Santamato, E., "Geometric derivation of the Schroedinger equation from classical mechanics in curved Weyl spaces", *Phys. Rev. D 29*, 216-22 (1984).

291. Santamato, E., "Statistical interpretation of the Klein-Gordon equation in terms of the space-time Weyl curvature", *Math. Phys. 25*, 2477-80 (1984).

292. Santilli, R.M., "Remarks on the problematic aspects of Heisenberg-Lie sumplectic formulations", *Hadronic J. 3*, 854-913 (1979/80).

293. Santilli, R.M., "Lie-isotopic lifting of unitary symmetries and of Wigner's theorem for extended, deformable particles", *Lett. Nuovo Cimento 38*, 509-521 (1983).

294. Scalera, G.C., "On a local hidden-variable model with unusual properties", *Lett. Nuovo Cimento 38*, 16-18 (1983).

295. Scalera, G.C., "Local models violating Bell's inequality by time delays", *Lett. Nuovo Cimento 40*, 353-361 (1984).

296. Schiavulli, L., "Generalized Bell's inequalities", *Lett. Nuovo Cimento 20*, 400-404 (1977).

297. Schiavulli, L., Selleri, F., "Further consequences of Einstein

locality", *Found. Phys. 9*, 339 (1979).

298. Selleri, F., "On the wave-function of quantum mechanics", *Lett. Nuovo Cimento 1*, 908-910 (1969).

299. Selleri, F., "Realism and the wave-function of quantum mechanics", in *Foundations of Quantum Mechanics*, (Academic Press, New York, 1971), pp. 398-406.

300. Selleri, F., "A stronger form of Bell's inequality", *Lett. Nuovo Cimento 3*, 581 (1972).

301. Selleri, F. "On the consequences of Einstein locality", *Found. Phys. 8*, 103 (1978).

302. Selleri, F., Tarozzi, G., "Is non-distributivity for microsystems empirically founded?", *Nuovo Cimento B 43*, 31 (1978).

303. Selleri, F., Tarozzi, G., "Nonlocal theories satisfying Bell's inequality", *Nuovo Cimento B 48, 120* (1978).

304. Selleri, F., Tarozzi, G., "Extension of the domain of validity of Bell's inequality", *Epist. Lett. 21*, 1 (1978).

305. Selleri, F., "Quantum correlation and separability", *Lett. Nuovo Cimento 27*, 1 (1980).

306. Selleri, F., "Photon coincidences with crossed polarizers", *Epistem. Lett. 25*, 39 (1980).

307. Selleri, F., "Einstein locality and the quantum mechanical long-distance", in Blaquiere, A., et al., eds. *Dynamical Systems and Microphysics* (Springer, New York, 1980).

308. Selleri, F., "Particules élémentaires et mécanique quantique", in Diner, S., Fargue, D., Lochak, G., eds., *La Pensée physique contemporaine*, (Fresnel, Paris, 1980)

309. Selleri, F., Tarozzi, G., "Is Clauser and Horne's factorability a necessary requirement for a probabilistic local theory?", *Lett. Nuovo Cimento 29*, 533 (1980).

310. Selleri, F., Tarozzi, G., "Quantum mechanics, reality and separability", *Riv. Nuovo Cimento 4*, 1 (1981).

311. Selleri, F., "Il problema del dualismo onda-particella", *La Nuova Critica 57-58*, 49-56 (1981).

312. Selleri, F., "Generalized EPR-paradox", *Found. Phys. 12*, 645 (1982).

313. Selleri, F., "Can an actual existence be granted to quantum waves?", *Ann. Fond. Louis de Broglie 7*, 45 (1982).

314. Selleri, F., "On the direct observability of quantum waves", *Found. Phys 12*, 1087 (1982).

315. Selleri, F., "L'altro labirinto", *La Nuova Critica 2*, 9 (1982).

316. Selleri, F., Tarozzi, G., "A probabilistic generalization of the concept of physical reality", *Spec. Sci. Tech. 6*, 55 (1983).

317. Selleri, F., Vigier, J.P., "Induced superfluorescence and the nature of the wave-particle duality", in Van der Merwe, A., ed., *Old and New Questions in Physics, Cosmology, Philosophy, and Theoretical Biology*, (Plenum, New York, 1983).

318. Selleri, F., "Einstein locality and the K'K'-system", *Lett. Nuovo Cimento 36*, 521 (1983).

319. Selleri, F., "Quantum reality as an empirical problem", in Bitsakis, E., ed., *The Concept of Reality*, (Zacharopoulos,

        Athens, 1983).
320.   Selleri, F., "Gespensterfelder", in Diner, S., et al., eds., The
        Wave-Particle Dualism (Reidel, Dordrecht, 1984).
321.   Selleri, F., "Questioni aperte in fisica quantistica", Paradigmi
        2, 105-116 (1984).
322.   Selleri, F., "Einstein locality for individual systems and for
        statistical ensembles", in Tarozzi, G., and Van der Merwe, A.,
        eds., Open Questions in Quantum Physics, (Reidel, Dordrecht,
        1985), pp. 153-170.
323.   Selleri, F., "Photon scattering with Doppler-shift in the atomic
        cascade experimental tests of Bell inequalities", Lett. Nuovo.
        Cimento 39, 252 (1984).
324.   Selleri, F., "A propos des expériences de type EPR", Ann. Fond.
        Louis de Broglie 9, 331 (1984).
325.   Selleri, F., "The Einstein-Podolsky-Rosen-Paradox fifty years
        later", in Kleinpoppen, H., Briggs, J.S., Lutz, H.O., eds.
        Fundamental Processes in Atomic Collision Physics (Plenum, New
        York, 1985)
326.   Selleri, F., "Local realistic photon models and EPR-type
        experiments", Phys. Lett. 108 A, 197-202 (1985).
327.   Selleri, F., "On the notion of probability for pairs of
        correlated systems", in Bitsakis, E.I., and Tambakis, N., eds.
        Determinism in Physics (Gutemberg, Athens, 1985)
328.   Selleri, F., "New variable detection probability model for
        EPR-type experimentis", in Lahti, P., and Mittelstaedt, P.,
        eds., Symposium on the Foundations of Modern Physics (World
        Scientific, New York, 1985).
329.   Spinelli, G., "Absolute synchronization: faster-than-light
        particles and causality violation", Nuovo Cimento B 75, 11-18
        (1983).
330.   Strocchi, F., "Difficulties of a quantum field theory of
        electrodynamics", Lett. Nuovo Cimento 1, 169-173 (1969).
331.   Strocchi, F., "Microscopic and macroscopic quantities in
        statistical mechanics", Nuovo Cimento 65 B, 239-264 (1970).
332.   Tagliagambe, S., "Il dibattito sull'interpretazione filosofica
        della meccanica quantistica", in Geymonat, L., Storia del
        pensiero filosofico e scientifico, vol. IX (Garzanti, Milano,
        1972).
333.   Tarozzi, G., "Realistic interpretation of physical theories",
        Mem. Accad. Naz. Sci., Lett. Arti di Modena 20, 49-62 (1978).
334.   Tarozzi, G., "The conceptual development of the E.P.R.
        argument", Mem. Accad. Naz. Sci., Lett. Arti di Modena 21, 353-
        373 (1979).
335.   Tarozzi, G., "Remarques sur les concepts de localité
        séparabilité et réalité physique", Proceedings of the Colloque
        "Indeterminisme Quantique et Variables Cachées, Genève (1979),
        pp. 1-8.
336.   Tarozzi, G., "The principle of empiricism and quantum theory",
        Epistemologia 3, 13-28 (1980).
337.   Tarozzi, G., "Realism as a meaningful philosophical hypothesis",
        Atti Accad. Sci. Ist. Bologna, Rendiconti 7, 89-98 (1980).

338. Tarozzi, G., "On the relevance of realistic assumption in the proof of Bell's inequality", *Atti Accad. Naz. Sci., Lett. Arti di Modena 22-23*, 81-86 (1980-1981).

339. Tarozzi, G., "Realisme d'Einstein et méchanique quantique: un cas de contradiction entre une théorie physique et une hypothèse philosophique clairement définie", *Revue de Synthèse 101-102*, 125-158 (1981).

340. Tarozzi, G., "On the essential role of the realist hypothesis in all derivations of E.P.R.-types paradoxes", *Epistemologia 4*, 407-422 (1981).

341. Tarozzi, G., "The theory of observations, Wigner paradox and the mind-body problem", in *Le mental et le corporel: Archives de l'Institut International des Sciences Théorique 24*, Bruxelles, 1982; *Epistemologia IV*, numero speciale, 37-52 (1981).

342. Tarozzi, G., "Local realism and Bell's theorem without the hidden-variable hypothesis", *Atti Accad. Sci. Torino 108*, 119-124 (1981).

343. Tarozzi, G., "Empirical realism and the foundations of quantum theory", to appear in *Actes du Congrés La Nature de la verité scientifique, Bruxelles, 1981*.

344. Tarozzi, G., "Chance and order's coexistence as an insuperable dualism in the logical structure of quantum physics", in *Chance and Order in Natural Sciences* (Proceedings of the International School of Logic and Scientific Methodology, Erice, 1981).

345. Tarozzi, G., "From ghost to real waves: a proposed solution to the wave-particle dilemma", in Diner, S. *et al.*, *The Wave-Particle Dualism*, (Reidel, Dordrecht, 1984).

346. Tarozzi, G., "Two proposals for testing physical properties of quantum waves", *Lett. Nuovo Cimento 35*, 53-59 (1982).

347. Tarozzi, G., "Physical reality: from the metaphysical notion to its empirical definition", in Bitsakis, E., ed., *The Concept of Physical Reality*, (Zacharopoulos, Athens, 1983), pp. 197-215.

348. Tarozzi, G., "Due tentativi di falsificazione del principio di Heisenberg: Popper come precursore e continuatore dell'argomento di EPR", in *Atti del IV Congresso di Storia della Fisica*, (CLUED, Milano, 1983), pp. 183-196.

349. Tarozzi, G., "Il teorema di Nobili e la natura dell'azione a distanza", Giornale di Fisica 3-4, 361-373 (1984).

350 Tarozzi, G., "A unified experiment for testing both the interpretation and the reduction postulate of the quantum mechanical wave function", in Tarozzi, G., and Van der Merwe, A., *Open Questions in Quantum Physics* (Reidel, Dordrecht, 1985).

351. Tarozzi, G., Teoria e strumento in microfisica", *Epistemologia 8*, 83-118 (1985).

352. Tarozzi, G., "Experimental tests of the properties of the quantum mechanical wave function", *Lett. Nuovo Cimento 42*, 439-442 (1985).

353. Tarozzi, G., "Critica alle interpretazioni del teorema di Bell", paper presented at the conference of the Italian Society of Logic and Philosophy of Science, held at S. Gimignano, December 7-11 1983, to be published in E. Agazzi et al. eds., *Logica e*

filosofia della scienza oggi, (Clueb, Bologna, 1986)

354.   Tarozzi, G., "Filosofia della fisica", lecture delivered at the
       conference of the Italian Philosophical Society, held at
       Bergamo, March 15-17, in E. Agazzi, ed., *La filosofia della*
       scienza in Italia nel '900, (Angeli, Milano, 1986).

355.   Tonini, V., "Nouvelles tendences réalistes dans l'interpretation
       des théories physiques", in *Philosophie de la Physique.*
       Colloque de l'Académie Internationale de Philosophie des
       Sciences. Paris, 16-18 Sept. 1961, (Office International de
       Librairie, Bruxelles, 1962)

356.   Tonini, V., "Il testamento scientifico di Einstein e la
       filosofia della fisica oggi", *La Nuova Critica*, 50-51 (1979).

357.   Tonini, V., "La realtà non è un sillogismo della ragione", *La*
       Nuova Critica 52 (1982).

358.   Toraldo di Francia, G., "Induction in Physics", *Riv. Nuovo*
       Cimento 4, 144-165 (1974).

359.   Toraldo di Francia, G., "What have physicists learned from
       experience about inductive inference?", in Przelecki, M., ed.,
       Formal Methods in the Methodology of Empirical Sciences,
       (Reidel, Dordrecht, 1976).

360.   Toraldo di Francia, G., "The concept of progress in Physics", in
       Dalla Chiara, M.L., ed., *Italian Studies in the Philosophy of*
       Science, (Reidel, Dordrecht, 1981).

361.   Zecca, A., "On the coupling of logic", *Math. Phys 19*, 1482-1485
       (1978).

362.   Zecca, A., "The superposition of the states and the logic
       approach to quantum mechanics", *Int. J. Theor. Phys. 20*, 191-230
       (1981).

BOOKS

1.   Agazzi, E., *Temi e problemi di filosofia della fisica*,
       (Manfredi, Milano, 1969).

2.   Bellone, E., *I modelli e la concezione del mondo nella filosofia*
       moderna da Laplace a Bohr, (Feltrinelli, Milano, 1974).

3.   Beltrametti, E., Cassirelli, G., *The Logic of Quantum Mechanics*,
       (Addison-Wesley, New York, 1982).

4.   Beltrametti, E., van Frassen, B., eds., *Current Issues in*
       Quantum Logic, (Plenum, New York, 1981).

5.   Blaquerie, A., Fer, F., Marzollo, A., eds., *Dynamical Systems*
       and Microphysics, (Springer, New York, 1980).

6.   Caldirola, P., *Dalla microfisica alla macrofisica*, (Mondadori,
       Milano, 1974).

7.   Caldirola, P., Cirelli, R., Prosperi, G.M., *Introduzione alla*
       fisica teorica, (UTET, Torino, 1982).

8.   Costa, S., Predazzi, E., *Origine classica della fisica moderna*,
       (Levrotto e Bella, Torino, 1984).

9.   Dalla Chiara, M.L., Toraldo di Francia, G., *La Teoria Fisica*,
       (Boringhieri, Torino, 1982).

10.   Diner, S., Fargue, D., Lochak, G., Selleri, F., eds., *The Wave-*

Particle Dualism: A Tribute to Louis de Broglie on his 90th Birthday, (Proceedings of an International Symposium held in Perugia, April 22-30, 1982) (Reidel, Dordrecht, 1984).

11. Donini, E., *Il caso dei quanti. Dibattito in fisica e ambiente storico 1900-1927*, (Clup-Clued, Milano, 1982).

12. Duimio, F., *Meccanica quantistica*, in R. Fieschi, ed., *Enciclopedia della Fisica*, (ISEDI, Milano, 1975).

13. Fano, G., *Mathematical Methods of Quantum Mechanics*, (Mc Graw Hill, New York, 1971).

14. Ferretti, B., *Le radici classiche della fisica quantica*, (Boringhieri, Torino, 1980).

15. Pignedoli, A., *Alcune teorie meccaniche superiori*, (CEDAM, Padova, 1969).

16. Recami, E., ed., *Tachyons, Monopoles and Related Topics*, (North Holland, Amsterdam, 1976).

17. Selleri, F., *Die Debatte um die Quantentheorie*, (Vieweg, Braunschweig, 1983).

18. Selleri, F., Tarozzi, G., van der Merwe, A., eds., *Microphysical Reality and Quantum Formalism*, Vols. 1 and 2, Proceedings of the 1985 Urbino conference (Reidel, Dordrecht, to appear).

19. Tagliagambe, S., *L'interpretazione materialistica della meccanica quantistica. Fisica e Filosofia in URSS*, (Feltrinelli, Milano, 1972).

20. Tarozzi, G., van der Merwe, A., eds., *Open Questions in Quantum Physics*, (Reidel, Dordrecht, 1985).

21. Toraldo di Francia, G., ed., *Problems in the Foundations of Physics*, (Proceedings of the International School of Physics "Enrico Fermi, Course LXXXII, held at Varenna, 25th July-6th August 1977), (North Holland, Amsterdam, 1979).

INDEX

Abbagnano, N., 388, 389, 395
Accardi, L., 17, 18, 20, 21
action at a distance, 38-45, 201, 266
additivity of probability, 16
Aerts, D., 272
Agazzi, E., 6, 7, 8, 26, 389
Alexandrov, A.D., 378
algebra
    of Heisenberg, 272
    of measurement, 272
    of observables, 346
    of von Neumann, 305
algebraic semantics, 335
Aliotta, A., 388, 394
Angelidis, T., D., 41
Aristotle, 164
    physics of, 70
arrow of time, 109
axiom(s)
    of classical probability theory, 13, 264
    of dynamics, 232-233
    of measure, 233-234
    of quantum mechanics, 172, 173, 178, 191
axiomatization, 60, 61-64

Bacon, F., 44
Baracca, A., 223, 224
Barchielli-Lanz-Lupieri-Prosperi construction, 295-298
Bayes analysis of probability, 277